High-Frequency Application of Semiconductor Devices

STUDIES IN ELECTRICAL AND ELECTRONIC ENGINEERING

Vol. 1 Solar Energy Conversion: The Solar Cell (Neville)
Vol. 2 Charge Storage, Charge Transport and Electrostatics with their Applications (edited by Wada, Perlman and Kokado)
Vol. 3 Integrated Functional Blocks (Novák)
Vol. 4 Operational Amplifiers (Dostál)
Vol. 5 High-Frequency Application of Semiconductor Devices (Kovács)

STUDIES IN ELECTRICAL AND
ELECTRONIC ENGINEERING 5

High-Frequency Application of Semiconductor Devices

FERENC KOVÁCS

Research Institute for Electronics,
Budapest, Hungary

ELSEVIER SCIENTIFIC PUBLISHING COMPANY

Amsterdam — Oxford — New York 1981

The original:
Félvezetők nagyfrekvenciás alkalmazása
was published by Műszaki Könyvkiadó, Budapest

Translated by
T. Sárkány

The distribution of this book is being handled by the following publishers
for the U.S.A. and Canada
Elsevier/North-Holland, Inc.
52 Vanderbilt Avenue
New York, New York 10017, U.S.A.

for the East European Countries, Democratic People's Republic of Korea, People's Republic of China, People's Republic of Mongolia, Republic of Cuba and Socialist Republic of Vietnam
Akadémiai Kiadó, The Publishing House of the
Hungarian Academy of Sciences, Budapest

for all remaining areas
Elsevier Scientific Publishing Company
P.O. Box 211, 1000 AE Amsterdam, The Netherlands

Library of Congress Cataloging in Publication Data

Kovács, Ferenc, Dr.-Ing.
   High-frequency application of semiconductor devices.

   (Studies in electrical and electronic engineering; v. 5)
   Rev. and updated translation of Félvezetők nagyfrekvenciás alkalmazása.
   Bibliography: p.
   Includes index.
   1. Semiconductors. 2. Amplifiers (Electronics)
I. Kovács, Ferenc, Dr.-Ing. Félvezetők nagyfrekvenciás alkalmazása. II. Title. III. Series.
TK7871.85. K69513   621.381'33   80–20023
ISBN 0–444–99756–3

ISBN 0–444–99756–3 (Vol. 5)
ISBN 0–444–41713–3 (Series)

© Akadémiai Kiadó, Budapest 1981

Joint edition published by
Elsevier Scientific Publishing Company, Amsterdam, The Netherlands and
Akadémiai Kiadó, The Publishing House of the Hungarian Academy of Sciences
Budapest, Hungary

Printed in Hungary

PREFACE

High-frequency application of semiconductor devices is so varied and complex that it is impossible to discuss all the applications in a single volume.

This particular field includes (in addition to circuit design in a limited sense) semiconductor theory, filter theory, and, among others, the theory of nonlinear circuits; all need to be dealt with thoroughly if detailed questions are to be answered adequately. Since this was not possible within the given framework, a practical method of treatment had to be used: one that emphasises the most important (in the author's view, indispensable) relationships of the field. Thus from semiconductor theory just a few of the more well known equations have been used.

Another interesting feature of the theme is its rapidly changing character. Almost every day new high-frequency semiconductor devices are created, and these in turn permit new solutions in circuits. For this reason, an effort has been made to include in the book the most up-to-date devices and applications.

The level of treatment chosen is intended to be acceptable to electrical engineers engaged in design of high-frequency circuits. Thus the analysis of circuits is presented in simple fashion and the more sophisticated methods are only briefly mentioned. At the same time, a series of real circuit applications are discussed.

I take this opportunity to express my thanks to Professor I. P. Valkó for his invaluable and patient assistance.

F. Kovács

CONTENTS

1. High-frequency applications of semiconductor diodes

1.1. Point-contact and junction diodes 11
1.2. PIN diodes 14
1.3. Hot-carrier diodes 20
1.4. Tunnel diodes 22
1.5. Varactor diodes 25
1.6. Microwave diodes 30

2. High-frequency properties and equivalent circuits of transistors

2.1. High-frequency properties of bipolar transistors 35
2.2. Parameters determining the transition frequency f_T 37
2.3. High-power, high-frequency bipolar transistors 39
2.4. Current gain and equivalent circuit of bipolar transistors 45
2.5. High-frequency operation of JFET and MESFET 49
2.6. High-frequency operation of MOS field-effect transistors 51
2.7. Large-signal transistor equivalent circuits 55

3. High-frequency properties of linear integrated circuits

3.1. Structure of monolithic integrated circuits 58
3.2. High-frequency properties of monolithic integrated circuits ... 61
3.3. Integrated-circuit delay lines 64

4. Characterization of solid-state devices by four-pole parameters

4.1. Four-pole equivalent circuit; four-pole parameters 68
4.2. Properties of terminated four-poles 72
4.3. Stability of active four-poles 73
4.4. Power transfer of unconditionally stable four-poles 75
4.5. Power transfer of potentially unstable four-poles 76
4.6. Connection of four-poles 77
4.7. Types of high-frequency four-pole parameters 80

5. Properties of wideband amplifiers

5.1. Analysis by the pole-zero method 85
5.2. Transfer functions and frequency properties 86
5.3. Transfer function with first-degree denominator 88
5.4. Transfer functions made up of n identical first-degree denominators 89

5.5. Transfer functions made up of n first-degree denominators having different cut-off frequencies 89
5.6. Transfer function with second-degree denominator 91
5.7. Practical expressions for transfer functions of higher degree 92

6. Frequency response of transistor configurations

6.1. The cut-off frequency of the common-emitter configuration 94
6.2. Input and output impedances of the common-emitter configuration 98
6.3. Frequency dependence of the common-base and common-collector configurations 99
6.4. Field-effect transistor configurations 102

7. Frequency response of multi-stage wideband amplifiers

7.1. Frequency response of a two-stage common-emitter amplifier 106
7.2. Frequency response of a common-emitter–common-base cascade connection 107
7.3. Frequency response of a common-collector–common-emitter cascade connection 108
7.4. Frequency response of a common-emitter three-stage configuration 110
7.5. Monolithic integrated multi-stage wideband amplifiers 111

8. Compensated wideband amplifiers

8.1. Common-emitter configuration with output parallel compensation 114
8.2. Common-emitter configuration with input parallel compensation 117
8.3. Common-emitter configuration with input series compensation 118
8.4. Common-emitter configuration with combined compensation 118
8.5. Stages with frequency-independent input impedances 121

9. Single-stage feedback amplifiers

9.1. General characteristics of feedback amplifiers 123
9.2. Stability of feedback amplifiers 125
9.3. Amplifier with emitter feedback 127
9.4. Emitter-feedback amplifier with compensation 129
9.5. Single-stage amplifier with current feedback 132
9.6. Single-stage compensated amplifier with current feedback 133

10. Multi-stage wideband amplifiers with feedback

10.1. Two-stage amplifier with current feedback 135
10.2. Multi-stage feedback amplifiers 139
10.3. Cascade connection of single-stage feedback amplifiers 141
10.4. Two-stage amplifier with stage-by-stage feedback 143
10.5. Multi-stage amplifiers with several feedback loops 144
10.6. Input impedance of feedback amplifiers 150

11. Wideband applications of feedback integrated amplifiers

11.1. Types of integrated amplifiers 151
11.2. Compensation of monolithic integrated amplifiers 152
11.3. Wideband monolithic amplifiers 156
11.4. Hybrid integrated wideband amplifiers 160
11.5. Output power frequency dependence 162

12. Wideband amplifiers with transmission-line coupling

- 12.1. Properties of transmission-line transformers 164
- 12.2. Amplifiers with transmission-line transformer coupling 168
- 12.3. Push-pull amplifiers with transmission-line transformers 169
- 12.4. Effect of mismatch on transmission flatness 172

13. Distributed amplifiers

- 13.1. Principle of operation 175
- 13.2. Distributed amplifiers using transistors 177

14. General design considerations of tuned amplifiers

- 14.1. Building blocks of tuned amplifiers 179
- 14.2. Design by the pole-zero method 180
- 14.3. Stability 182
- 14.4. Neutralization 184
- 14.5. Design based on the loop-gain limit. Linvill design 187
- 14.6. Stability calculation by scattering parameters 189

15. Single-tuned amplifiers

- 15.1. Power gain 193
- 15.2. Stability 196
- 15.3. Design for maximum power gain in the case of ideal neutralization . 200
- 15.4. Design for given stability factor 201
- 15.5. Frequency-response distortion due to internal feedback 202
- 15.6. Stagger tuning 207

16. Double-tuned amplifiers

- 16.1. Voltage transfer 209
- 16.2. Power gain at resonance 211
- 16.3. Stability 213
- 16.4. Design procedure with ideal neutralization 216

17. Integrated-circuit tuned amplifiers

- 17.1. General considerations 220
- 17.2. Application of monolithic differential amplifiers 222
- 17.3. Application of high-complexity monolithic circuits 226
- 17.4. Application of hybrid integrated circuits 235

18. UHF and microwave amplifiers

- 18.1. Wideband amplifiers with reactive filters 238
- 18.2. Wideband amplifiers with transformers 243
- 18.3. Narrow-band amplifiers with transmission lines 247
- 18.4. Wideband amplifiers with transmission lines 252
- 18.5. Wideband amplifiers with directional couplers 258

19. High-frequency power amplifiers. Theory and design principles

- 19.1. Factors limiting high-level operation 262
- 19.2. Instability of power amplifiers 266
- 19.3. Classification of power amplifiers. Design of class A amplifiers .. 268

19.4. Design of class B and class AB amplifiers 273
19.5. Design of class C amplifiers 279
19.6. Push-pull amplifiers 282
19.7. Nonlinear equivalent circuit of power amplifiers 285
19.8. Approximate relations of power amplifiers 290

20. High-frequency power amplifiers. Practical design

20.1. Power-amplifier matching four-poles 293
20.2. High-frequency power amplifiers with hybrid coupling 299
20.3. Biasing of power amplifiers 302
20.4. High-frequency power amplifier circuits 308

21. High-frequency converters

21.1. Theoretical fundamentals of conversion 313
21.2. Determination of conversion parameters 318
21.3. High-frequency mixer circuits 323

22. Frequency multipliers

22.1. Basics of operation 329
22.2. Frequency-multiplier circuits 334

23. High-frequency oscillators

23.1. Operation and design methods 342
23.2. Circuit principles 343
23.3. Amplitude and frequency stability 349
23.4. Circuit implementation 352

24. High-frequency noise of solid-state devices

24.1. The concept of noise figure and noise four-pole 363
24.2. Noise figure of solid-state devices 366
24.3. Design of input stages 370

References 373

Subject index 389

1. HIGH-FREQUENCY APPLICATIONS OF SEMICONDUCTOR DIODES

1.1. Point-contact and junction diodes

There are two basic types of semiconductor diodes, the point-contact diode and the junction diode. From these basic types, several versions of modern semiconductor diodes have been developed to give favourable high-frequency properties by employing special structures, and to reduce the high-frequency limiting factors which are present in conventional diode types. For this reason, high-frequency properties will be investigated first on conventional diode types.

The two conventional diode types, point-contact and junction, have much in common, including high-frequency limiting factors which only differ in

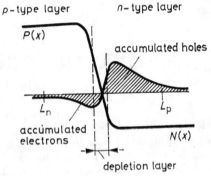

Fig. 1.1. Layers of the junction diode and the charge distributions

numerical values. It should be stated that the theory of operation of point-contact diodes is not quite resolved yet. However, it is well established that similar phenomena occur at the tip of the needle contacting the semiconductor crystal and at the junction of two semiconductor layers having opposite dopings. The behaviour of the junction diode will be surveyed in the following, but our investigations will also apply — within limits — to point-contact diodes [1.18].

The layers of the junction diode and the charge distribution within these layers are shown schematically in Fig. 1.1. The diode effect (rectification property) is produced at the junction of the two sides having opposite impurities. The electrons flowing from the n-type layer into the p-type layer accumulate in the latter near the junction, and a similar phenomenon occurs with the holes i.e. defects of electrons flowing into the n-type layer in the opposite direction. Charge accumulation takes place within regions L_n and L_p, which are the diffusion lengths pertaining to the electrons and holes. Between the two accumulated charge carriers of opposite polarities is the depletion layer, in which there are practically no charge carriers.

The magnitude of the accumulated charges in the two layers depends on the diode current. In order to increase the diode current, the charge already present must be increased by excess charge, i.e. the two diode layers "have to be charged".

The structure thus behaves like a capacitance, and the so-called diffusion capacitance which is formed is given by

$$C_d = \text{constant} \cdot I_d, \qquad (1.1.1)$$

i.e. it is proportional to the instantaneous diode current. This diffusion capacitance will limit the speed of diode operation.

The diffusion capacitance is primarily effective in the conducting state of the diode. In the non-conducting state, the junction diode is also capacitive. The width of the depletion layer depends on the magnitude of the reverse diode voltage, as does the overall quantity of charges within the layer. This change in charges represents another capacitance which has a value of

$$C_T = \frac{C_{T0}}{(1 - V/V_0)^{1/n}}, \qquad (1.1.2)$$

where C_{T0} is the transition capacitance at the voltage $V = 0$, and V_0 and n are constants characterizing the pn junction. This capacitance becomes significant in the reverse-bias state, while in the forward-bias state, the diffusion capacitance according to (1.1.1) dominates.

As a result of the charge-storage effect of the pn junction, the static diode characteristic is no longer valid above a given frequency. Below this frequency, ideal diode operation may be assumed, implying an open-circuit in the reverse direction and a short-circuit in the forward direction. The diode current with sinusoidal diode voltage is shown in Fig. 1.2. Figure (a) shows the current for an ideal diode, and Fig. 1.2(b) the current for a real diode with stored charge. Owing to the charge stored during the forward-biased period (positive half-wave), the diode can not assume the reverse-voltage state since the reverse voltage (negative half-wave) must first remove the stored charge. Accordingly, the clipping of the negative half-wave will not be perfect, and the remainder of the negative signal at the beginning of the negative half-cycle will be proportional to the signal frequency. This effect is characterized by the so-called recovery time, which is the time needed by the reverse-direction discharge current to decrease to a given value, when the diode switches over from forward-biased to reverse-biased state.

One of the most important consequences of the pn-junction frequency dependence is the decrease in rectification efficiency. Another consequence is the difficulty of wideband detection with frequency-dependent diodes, though wideband detection is a frequent requirement in instrumentation.

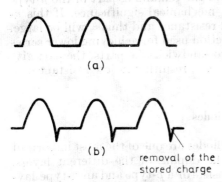

Fig. 1.2. Rectified signal in the case of (a) an ideal diode and (b) a real diode with stored charge

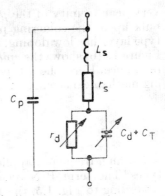

Fig. 1.3. High-frequency equivalent circuit of semiconductor diode

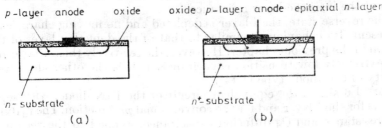

Fig. 1.4. Cross-section of (a) planar diode and (b) an epitaxial planar diode

For rectification of low currents, point-contact diodes are more suitable, especially tungsten diodes which have lower capacitance compared with gold-contact diodes. In the microwave region, special point-contact diodes placed in cartridges are used for easy insertion into the transmission line.

In addition to the frequency-limiting factors arising from the properties of the *pn* junction, high-frequency operation is determined by further parameters. These are shown in Fig. 1.3, which gives equivalent circuit of the complete semiconductor diode. The *pn* junction is characterized by the parallel RC network, where r_d is the dynamic resistance of the diode, and the capacitance is the sum of diffusion and transition capacitances. Further parts of the crystal block and connecting leads are accounted for by series resistance r_s, and lead inductance is denoted by L_s. The case containing the crystal has a parallel capacitance C_p. As seen from Fig. 1.3, parasitic elements are practically independent of the working point, with the exception of the parallel RC network.

The significance of series resistance r_s is demonstrated by Fig. 1.4(a) which shows a *pn* diode of planar construction. Rectification takes place only in a very near vicinity of the *pn* junction, and the remaining part of the *n*-type bulk up to the soldering plate has only mechanical significance. If this *n*-type layer has low doping, then the bulk resistance and thus r_s will be large. Figure 1.4(b) shows the *epitaxial* construction used for achieving lower series resistance. Here, the *n*-type layer has two sandwich-like parts; the part giving mechanical strength is highly doped (n^+), resulting in low resistance.

1.2. PIN diodes

For high-frequency applications, PIN diodes are one of the most important types. The term PIN diode refers to the doping of the different layers. According to Fig. 1.5, the PIN diode consists of a *p*-type and an *n*-type layer, and between these two an *i* (intrinsic) layer, with very low doping. In the conducting state, charge is stored in the intrinsic layer. This has the effect of decreasing the resistance of this layer which is otherwise high. This is also called conductance modulation: the presence of the stored charge allows a "modulation" by the forward current of the diode resistance, by means of the intrinsic-layer resistance.

In the reverse state, the *i*-layer is depleted and no moving charge carriers are present. Its behaviour is similar to that of the depletion layer of the *pn* junction in the previous chapter. However, the relatively thick *i*-layer results in an extremely low transition capacitance. This is the other characteristic property of this diode type [1.14].

Figure 1.6 shows the equivalent circuit of the PIN diode, with separate elements for the *i*-layer and for the conventional *pn* junction. The r_d dynamic diode resistance and C_d diffusion capacitance, applied to the *pn* junction, denote the same quantities as before. Elements r_i and C_i, in series with the former, characterize the resistance and capacitance of the *i*-layer.

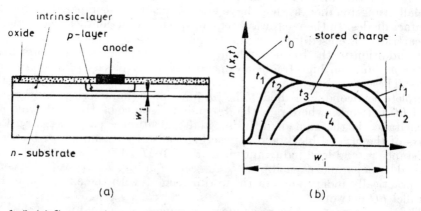

Fig. 1. 5. (a) Cross-section of a PIN diode, (b) time dependence of charge-carrier concentration in the intrinsic region after switching into the reverse direction

1.2. PIN diodes

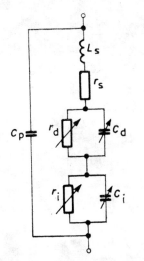 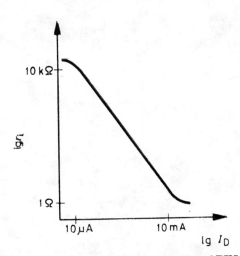

Fig. 1.6. High-frequency equivalent circuit of PIN diode

Fig. 1.7. The current dependence of PIN diode resistance

Resistance r_i depends on the forward current as a consequence of the conductance modulation, and with good approximation,

$$r_i = K I_D^n,\qquad(1.2.1)$$

where K and n are constants. As shown in Fig. 1.7, this represents a straight-line section over a wide current range using logarithmic scales. The resistance at high currents is extremely low (nearly short-circuit), and at low currents extremely high (nearly open-circuit) [1.21], [1.22].

In the reverse-bias state, the capacitance C_i in parallel with the high resistance becomes significant. This has a value of about 0.1 pF, so the high-frequency impedance in the reverse-bias state is nearly infinite.

The PIN diode may thus be considered as a hardly ideal electronic switch having a resistance which is nearly short-circuit at high forward currents and nearly open-circuit at reverse bias, even at high frequencies. The situa-

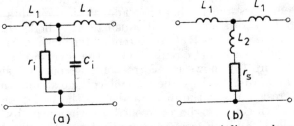

Fig. 1.8. Equivalent circuit of a PIN diode placed in stripline package in (a) reverse-bias and (b) in forward-bias state

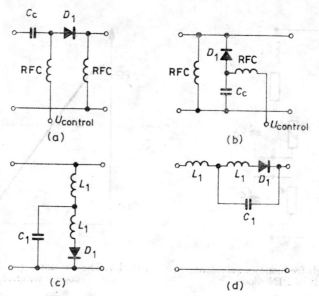

Fig. 1.9. PIN diode switching circuits: (a) series diode, (b) parallel diode, (c) parallel diode tuned to resonance, (d) series diode tuned to resonance

tion is impaired by the series inductance L_s and series resistance r_s, and further by the parallel capacitance C_p arising from the diode package. The package is designed to minimize these parameters.

Figure 1.8 shows the equivalent circuit of a diode in stripline package. Figure (a) in the reverse-bias state and (b) is valid in the forward-bias state. The diode is placed parallel in a short stripline section which is a transmission line made up of two parallel striplines also serving as output leads. With correct choice of stripline characteristic impedance, the capacitance C_i in the reverse-bias state and the lead inductance L_2 in the conducting state can be made parts of the transmission line. As the impedance of the stripline section is determined by resistances r_i and r_s, it appears as nearly ideal open circuit and short circuit, respectively, without reactive components.

The PIN diode can be applied advantageously in high-frequency switches, adjustable attenuators and modulators. Four typical circuits are shown in Fig. 1.9. Figure (a) shows a series switch using series diode D_1. Capacitance C_c serves for coupling, chokes RFC are intended for biasing. At high forward currents, the diode is a short-circuit, and in the reverse-bias state it is an open-circuit. Figure (b) shows the corresponding parallel switch, where the diode D_1, forward-biased by the control voltage shunts the output. In both cases the resistance of the diode and thus the series or shunt resistance of the switch can also be controlled continuously by the forward current of the diode. In high-frequency applications, the parallel circuit of Fig. (b) is more frequently used. If the circuit is inserted into a transmis-

1.2. PIN diodes

sion line of characteristic impedance Z_0, then the insertion loss will be given by the following expression:

$$a_1 = 10 \lg \left[(1 + GZ_0/2)^2 + (BZ_0/2)^2\right], \qquad (1.2.2)$$

where G and B are real and imaginary parts of the diode admittance in the reverse-bias state. For applications in a narrow frequency range, diode reactance B may be compensated by a parallel inductance.

Reverse attenuation is given by the following expression:

$$a_r = 10 \lg \left\{\left[1 + \frac{RZ_0}{2(R^2 + X^2)}\right]^2 + \left[\frac{XZ_0}{2(R^2 + X^2)}\right]^2\right\}, \qquad (1.2.3)$$

where $Z = R + jX$ is the series impedance of the forward-biased diode. Series reactance X may also be compensated in a narrow frequency range.

Figures (c) and (d) show resonant switches in parallel and series configurations, respectively. Depending on the diode impedance, either a series or a parallel resonant circuit appears between the connection points. The impedance depends on the Q factor of the resonant circuit, and this, in turn, depends on how closely the diode impedance approximates the short-circuit and open-circuit condition.

Circuits according to Figs. 1.9 (a) and 1.9 (b) have the drawback of producing reflections in the transmission line of characteristic impedance Z_0 as a result of mismatches. Figure 1.10 shows adjustable attenuators having

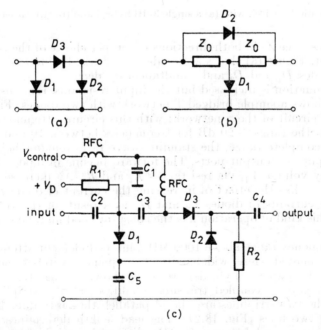

Fig. 1.10. PIN diode attenuator with (a) π network, (b) bridged-T network, (c) π network and DC control circuit

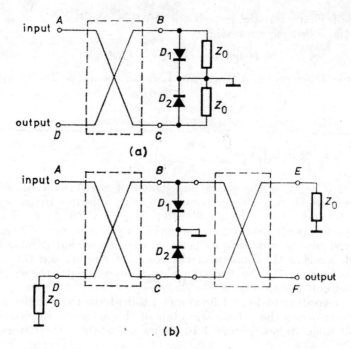

Fig. 1.11. PIN diode switch with (a) a single 3-dB hybrid and (b) two cascaded hybrids

constant resistance Z_0 in both directions by proper choice of the control forward current. Figure (a) shows a 3-diode π network: by increasing the currents of diodes D_1 and D_2 and simultaneously decreasing the current of D_3, the attenuation is increased but the input resistance remains constant. Figure (b) shows a simple bridged-T network with two diodes. Figure (c) is the detailed circuit of the π network; with this circuit, attenuation may be controlled in the range 1–20 dB for frequencies between 10 and 100 MHz. Within the complete range, the standing-wave ratio remains below two at both the input and output ports. The forward-biasing of diodes D_1 and D_2 is assured by voltage V_D via resistances R_1 and R_2. An increase in control voltage $V_{control}$ has the effect of increasing the current of diode D_3 and decreasing the currents of diodes D_1 and D_2, as a result of the voltage drop across R_2. This kind of operation has the constant good matching mentioned earlier.

In the frequency range above 1000 MHz, lumped-element attenuators can not be implemented. Here, wideband high-frequency switches and attenuators with 3-dB couplers (hybrids) are used, shown in Fig. 1.11 (a): this is essentially a pair of coupled transmission lines, usually striplines. Long sections of the two striplines are placed parallel, thus introducing coupling between the two lines (Fig. 18.27). The lead length determines the lower cut-off frequency, and below 100 MHz difficulties result from the large lengths needed.

If the points B and C are matched (i.e. the resistances of the diodes are infinite), the signal power P_A given at point A will be evenly divided between points B and C, with a phase shift $90°$ between the two signals. No signal reaches point D, so the attenuation towards the output port is theoretically infinite.

If the resistances of diodes D_1 and D_2 are zero, then power is reflected from points B and C, and because of the phase relations, all input power reaches output port D.

In intermediate cases, the output power depends on the resistances of diodes D_1 and D_2, i.e. the forward currents. In practice, an attenuation range of 1–20 dB may be obtained. If the two diodes have equal resistances, the standing-wave ratio at input port A and output port D will be unity. Thus diode symmetry is important. A drawback is the rippled frequency response of the attenuation which is not generally tolerable.

The circuit shown in Fig. 1.11 (b) has better performance since two 3-dB couplers are applied. Operation is now described briefly.

If the two diode resistances are infinite, powers reaching points B and C are added at point F, thus the whole input power theoretically reaches output port F, and there are no signals at points E and D. If the two diode resistances are zero, then the signal is reflected from points B and C, and is totally dissipated at point D. Output power is then theoretically zero. In intermediate cases, output power depends on the diode resistances and thus may be controlled by the DC bias of the diodes. The attenuation range is equal to that in the previous case, but ripple and standing-wave ratio are substantially better.

The switch shown in Fig. 1.11 (b) may be used as a duplexer if the antenna is connected to point A, the transmitter to point D and the receiver to point F. In the case of infinite diode resistances, the antenna signal reaches the receiver, and the transmitter signal is dissipated in the termination at point E. In the case of zero diode resistances, the transmitter signal is passed to the antenna.

In Ref. [1.30], a gain-control circuit applying series PIN diodes at a receiver input is considered. Wideband microwave switches are investigated in Refs. [1.38] and [1.32]. Further switching and control circuits are presented in Refs. [1.23], [1.24], [1.33] and [20.17], and a limiter is investigated in Ref. [1.25]. Switching-time problems are analyzed in Ref. [1.39].

High-power PIN diodes have made the switching of extremely high powers [1.34] possible. Reference [1.20] presents a special configuration for switching 100 kW of transmitter power in the frequency range 6–40 MHz by applying water-cooled PIN diodes. Open-switch attenuation is 10.1 dB, and closed switch attenuation is 70 dB. Reference [1.43] reports on a 1.75 MW, 224 MHz switch.

Harmonics have special significance when switching high powers. Intermodulation distortions for different circuit configurations and signal levels are treated extensively in the literature [1.8], [1.12]. An important application field of PIN diodes is phased-array radar, where diodes are used as phase shifters.

1.3. Hot-carrier diodes

The hot-carrier diode or Schottky diode is intermediate between point-contact diodes and junctions diodes, but has favourable high-frequency properties. It has been made possible by the process for fabricating metal–semiconductor junctions. As early as 1942, the well known physicist, W. Schottky mentioned the rectification properties of metal–semiconductor junctions in his paper, but many years passed before the practical implementation of this effect. This is due to the fact that the slightest contamination at the metal–semiconductor interface or distortion in the crystal-lattice results in a decreased or lack of rectification effect, so junction fabrication required highly developed technology.

In hot-carrier diodes, there are no minority charge carriers in contrast with *pn* diodes, and thus no charge storage takes place, which is an advantage. In connection with Fig. 1.2, the recovery time due to charge storage and its effect on the switching properties of diodes were mentioned. The charge-storage time of hot-carrier diodes is practically negligible and can therefore be used advantageously for clipping half periods of sinusoidal signals; the waveform will approximate the ideal waveform shown in Fig. 1.2 (a). Apart from this advantage, further features of the hot-carrier diode are the low series resistance, low noise figure, low threshold voltage, and ability to handle high AC powers.

The operation of the hot-carrier diode may be explained using the energy band theory of metals and semiconductors. Without going into detail, the most important aspects of operation are summed up as follows. Similarly to the *pn* junction, a potential barrier is present at the metal–semiconductor junction which can be attributed to the difference between energy levels in the two materials. Applying a negative voltage to the *n*-type semiconductor, this external voltage decreases the potential barrier producing a flow of current. This forward current means that electrons are drawn from the *n*-type semiconductor into the metal. These electrons are injected into the metal with high energy, and are, therefore, called "hot" electrons (this explains the name of the diode). In contrast with the *pn* junction, there is no flow of holes as these are not present in the metal. Thus minority carriers are not stored in the *n*-type layer, resulting in a much lower recovery time. The DC diode characteristic is exponential, as for the *pn* junction. It should be noted, however, that this diode type can very easily burn out. The equivalent circuit of the hot-carrier diode is shown in Fig. 1.12, where R_j and C_j are the bias-dependent elements of the metal–semiconductor junction, L_s and r_s are the series, and C_p is the parallel parasitic element arising mainly from the package. It is customary to define a cut-off frequency as given by

$$\omega_H = \frac{\sqrt{1 + r_s/R_j}}{C_j \sqrt{r_s/R_j}}, \qquad (1.3.1)$$

(see Ref. [1.5]). As the capacity of the forward-biased diode is very low ($C_j < 1$ pF), the cut-off frequency is extremely high.

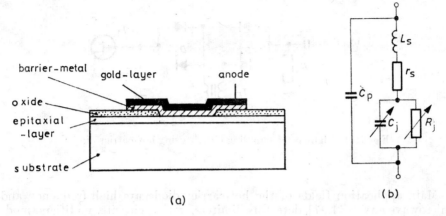

Fig. 1.12. (a) Cross-section and (b) high-frequency equivalent circuit of hot-carrier diode

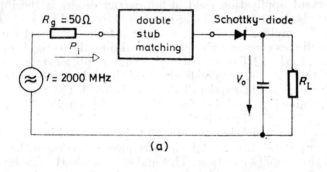

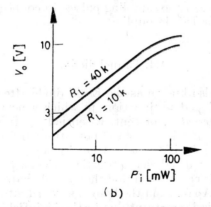

Fig. 1.13. (a) Hot-carrier diode rectifier circuit and (b) output voltage as a function of input power

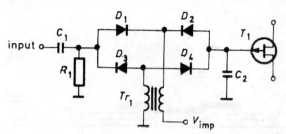

Fig. 1.14. High-speed sampling circuit using hot-carrier diodes

Main application fields of the hot-carrier diode are high-frequency and microwave mixers [1.27], detectors, limiters, gating circuits, rectifiers, modulators [1.30] and switches [1.22]. The application for low-level detection is justified by favourable noise properties. High-level detection is illustrated in Fig. 1.13. Figure (a) shows a circuit of a rectifier, operating at 2000 MHz, matched with a double stub, and having a high efficiency as shown in (b).

An important application field of hot-carrier diodes is the input gate of sampling heads, built from a diode-quad as shown in Fig. 1.14. This circuit operates by "sampling" the input waveform for a very short time interval: opening the diodes connects the input signal directly to the input amplifier (in the figure, to the FET representing the first amplifier stage). As the opening time of the gate is extremely short, of the order of 100 ps, only high-speed diodes such as hot-carrier diodes may be used. The opening is initiated by a needle-pulse V_{imp} arriving from the transformer Tr_1. The input signal charges capacitor C_2 through the turned-on diodes and feeds a special feedback amplifier which has the effect of signal stretching. By correct timing of the sampling instants, a true low-frequency version of the input signal appears at the amplifier output. This makes possible the display and measurement of fast pulses and high-frequency signals. A special circuit is re, sponsible for the timing of the sampling pulses. In connection with Fig. 1.14- capacitor C_1 serves for DC decoupling.

1.4. Tunnel diodes

The tunnel diode, also known as the Esaki diode after its inventor [1.1], consists of a highly doped *pn* junction, in which a new kind of additional current, the *tunnel current*, is present. Essentially, a fraction of the charge carriers are capable of getting over the potential barrier, as a result of the high degree of doping. This tunnel current has a characteristic voltage dependence and alters the usual characteristic of the *pn* junction significantly as shown in Fig. 1.15. It can be seen that the overall characteristic is made up of three components. Apart from the known forward diffusion current, there is significant field-emission current due to the high field strength, resulting from the high level of doping in the transition layer, similarly to the Zener

1.4. Tunnel diodes

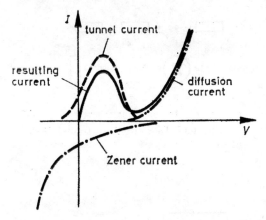

Fig. 1.15. Current components and overall characteristic of tunnel diodes

diodes. The third component is the tunnel current which has a maximum, i.e. the tunnel current decreases with increased forward biasing. These three components result in an overall characteristic which has a falling, i.e. negative resistance part. It is known that all characteristics with negative resistance parts are suitable for amplification, and this is utilized in tunnel diodes. Since both the Zener current and the tunnel current are generated much faster than the diffusion current, the tunnel diodes have faster operation than conventional *pn* junction diodes (in the low forward-voltage region). Locations and values of the current maximums and minimums are important parameters of the tunnel-diode DC characteristic.

A small-signal equivalent circuit of tunnel diodes is shown in Fig. 1.16. As in previous diode types, the junction is characterized by a parallel *RC* network where the resistance has a negative value at the falling part of the

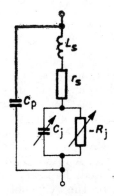

Fig. 1.16. High-frequency equivalent circuit of tunnel diode

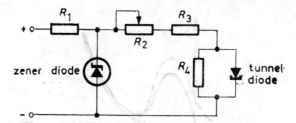

Fig. 1.17. Bias circuit of tunnel diode

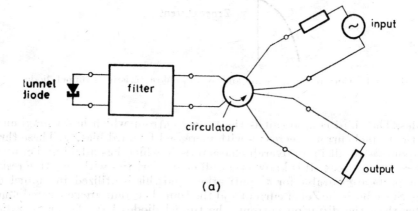

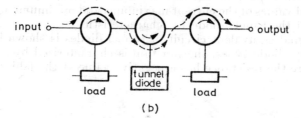

Fig. 1.18. (a) Tunnel diode amplifier with a single circulator and (b) with three circulators and low reflection

characteristic. The series resistance r_s is given by the bulk, and L_s/C_p are the series/parallel parasitic elements given by the case.

In addition to applications in fast-pulse circuits, tunnel diodes are applied widely in the high-frequency range, primarily as microwave amplifiers and oscillators. In tunnel-diode amplifiers, the diode working point should be maintained within the falling range of the characteristic, requiring a DC source of low internal resistance. A bias circuit is shown in Fig. 1.17, where a low-valued resistance R_4 is used. Typical working-point voltage is within the range 110–130 mV, showing the extremely limited high-level operating range. In most cases, the working point is determined by optimum noise

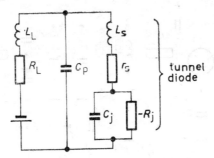

Fig. 1.19. Simplified circuit of tunnel-diode oscillator

figure and available gain. As the former requires a higher, and the latter a lower DC voltage, a compromise is chosen. A typical amplifier circuit is shown in Fig. 1.18 (a). Input and output ports are separated by a circulator in which the signal only propagates in the direction shown by the arrow. The generator signal is passed through the filter, intended for matching and tuning, to the tunnel diode. Reflected from the negative resistance of the tunnel diode, the signal propagates further in the direction of the arrow and is passed to the output termination. Input and output reflections may be substantially reduced by the circuit shown in Fig. 1.18 (b). This uses three circulators, and signal path is denoted by a dashed line. The resistors serve for dissipation of the reflected signals. With this solution, high quality, relatively wideband amplifiers having flat responses are built in the GHz frequency range [1.11].

A high-frequency tunnel-diode oscillator is shown in Fig. 1.19. The condition for oscillation is given by zero dissipated power in the resistive components, i.e. the positive and negative resistances should cancel each other. As the negative resistance of the tunnel diode is dependent on the working point, this condition also determines the oscillation amplitude. The actual oscillation amplitude allows calculation of a so-called average negative diode resistance which has to be compensated.

1.5. Varactor diodes

The capacity of the *pn* junction depends on the reverse voltage. This property is made use of in the case of *varicap diodes* and *varactor diodes*. These have special *pn* junctions with extremely high capacity change as a function of reverse voltage. A typical application is a tuning diode for automatic tuning correction, but varicap diodes are equally widely used for frequency multiplication and parametric amplification. The most important parameters of both diode types are the capacity/voltage curve and the ratio of maximum to minimum capacity.

According to eqn. (1.1.2), the junction capacity is proportional to the square-root reverse voltage for abrupt impurity profile, and to the cube-

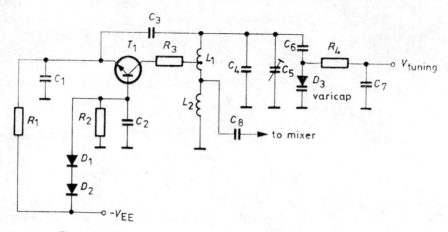

Fig. 1.20. High-frequency oscillator with varicap diode tuning

root reverse voltage for linear impurity profile. For other impurity profiles of more complicated form, similar voltage dependences of limited slopes prevail. For a high capacity change, wide voltage swings are necessary, limited by the forward voltage and by the reverse breakdown voltage. The impurity profile of the junction has thus to be chosen to provide optimum capacity voltage dependence and maximum breakdown voltage.

Another problem is presented by the diode series resistance which is a series loss resistance and impairs the quality factor. A low series resistance is provided by using a highly doped and thus low resistivity epitaxial layer.

Varicap diodes are operated at low signal levels in tuning applications, and the voltage which appears is always reverse. At low reverse voltages, especially with commensurable signal voltages, harmonics are produced which impair reception. In this case, a *back-to-back diode pair* is applied, which results in opposite capacity changes for the two diodes at high input levels, which cancel each other.

Figure 1.20 shows a typical oscillator with varicap diode tuning. The diode acts as a variable parallel capacity in the resonant circuit of the oscillator.

The main application field of *varactor diodes* is frequency multiplication, primarily at high levels. Varactor diodes may be classified into three types (Fig. 1.21):

(a) The *tuning varactor* (C-swing varactor) has a capacity which is highly dependent on reverse voltage and is used for frequency multiplication.

(b) The *storage-charge varactor* (snap-off varactor, step-recovery diode) utilizes the fast current change occurring at cut-off. These diodes have capacities which are less dependent on the reverse voltage.

(c) In *dual-mode varactors*, both of the above effects are used.

The cross-section of high-power varactor with *mesa* construction is shown in Fig. 1.22 (a). The static Q factor is expressed from the equivalent circuit

1.5. Varactor diodes

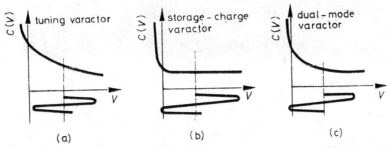

Fig. 1.21. Varactor diode characteristics: (a) tuning varactor, (b) storage-charge varactor, (c) dual-mode varactor

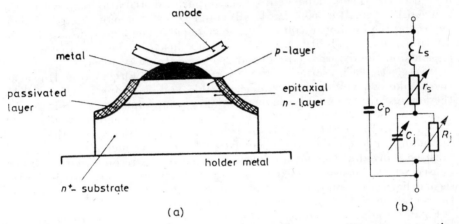

Fig. 1.22. (a) Cross-section and (b) equivalent circuit of varactor with mesa construction

of Fig. 1.22 (b) as

$$Q_{\text{stat}} = 1/\omega C_j r_s, \tag{1.5.1}$$

where ω is the operating frequency, r_s is the bulk resistance of the crystal and C_j is the diode capacity at a given reverse voltage. Other elements of the equivalent circuit are known from the previous diode types, and the effect of parallel resistance R_j may be neglected in the reverse direction.

The static Q factor has the drawback of expressing the properties of a single working point alone. Dynamic properties are expressed by the dynamic Q factor given by

$$Q_{\text{dyn}} = S_1/\omega r_s, \tag{1.5.2}$$

where r_s is again the bulk resistance, ω is the operating frequency, and S_1 is the first Fourier component amplitude of the time-dependent capacity reciprocal value. Applying a sine-wave signal to the diode, the diode capacity-time function may be expanded into a Fourier series:

$$C_j(t) = \sum_{n=-\infty}^{\infty} C_n e^{jn\omega t}. \qquad (1.5.3)$$

Eliminating the r_s series resistance by using eqn. (1.5.1) from the dynamic Q factor we have the usual expression:

$$Q_{dyn} = \Gamma Q_{stat} \frac{C_{j0}}{C_j}, \qquad (1.5.4)$$

where Q_{stat} is the static Q factor at zero voltage, C_{j0} is the junction capacity at zero voltage, C_j is the junction capacity at the working bias, and Γ is a Fourier coefficient dependent on the impurity distribution and within the range 0.17–0.25. It is seen that the dynamic Q factor is more suited to characterizing the nonlinear voltage capacity dependence of the device, i.e. the multiplication capability, but is not an exclusive device parameter as it depends on the external circuit and the generator waveform.

It is seen that a low series resistance is required for high Q factors. However, the breakdown voltage and thus the usable voltage swing is decreased by the low-resistivity epitaxial layer, forcing a compromise.

A frequency tripler circuit using tuning varactor is shown in Fig. 1.23. According to Fig. 1.21 (a) it works only in reverse direction, and so the unused harmonics of low order are produced with a relative high amplitude, these must be shorted at the output for good efficiency. A resonant circuit (idler) is used which consists of elements $L_2 C_3$; it provides a short circuit for the second harmonic component. Elements L_1, C_1, C_2 and L_3, C_4, C_5 are used for the matching to the input and output respectively.

Finally, the *storage-charge varactor* or *step-recovery diode (snap-back diode)* should be mentioned [1.3], [1.16]. The storage-charge property of the pn junction has been investigated before; in contrast with those above, the charge is stored for a long time in these diodes, i.e. changing the voltage from forward to reverse direction, a relatively long time has to elapse for complete cut-off. The time dependence of the diode current is shown in Fig. 1.24 (a). The storage time t_s in the reverse-biased period is extremely long, but after this, the transition time t_t required for voltage fly-back is extremely short. This steep voltage change is suitable for harmonic generation, so the storage-charge diode is often applied for multiplication and harmonic generation, as for instance in comb generators producing a series of harmonics [1.9]; this application will be discussed in detail in Section 22.2.

The storage of charges in this diode may be explained by a special kind of impurity distribution. There are two requirements: first, the recombination of the charge carriers must be slow, and second, they should neither diffuse nor accumulate far from the junction, thus implementing a fast fly-back (i.e. a fast depletion of the charge carriers). In order to meet the second requirement, a steeply rising impurity profile is obtained in the n-type material. This acts as a drift region, forcing the holes coming from the p-type material to fly-back near the junction. The charge, thus concentrated into

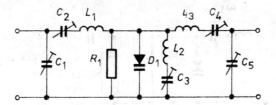

Fig. 1.23. Frequency tripler circuit using tuning varactor

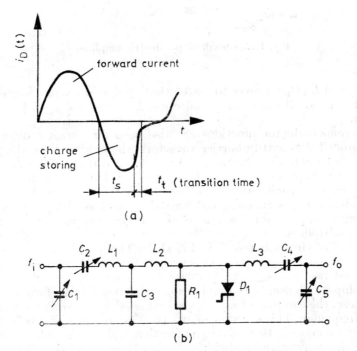

Fig. 1.24. Storage-charge varactor (a) current–time dependence, (b) application of varactor in a frequency-multiplier circuit

the immediate vicinity of the junction, disappears in an extremely short time (about 100 ps) after the discharge.

A frequency-multiplying circuit is shown in Fig. 1.24 (b) where the elements L_2 and C_3 are used for utilizing the extremely fast fly-back of the diode current. The voltage appeared on the inductance L_2 may be expressed as $v_2 = L_2 di/dt$. The value of resistance is $R_1 \approx 5t_s/N^2 C_{j0}$, where N is the multiplication factor and C_{j0} gives the junction capacity at zero bias. Since the circuit produces also high-order harmonics with relative high amplitude, idlers are not required to select the low-order harmonics. The elements

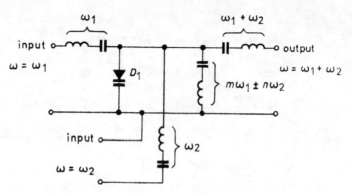

Fig. 1.25. Circuit of parametric amplifier

L_1, C_1, C_2 and L_3, C_4, C_5 serve to match the input and output ports, respectively. For good efficiency the conditions $f_i > 10 t_s$ and $f_0 \leq 1/t_t$ must be satisfied.

Metal–semiconductor junctions are also used for varactor diodes; these are the so-called Schottky-barrier varactors which differ from the previous diodes in both capacity change and breakdown voltage. The junction may be taken as abrupt, resulting in a high capacity change, but the lack of forward diffusion capacity results in a somewhat lower capacity ratio. The series resistance r_s is smaller, and breakdown is less abrupt. These diodes are therefore applied for lower level signals, i.e. parametric amplifiers instead of multipliers.

This kind of circuit is shown in Fig. 1.25. The signal source frequency is ω_1, and the so-called pump frequency is ω_2. Both generators are applied via series tuned resonant circuits. Varactor D_1 has a nonlinear voltage-capacity relationship which results in combination frequencies of the form $m\omega_1 + n\omega_2$. From these, the component $\omega_1 + \omega_2$ is extracted by a resonant circuit tuned to this frequency. This solution allows the design of amplifiers in the GHz frequency range with 10–20 dB amplification and 2–5 dB noise figure, naturally with low relative bandwidth. It should be noted that Schottky-barrier varactors utilize gallium arsenide instead of silicon to produce more favourable properties.

1.6. Microwave diodes

Owing to the high reverse voltage, charge carriers are accelerated in the junction region, and further charge carriers are generated by collision. This is called the avalanche effect and the multiplication current thus produced is called the avalanche current, because of the rapid growth of the multiplication of charge carriers. Investigations show that in this operating region, a peculiar high-frequency oscillation is produced which was first explained by W. Read [1.3]. These kinds of diodes are called IMPATT diodes (Impact Avalanche Tranist Time).

1.6. Microwave diodes

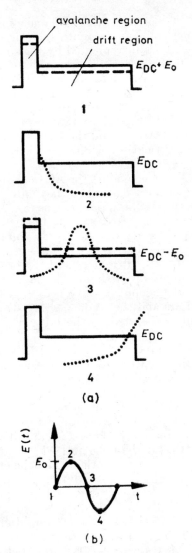

Fig. 1.26. Illustration of IMPATT diode operation. (a) Field strength and number of charge carriers as a function of displacement, (b) field strength as a function of time

Diode operation is investigated in Fig. 1.26 [1.18]. This figure shows the electric field strength as a function of displacement. At the junction (avalanche region) the field strength is high, while outside it (drift region) only a low DC field E_{DC} is present. Let us investigate the charge-carrier density, denoted by the dashed line in Fig. (b) at four different time instants of the sinusoidal field strength.

Let us assume that at instant *1*, the avalanche effect has not yet been produced, there is practically no current, and carrier density is low. At instant *2*, the field strength is incremented by the sinusoidal amplitude and affects the avalanche multiplication: current flow builds up, and charge carriers are transferred from the junction region into the drift region. At instant *3*, field strength is once more at the initial level, but the charge carriers, transferred earlier into the drift region, are transferred further to the left owing to the DC field E_{DC}. At instant *4*, the polarity of the AC field is reversed and is thus subtracted from the DC field, but the charge carriers are transferred by the DC field further to the right. This is the essence of operation.

Since the carriers are moving against the AC field, energy is transferred to the field (this contrasts with the case when they move in the direction of the field, thus receiving energy from the field). Thus in this period, the energy of the charge carriers, taken earlier from the DC field, is transferred to the AC field, i.e. AC voltage is produced from the supply voltage, accounting for the oscillations. After this the period comes to an end, the charge carriers are transferred to the end of the drift region, and the process starts again.

Although it involves many approximations, the above explanation allows insight into the physical phenomena involved. It is seen that the transit time of the charge carriers and the frequency are strongly related, and the oscillations are normally generated in the GHz region, in many cases at fairly high power levels.

IMPATT diodes can be produced from either silicon [1.37] or GaAs [1.31] material. Using the negative resistance of the diode, pulse amplifiers of several watts output power can be designed in the GHz frequency range [1.41].

In contrast with IMPATT diode consisting of avalanching *pn* junction, the BARITT diode uses a *pnp* structure operating in punch-through. Due to the injecting forward-biased junction, a capacitive delay appears and makes it usually less efficient than the IMPATT diode.

TRAPATT diodes are also based on the avalanche effect and on the finite transit time of charge carriers; however, during part of the oscillation cycle, an electron–hole *plasma* is also generated, and interaction takes place with the high-frequency field. TRAPATT diodes are used for both oscillators and amplifiers [1.40]. Figure 1.27 shows a class C amplifier utilizing a circulator with a 5-dB gain and 70-watt output power at 3.6 GHz frequency [1.28].

Operation of *Gunn diode* (called also transferred-electron diode, TED) is based on the field-strength dependence on the charge-carrier mobility. At high field strengths, the charge carriers are transferred to a high-energy state in which mobility is lower. As a result of the presence of these lower mobility *domains*, current is decreased, and a negative-resistance region is produced, which may be used to generate high-power microwave oscillations.

The cross-section of the diode is shown in Fig. 1.28. Width L of the *n*-type epitaxial layer (active region) is related with the oscillation frequency $L = v/f_0$, where v is the average electron velocity. In this case due to the finite transit time the current at the anode follows the voltage with 180°. The oscillation frequency is determined primarily by the width of the *n* layer

1.6 Microwave diodes

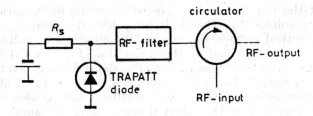

Fig. 1.27. Microwave amplifier using a TRAPATT diode

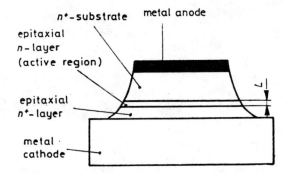

Fig. 1.28. Cross-sectional view of *Gunn* diode

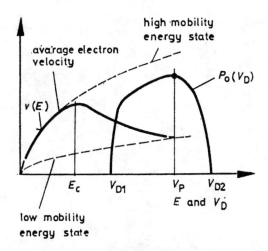

Fig. 1.29. Dependence of average electron velocity on the field strength and the available power as a function of the operating voltage in a *Gunn* diode

3 Kovács: High-Frequency

($f_0 = 4$–50 GHz) but it can be changed according to the ratio 2 : 1 with the operating voltage V_D. The field strength of the operating voltage V_D exceeds the critical voltage E_C. The oscillation starts at voltage V_{D1}, the available power is maximal at the value V_P ($P_{out} = 0.1$–1 W); at the value V_{D2} the oscillation breaks down. The dependence of available power on the field strength and operating voltage is shown in Fig. 1.29.

The transferred-electron diode as an oscillator can provide much higher pulse power operating in the "limited space-charge accumulation" (LSA) mode, where the time for exceeding the critical voltage will be properly limited by the circuit in each cycle of the oscillation. The thickness of LSA diodes exceeds the value obtained from the nominal transit time.

2. HIGH-FREQUENCY PROPERTIES AND EQUIVALENT CIRCUITS OF TRANSISTORS

2.1. High-frequency properties of bipolar transistors

In Fig. 2.1, the principal parameters of microwave transistors are plotted as a function of frequency. The *gain* between matched terminations can not be defined above a given frequency limit which is the lower boundary of the *potentially unstable* frequency range of the transistor. Further increase in frequency has the effect of lowering the gain, again to a stable operating range for which the 6 dB/octave drop will be characteristic:

$$G(f) = f_{max}^2/f^2. \qquad (2.1.1)$$

The gain reaches unity at f_{max}, which is the upper frequency limit of transistor operation.

Another relevant transistor parameter is the *noise figure* which increases abruptly above the frequency $f_N \approx f_T/\sqrt{h_{FE}}$. Finally, high-frequency behaviour is characterized by the frequency dependence of the *available power* at which the low-level gain is decreased by 1 dB (1-dB gain compression). Higher output power results in decreased gain and also rapidly increased distortion.

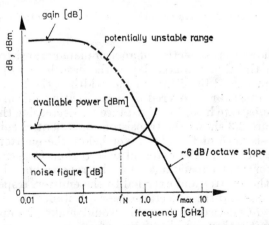

Fig. 2.1. Frequency dependence of microwave transistor parameters

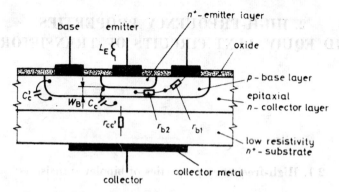

Fig. 2.2. Cross-sectional view of an *npn* planar transistor

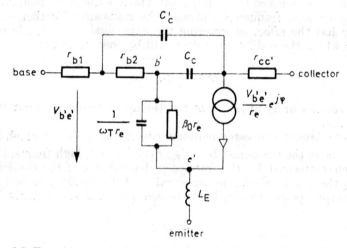

Fig. 2.3. Transistor equivalent circuit by using the notations of Fig. 2.2

Figure 2.2 shows a cross-section of an *npn* planar transistor. The transistor itself is within the thin epitaxial layer, the base layer (p) and the emitter layer (n^+) are produced by diffusion. The width of the base layer is w_B. The surface of the transistor is covered by a passivating silicon dioxide protecting glass. The substrate has a low resistance for decreasing the series collector resistance. Figure 2.3 shows the transistor equivalent circuit. Resistance r_{b2} is the series resistance of the base layer below the emitter region, and r_{b1} is the series resistance of the external base region. Capacitances C_c and C_c' are the collector–base capacitances for the reverse-biased junction in the regions below the emitter and external to the emitter, respectively. In fact, the base resistance and the collector–base capacitance should be represented by distributed RC networks, but this would make the application of the equivalent circuit rather complicated.

Emitter capacitance C_e is the sum of the emitter–base junction capacitance and of the base-layer capacitance, and may be calculated from the relation $C_e = 1/\omega_T r_e$ where ω_T denotes the current-gain transition frequency pertaining to unity current gain, and r_e is the dynamic resistance of the emitter–base diode. The emitter capacitance is shunted by a resistance which has a value of $\beta_0 r_e$ where β_0 is the low-frequency common-emitter current gain. The current of the output current generator is proportional to the voltage $V_{b'e'}$ across the emitter capacitance and has a phase which lags behind the phase of the voltage $V_{b'e'}$ by an angle φ.

The series resistance of the collector layer $r_{cc'}$ arises from the substrate layer and that region of the epitaxial layer outside the collector–base junction. Inductance L_E arises from the emitter lead.

An important basic relation is deduced by neglecting some elements of the equivalent circuit shown in Fig. 2.3. Let us assume that $C'_c = 0$, $r_{b2} = r_{cc'} = 0$, $L_E = 0$, and let us introduce the simplified notation $r_{b2} = r_b$. The frequency at unity power gain is thus calculated as

$$f_{max} = \sqrt{\frac{f_T}{8\pi r_b C_c}}, \qquad (2.1.2)$$

where $f_T = \omega_T/2\pi$ is the frequency at unity current gain in the common-emitter circuit. The parameters determining the frequency f_T are dealt with in Section 2.2.

2.2. Parameters determining the transition frequency f_T

The transition frequency f_T of the bipolar transistor can be expressed by the overall transit time τ_T of the minority carriers according to the relation $\tau_T = 1/\omega_T = 1/2\pi f_T$, and is comprised of the components within the different regions [2.45] as shown in Fig. 2.4:

$$\tau_T = \tau_e + \tau_E + \tau_B + \tau_x + \tau_C. \qquad (2.2.1)$$

We shall examine these components further. The transit time within the emitter region is well approximated by $\tau_e \simeq 0$ and may thus be neglected. The transit time within the emitter–base junction τ_E is dependent on the dynamic resistance r_e of the emitter–base diode, and the junction capacitance C_{Te} of this diode:

$$\tau_E = r_e C_{Te}. \qquad (2.2.2)$$

The transit time within the base region τ_B is given, for *npn* transistors, by

$$\tau_B = w_B^2 / D_n m(E), \qquad (2.2.3)$$

where w_B is the base width, D_n is the mobility of the electrons and $m(E)$ is a factor dependent on the drift field [2.3], [2.4]. The drift field depends on the base impurity profile, and τ_B decreases with increasing drift field.

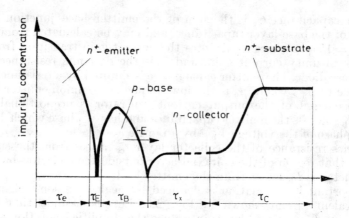

Fig. 2.4. Transit times in the different regions of the transistor

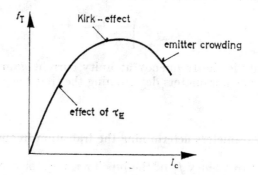

Fig. 2.5. Dependence of transition frequency f_T on the collector current

The transit time within the reverse-biased collector–base junction can be calculated from the width x_C of the junction region and the space-charge velocity of the minority carriers v_{SC} according to the relation $\tau_x = x_C/2v_{SC}$.

The transit time within the collector layer is given by $\tau_C = r_{cc'} C_c$ where $r_{cc'}$ is the series collector resistance and $C_c = C_{Tc}$ is the collector–base junction capacitance. Resistance $r_{cc'}$ arises from the fact that electrical contact to the collector layer is established through the rear side of the crystal chip by means of soldering. Decreasing the specific resistance of the collector layer is not possible because of the breakdown voltage of the collector. This problem is overcome by applying two collector layers in a sandwich-type arrangement. The actual collector is placed within the very thin layer which has high resistance, and the thick supporting layer of low resistance has the single purpose of ensuring the mechanical stability of the crystal. Several methods exist for producing such structures.

Using the *epitaxial procedure*, the high-resistance layer is grown onto the low-resistance substrate, and subsequently, the active transistor is built into this layer.

In the *triple-diffusion procedure*, the resistance of the substrate layer is decreased by adding impurity from the rear side.

Figure 2.5 shows the dependence of transition frequency f_T on collector current. At low collector currents, the approximately linear dependence is given by the component τ_E since r_e is inversely proportional to collector current ($r_e = 26$ mV/I_C). Above a given collector current, there is no further increase in f_T because of the so-called *Kirk effect*. This results principally from the fact that owing to the increasing minority carrier concentration, the base-layer becomes *wider* and the base–collector junction layer becomes *narrower*, thus increasing the effective value of w_B and thence also τ_B. The further increase of collector current introduces the so-called *emitter-crowding effect* which decreases f_T. This is treated in Section 2.3.

2.3. High-power, high-frequency bipolar transistors

To obtain a high cut-off frequency f_{max}, we require a high f_T value, a low value of base resistance r_b and a low value of collector capacitance C_c, according to eqn. (2.1.2). Let us now investigate the different transistor structures which can provide these conditions.

According to eqn. (2.2.3), transition frequency f_T is proportional to the square of the base width w_B which is, however, unfavourably influenced by the so-called *emitter dip effect* (EDE). This is caused principally by the phenomenon where the base region below the emitter is shifted during the diffusion process producing the emitter layer, thus preventing production of a small enough value of w_B (see Fig. 2.6). This problem is overcome by producing the emitter layer by arsenic diffusion instead of phosphoric diffusion [2.36], [2.46], or using *ion implantation* instead of diffusion. In the latter case extremely flat diffusion profiles with good reproducibility may be established, so this process is widely used for high-frequency transistor structures.

In order to decrease the base resistance r_{b1}, a higher contamination is introduced into the base region outside the emitter, which has the effect of decreasing the specific resistance (see Fig. 2.2).

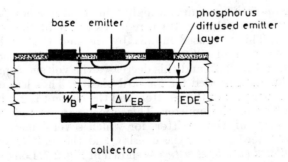

Fig. 2.6. Illustration of the EDE effect and the emitter-crowding effect

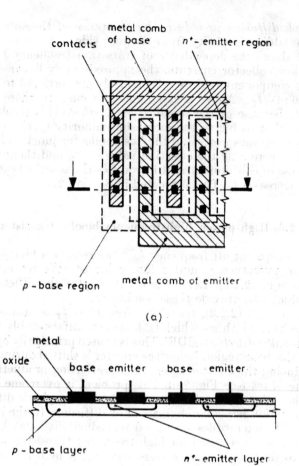

Fig. 2.7. Comb structure (a) top-view and (b) cross-sectional view

High-power operation requires a high value of frequency f_T even at high collector currents, and this, in turn, requires the reduction of the emitter-crowding effect. This is caused mainly by the fact that at high values of I_C, the high base current introduces a voltage drop ΔV_{EB} across the base resistance r_{b2} (see Fig. 2.6), and this decreases the voltage component at the emitter–base junction needed for forward-biasing. Thus less current is conducted by the inner part of the emitter, and the current is gradually crowded into the periphery of the emitter. This effect can be reduced by increasing the periphery of the emitter, for which several methods have been developed: the interdigitated, the overlay and the emitter-grid structures.

The *interdigitated (comb) structure* is shown in Fig. 2.7. Long emitter stripes are inserted within the base region, and both regions are covered by inter-

2.3. High-power bipolar transistors

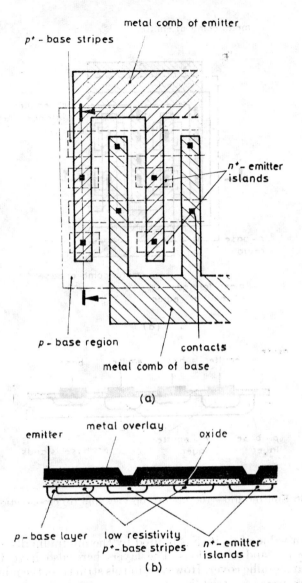

Fig. 2.8. Overlay structure (a) top-view and (b) cross-sectional view

digitated metallic layers. The periphery of the emitter is thus substantially increased, but this increase is limited by the longitudinal voltage drop introduced by excessively long digits.

The *overlay structure* is shown in Fig. 2.8. The emitter periphery is increased further by using separate emitter islands [2.24]. An additional low-resistance p^+-base layer is placed between the array of emitter islands, so

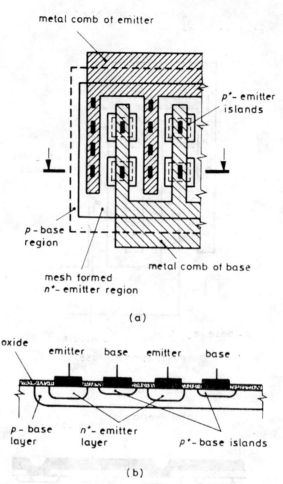

Fig. 2.9. Emitter-grid structure (a) top-view and (b) cross-sectional view

the minimum value of r_{b2} is thus limited by the minimum distance available between the p^+- and n^+-regions. The layers here also have interdigitated structure and metallic cover. However, in this structure, the mutual coverage of the layers imposes strict isolation requirements because of the metal overlay.

The *emitter-grid structure* is shown in Fig. 2.9. This is a further development of the previous one and has the concept of producing a better current distribution by placing base segments into the emitter region [2.38], [19.28], instead of emitter segments into the base region. An emitter grid is therefore built into the base region of higher resistance, and p^+-base segments are placed into the blank squares of the grid. The layers here also have interdigital form with metallic cover, similarly to the overlay structure.

2.3. High-power bipolar transistors

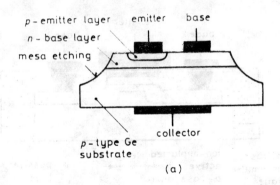

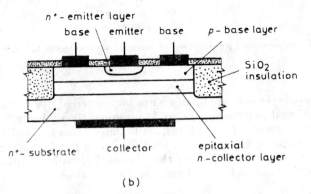

Fig. 2.10. (a) Mesa structure and (b) LOCOS structure, having low C'_c collector capacitances

To obtain a high value of cut-off frequency f_{max}, the collector capacitance C'_c must also be reduced to a low value. This may be set up by using either the mesa structure or the LOCOS structure. The *mesa structure* is shown in Fig. 2.10 (a). This is primarily applied to germanium transistors, and essentially consists of *etching* that part of the base layer external to the emitter. This introduces a form resembling a table (mesa), hence the name. The removal of the outer part of the base layer also reduces the area of the collector–base junction and thus C'_c.

The *LOCOS structure*, shown in Fig. 2.10 (b), makes possible to have an oxide layer at the inactive part of the chip area, for obtaining transistor structures with oxide insulation on the side wall. These transistors do not have collector–base junction at the edge, and consequently C'_c has a low value.

The *microwave transistor structure* is shown in Fig. 2.11 [2.33]. This is a combination of the LOCOS structure and the ion implantation technology and has the following merits:

(a) The side wall does not contribute to capacitance C'_c because of an insulating oxide layer.

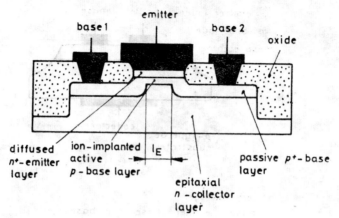

Fig. 2.11. Structure of ion-implanted microwave transistor fabricated using LOCOS technology

(b) The base layer is made up of an ion-implanted active region and a low-resistance diffused passive region, and consequently r_{b1} has a low value.

(c) The emitter width l_E which primarily determines the emitter-crowding effect has, because of the sideways p^+-diffusion, a much smaller value than the minimum resolution width of the photolithographical procedure, thus making possible extremely small emitter width values.

In microwave transistor structures, the aluminium leads of the emitter and base regions are inductively coupled, and special effects may be produced if the length of these leads is of the same order of magnitude as the wavelength. The theoretical and technological aspects of these effects are treated in Refs. [2.40] and [2.41].

Microwave transistor pairs having a common collector or a common emitter can be conveniently applied to fast flip-flop circuits [2.47].

High-frequency properties are basically influenced by the *transistor package*, by introducing further parasitic elements [2.28]. Connections to the external circuit are facilitated by the *stripline package* (see Fig. 2.12) in which striplines of characteristic impedance Z_{01} and Z_{02} are used for the base and collector leads [2.11], [2.16], [2.39]. Series inductances L_E and L_B are formed by the connecting leads, capacitances C_{BE} and C_{CE} exist between the striplines and the emitter line, and capacitance C_{CB} is formed by the striplines themselves.

High-frequency properties may be improved by the so-called *internal matching* technique, according to which elements L and C within the transistor package are used to obtain suitable frequency responses and improved circuit-device interface. This procedure may lead to higher bandwidth, higher output power and higher efficiency. Usually MOS capacitances and thin-film or thick-film inductances are mostly applied as reactance elements. Usually, the collector capacitance of the transistor is tuned out by a parallel

2.4. Equivalent circuit of bipolar transistors

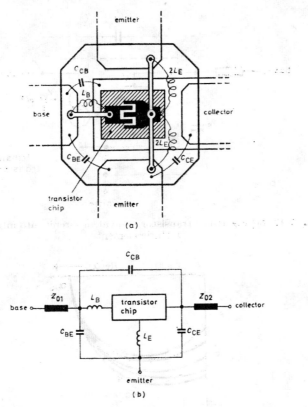

Fig. 2.12. (a) Stripline package transistor view and (b) equivalent circuit

inductance with series DC blocking capacitors; in common-base configuration the emitter-lead inductance is matched by a parallel capacitance to the input stripline. High-power transistors may have several sections, and by matching these separately, the deviation between them may be minimized — this is important from the stability viewpoint as shown in Section 19.2 [2.48], [2.49], [2.51].

The *beam-lead transistors* may be considered as a special family of high-frequency transistors in which the layer structure is not different from the conventional case. These will be treated in connection with integrated circuits which is their main field of application.

2.4. Current gain and equivalent circuit of bipolar transistors

To investigate current-gain factors, let us separate the equivalent circuit shown in Fig. 2.3 into parts consisting of *intrinsic* and *extrinsic* elements [2.2]. The following parts are within the intrinsic transistor: the base region

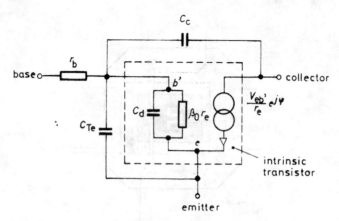

Fig. 2.13. Separation of transistor equivalent circuit into intrinsic and extrinsic elements

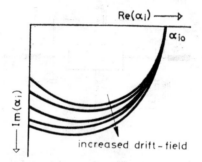

Fig. 2.14. Polar diagram of the current-gain factor in the complex plane, for various drift fields

in which the minority carriers proceed from the emitter to the collector, the ideal emitter responsible for the injection, and the collector collecting the minority carriers. According to Fig. 2.13, this fictive *intrinsic* transistor is made up of the base diffusion capacitance C_d, the input resistance $\beta_0 r_e$ and the output current source.

Of the *extrinsic* elements connected externally to the intrinsic transistor, the collector–base junction capacitance C_c, the base resistance $r_{b2} = r_b$ and the emitter–base junction capacitance C_{Te} will be used. According to the notation of Fig. 2.3, $C_e = 1/\omega_T r_e = C_d + C_{Te}$. Further extrinsic elements are neglected because of the simplifying assumption of $L_E = 0$, $r_{cc} = r_{b1} = 0$, and $C_c' = 0$.

The current gain frequency response of the common-base *intrinsic* transistor shown in Fig. 2.13 is given by

$$\alpha_i(\omega) = \alpha_{i0} \frac{e^{-j\varphi(E)\omega/\omega_{\alpha i}}}{1 + j\omega/\omega_{\alpha i}}, \qquad (2.4.1)$$

2.4. Equivalent circuit of bipolar transistors

where $\alpha_{i0} \approx 1$ is the low-frequency current gain, φ is the so-called excess phase shift depending on the accelerating field, and $\omega_{\alpha i}$ is the intrinsic cut-off frequency for the common-base configuration, $\omega_{\alpha i} \approx 1.2/\tau_B$.

The polar diagram of the current-gain factor α_i in the complex plane is shown in Fig. 2.14 for various accelerating (drift) field values. Increasing the drift field and thus the excess phase has the effect of shifting the diagrams downwards, in the direction of higher negative imaginary values.

Disregarding the exponential factor, eqn. (2.4.1) gives the frequency response of an RC voltage divider with a time constant $1/\omega_{\alpha i}$ and with a polar plot which is a semicircle on the complex plane, passing through the origin. Figure 2.14 shows how the plot is modified by the excess phase.

The extrinsic elements have the effect of decreasing the current gain. Several kinds of current-gain factors and thus several kinds of cut-off frequencies may be defined, depending on which extrinsic elements are taken into account.

One of the most significant extrinsic elements (especially at low currents) is the emitter–base transition capacitance C_{Te} by which the transit time τ_E is defined according to eqn. (2.2.2). If this capacitance alone is taken from the extrinsic elements, then the current-gain factor of the common-base configuration pertaining to the transistor may be determined from eqn. (2.4.1), by substituting the value

$$\omega'_\alpha = \omega_{\alpha i}/(1 + \omega_{\alpha i} r_e C_{Te}) \tag{2.4.2}$$

for $\omega_{\alpha i}$.

By also taking into account the base resistance r_b and the collector–base capacitance C_c, another current-gain factor may be defined as follows:

$$\alpha \cong \frac{\alpha_i + j\omega r_b C_c}{1 + j\omega r_b C_c}, \tag{2.4.3}$$

where α_i is the value according to eqn. (2.4.1) corrected by eqn. (2.4.2).

Further current-gain factors and cut-off frequencies could be defined by taking into account still more extrinsic elements. However, these will not be considered since they are much too complicated for practical applications.

The current-gain factor of the common-emitter configuration is given by

$$\beta = \alpha/(1-\alpha), \tag{2.4.4}$$

which is the result of the fact that the base current is given by the difference of the emitter current and the collector current. The current-gain factor β is largely dependent on the phase-angle of the current gain α. For increasing phase-angle values, $|1-\alpha|$ is rapidly increasing and thus β is rapidly reducing.

The frequency response of the current-gain factor pertaining to the common-emitter configuration may be expressed in the following form (independent of whether it is related to the extrinsic or intrinsic transistor):

$$h_{21e} = \beta(\omega) = \frac{\beta_0}{1 + j\omega/\omega_\beta}, \tag{2.4.5}$$

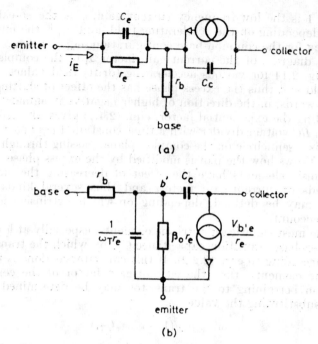

Fig. 2.15. Simplified equivalent circuit of bipolar transistors for (a) common-base configuration and for (b) common-emitter configuration

where $\beta_0 = \alpha_0/(1 - \alpha_0)$ is the low-frequency current gain, and the cut-off frequency is given by

$$\omega_\beta \simeq \omega_{\alpha i}/(1 + \beta_0)(1 + \varphi), \qquad (2.4.6)$$

where φ is the excess phase. This means that the merits of the drift field in the common-emitter configuration are less significant than in the common-base configuration. It should be noted here that the transition frequency f_T considered in Section 2.2 is defined by the equation $|h_{21e}(f_T)| = 1$, whence we obtain:

$$2\pi f_T = \omega_T \simeq \beta_0 \omega_\beta. \qquad (2.4.7)$$

The cut-off frequency "f_s" of the transistor is expressed by the scattering parameters; at this frequency, $|s_{21}|^2 = 1$.

Two simplified equivalent circuits of bipolar transistors frequently used are shown in Fig. 2.15. Figure (a) shows the T-equivalent circuit of the common-base configuration, where $C_e = C_d + C_{Te}$ and $r_e = 26$ mV/I_C. The hybrid-π equivalent circuit of the common-emitter configuration is shown in Fig. 2.15 (b). This may be extended by including further parasitic reactances according to the transistor structure and to the package.

The transistor models with distributed elements may only be handled by computers. These models have been used for computer-aided design of high-frequency circuits and to develop the optimal structure of new microwave transistors.

2.5. High-frequency operation of JFET and MESFET

The principle of field-effect transistors with pn junctions is based on the phenomenon that the width of the channel below the gate may be controlled by the gate voltage. The transistor current flows through this channel between the source and the drain. According to Fig. 2.16, the channel is formed by the p-doped layer. The depletion region, formed between the n-type gate layer and the channel, wides out with reverse voltage, decreasing hereby the channel width. The channel may also be cut-off completely by applying a high enough gate voltage to prevent current flow between the source and the drain [2.5], [2.8], [2.15].

Two regions are distinguished in the operation of the field-effect transistor. In the so-called triode range, the drain current I_D is dependent on both the gate voltage V_{GS} and the drain voltage V_{DS}. In the other region which is called pinch-off, the channel is already cut-off at one point. In this region, the current I_D no long depends on voltage V_{DS}, but only on the gate voltage. This means that the characteristic will be similar to the pentode characteristic. In most cases, the latter region is a more important one from the view of applications.

The high-frequency equivalent circuit of the field-effect transistor is shown in Fig. 2.17 where g_m is the transistor transconductance, $r_{ss'}$ and $r_{dd'}$ are the source and drain series resistances, respectively; r and C denote parallel resistances and capacitances, respectively between points denoted by the subscript letters, C_{gc} is the capacitance between the gate and the channel; and r_c is the channel resistance. In high-frequency operation, the latter two, both of which are functions of the gate voltage, have special significance. According to Ref. [2.7], there is an approximate relation between resistance r_c and the low-frequency value of the transconductance g_{m0} as follows:

$$1/r_c \approx k g_{m0}, \qquad (2.5.1)$$

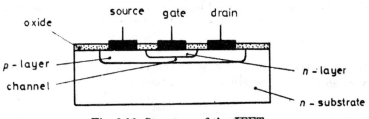

Fig. 2.16. Structure of the JFET

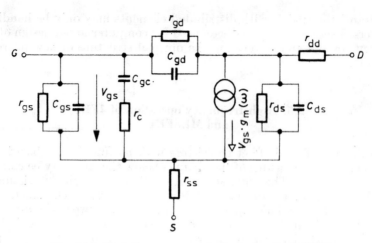

Fig. 2.17. High-frequency equivalent circuit of JFET

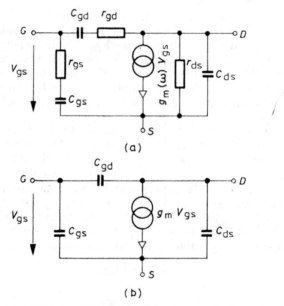

Fig. 2.18. (a) JFET simplified high-frequency equivalent circuit and (b) equivalent circuit for the frequency range $\omega \ll \omega_m$

where the proportionality factor is in the range $k = 5\text{--}10$. The frequency dependence of the transconductance is given by

$$g_m(\omega) = \frac{g_{m0}}{1 + j\omega/\omega_m}, \tag{2.5.2}$$

where the cut-off frequency ω_m is given approximately by

$$\omega_m \cong 1/2 r_c C_{gc}. \qquad (2.5.3)$$

At extremely high frequencies, the series lead inductances and stray capacitances should be included in the equivalent circuit of Fig. 2.17.

Figure 2.18 (a) shows the simplified high-frequency equivalent circuit, in which the impedances between gate and drain as well as between gate and source are represented by series RC networks.

Figure 2.18 (b) shows a low-frequency equivalent circuit for the frequency range $\omega \ll \omega_m$ which is mainly applied for broadband applications. The transconductance is frequency-independent and the capacitances between electrodes are lossless.

Extremely high cut-off frequencies can be attained by MESFET's (*Schottky-gate* field-effect transistors) with silicon substrate ($f_{max} > 10$ GHz) and with GaAs substrate ($f_{max} > 50$ GHz). In these transistors, the gate is formed by a metal–semiconductor junction instead of a *pn* junction [2.6], [2.13], [2.18], [2.23], [2.37], [2.50], [2.53]. In GaAs transistors, the epitaxial layer of less than 1 μm width is grown on the substrate onto which the metallization of the source, gate and drain electrodes is deposited. The gate electrode is fabricated by forming a metal–semiconductor junction, the other two contacts must be ohmic contacts. The width of the thin strip-like gate and its distance from the source- and drain-metallization is not more than 1 μm.

High-power transistors are fabricated by the parallel connection of MESFET's within a single package. Such a transistor has an output power of 25 W at a frequency of 30 MHz, and more than 10-dB gain at an intermodulation distortion of -35 dB [20.9].

The equivalent circuit of the GaAs MESFET is investigated in Refs. [2.39] and [19.32]. The cross-modulation is computed in Ref. [21.12]. A wideband internal matching of power GaAs MESFET's is presented in Refs. [2.52] and [2.54].

2.6. High-frequency operation of MOS field-effect transistors

Figure 2.19 shows that in MOS transistors, a thin insulating layer (mostly an oxide layer) exists between the gate and the channel. The drain current depends on the gate voltage V_G. Enhancement-type MOS transistors are off at $|V_G| < |V_T|$, and the drain current starts at gate voltage $V_G = V_T$, where the latter is the threshold or pinch-off voltage.

The limiting factors of high-frequency operation of MOS transistors are: the capacitance C_{gd}, the transit time of the carriers through the channel length L, and the gate-channel capacitance C_{gc}. The capacitance C_{gd} between gate and drain introduces feedback and thus decreases the gain. Its value is higher for larger overlap, i.e. for larger coverage of the drain electrode by the gate metallization.

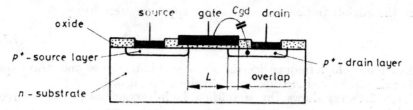

Fig. 2.19. Structure of MOS transistor

The channel length L is limited by the resolution of the photolithographic process. It determines the transit time which is the reciprocal value of the cut-off frequency:

$$\tau_T = 1/\omega_{max} = \frac{L^2}{\mu|V_G - V_T|}, \qquad (2.6.1)$$

where μ is the mobility of the minority carriers.

Figure 2.20 shows the high-frequency equivalent circuit of the MOS transistor, where $r_{ox} > 10^{10}\,\Omega$ is the insulation resistance of the oxide layer. All other elements were explained in connection with Fig. 2.16. The cut-off frequency of gain according to eqn. (2.6.1) is given by $\omega_{max} = g_m/C_{gc}$ where g_m is the transconductance assumed to be frequency-independent. In fact g_m is also frequency-dependent:

$$g_m(\omega) = \frac{g_m}{1 + j\omega/\omega_m}, \qquad (2.6.2)$$

where, however, the cut-off frequency $\omega_m \gg \omega_{max}$. The channel resistance is given by $r_c \approx 0.2/g_m$.

The capacitance C_{gd} due to overlap may be reduced by the *self-aligning* (e.g. silicon-gate) process. According to this process, the metallization of the

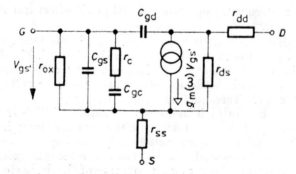

Fig. 2.20. High-frequency equivalent circuit of MOS transistor

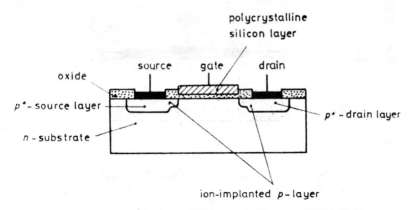

Fig. 2.21. Structure of ion-implanted silicon-gate MOS transistor

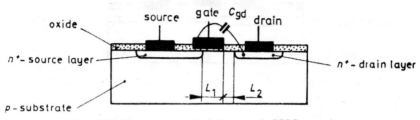

Fig. 2.22. Structure of depletion-mode MOS transistor

gate is not subsequently deposited between the source and drain (this is why the overlap is large), but the drain edge is masked during the diffusion of the p^+-layers by the previously deposited polysilicon gate itself. The overlap is thus much less, and only results from the lateral diffusion. This effect can be diminished further by fabrication of the shallow p-layers at the two sides of the gate by *ion implantation* (Fig. 2.21), subsequent to the deposition of the polysilicon gate [2.17].

The overlap can be eliminated completely in *depletion-mode* transistors, in which there is current flow even at $V_G = 0$ (normally-on transistor) because of the built-in charge of the gate-oxide layer which forms a channel. Increasing the voltage V_G will pinch off this channel, and for this purpose, only a section L_1 of the channel need to be controlled (Fig. 2.22). The remaining length L_2 is not covered by gate metallization, and consequently C_{gd} has a low value.

According to eqn. (2.6.1), the cut-off frequency of gain of the n-channel MOS transistors is higher because of the higher electron mobility μ_n.

The layer-to-substrate capacitances may be reduced considerably by the SOS (Silicon-On-Sapphire) technology, applying insulator as substrate (Fig. 2.23). The transistor is built into a thin silicon layer which is grown epitaxially onto a crystalline aluminium-oxide insulating layer (sapphire) [2.25], [2.42].

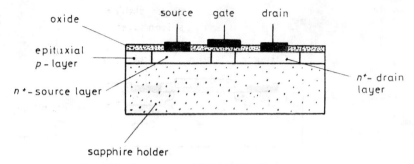

Fig. 2.23. Structure of SOS field-effect transistor

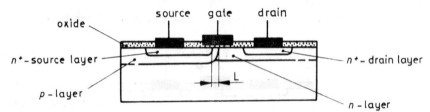

Fig. 2.24. Structure of double-diffused DMOS field-effect transistor

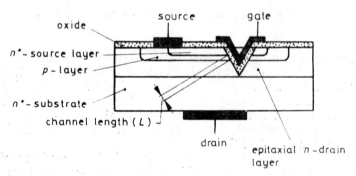

Fig. 2.25. Structure of V-MOS field-effect transistor

A new family of high-frequency, high-power devices is the *static-induction transistor* (SIT). In contrast with JFET's, the channel located in a thin epitaxial n-layer is perpendicular to the surface, thus the channel current flows down to the n-type bulk (using it as a drain). The channel will be controlled on the two sides by reversed-biased pn junctions, formed by p-type diffusion. The main difficulty with this structure is how to bring the diffusion layers very close (1.5—2 μm) to each other at a high production yield. Using a gate width of a few millimeters the output power exceeds 10W at 2GHz.

In the above types, the minimum value of the channel length L is limited by the resolution capability of the photolithographic process. But at the types DMOS and V-MOS investigated in the following, the channel length is given by the difference between the diffusion depths, and is thus very small. In the *double-diffused MOS transistor*, (Fig. 2.24), the diffusion of the p-layer, which will contain the channel, and the diffusion of the n-layer forming the source region are simultaneously performed through a common oxide window. The length of the channel is given by the excess penetration of the p-diffusion compared with the depth of the n-diffusion [2.30].

In the V-MOS structure, the channel is placed vertically along a V-shaped etching (Fig. 2.25). The length of the channel is given by the difference of the diffusion depths between the p-diffusion and the n-diffusion, similarly to the base width of bipolar transistors.

The *double* MOS transistors (dual-gate FET's) have extremely favourable high-frequency properties. These transistors, in which the drain of the first transistor is at the same time the source of the second transistor, are dealt with in Section 21.2 [2.25], [2.42], [14.10].

2.7. Large-signal transistor equivalent circuits

The equivalent circuits shown previously are only suitable for small-signal characterization of the device when the nonlinearity of the characteristics may be neglected. However, this is not the case for power amplifiers and oscillators, so these equivalent circuits may not be used, or have limited value. Linearity problems show up primarily with bipolar transistors in which the emitter–base diode has an exponentially shaped characteristic in the forward direction, and thus nonlinearity is present even at driving voltages of a few millivolts.

The large-signal characterization of bipolar transistors is given by the so-called Ebers–Moll symmetrical equations [2.1]:

$$i_e = -\frac{i_{e0}}{1-\alpha_f\alpha_r}(e^{\frac{qV_{eb}}{kT}}-1) + \frac{\alpha_r i_{c0}}{1-\alpha_f\alpha_r}(e^{\frac{qV_{cb}}{kT}}-1), \qquad (2.7.1)$$

$$i_c = \frac{\alpha_f i_{e0}}{1-\alpha_f\alpha_r}(e^{\frac{qV_{eb}}{kT}}-1) - \frac{i_{c0}}{1-\alpha_f\alpha_r}(e^{\frac{qV_{cb}}{kT}}-1). \qquad (2.7.2)$$

Additionally, the relation $\alpha_f i_{e0} = \alpha_r i_{c0}$ is valid where i_{e0} and i_{c0} are the reverse saturation currents; V_{eb} and V_{cb} are the emitter–base and collector–base voltages, respectively; finally α_f and α_r are the forward- and reverse-direction current gains, respectively, and have the following frequency dependence:

$$\alpha_f = \frac{\alpha_{f0}}{1+j\omega/\omega_f}; \qquad (2.7.3)$$

$$\alpha_r = \frac{\alpha_{r0}}{1+j\omega/\omega_r}. \qquad (2.7.4)$$

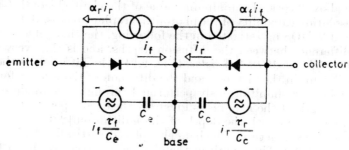

Fig. 2.26. Nonlinear equivalent circuit of bipolar transistor based on the symmetrical equations

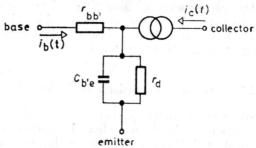

Fig. 2.27. Approximate nonlinear equivalent circuit of bipolar transistor

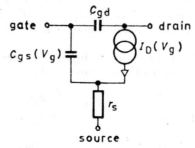

Fig. 2.28. Nonlinear equivalent circuit of the JFET

A shortcoming of the symmetrical equations is the poor characterization of the transistor frequency dependence due to neglection of the transition capacitances. This shortcoming is eliminated by the equivalent circuit shown in Fig. 2.26 in which the DC and AC parameters are separated. The currents of the current sources are given by

$$i_f = i_{f0}(e^{\frac{qV_{eb}}{kT}} - 1); \qquad (2.7.5)$$

$$i_r = i_{r0}(e^{\frac{qV_{cb}}{kT}} - 1), \qquad (2.7.6)$$

2.7. Large-signal equivalent circuits

and further the relation $\alpha_f\, i_{f0} = \alpha_r\, i_{r0}$ is valid. The transistor is characterized by eight parameters as follows: two reverse saturation currents, two current-gain factors, two transition capacitances, and two time constants. This equivalent circuit may be put to use in computer analysis of large-signal circuits.

An equivalent circuit particularly suitable for high-frequency applications is shown in Fig. 2.27, where $r_b = r_{b'b}$, the impedance of the emitter–base diode is represented by the RC network and the time function of the currents are defined by the following equation:

$$i_b(t) = \frac{1}{\omega_T} \frac{di_c}{dt} + \frac{i_c}{\beta_0}. \qquad (2.7.7)$$

This simplified equivalent circuit allows an approximate investigation of the large-signal operation of bipolar transistors.

The nonlinear behaviour of high-frequency circuits is treated in Ref. [21.17] based on the Gummel–Poon model which uses 21 parameters to characterize the transistor and thus also takes into account secondary effects. The simulation of nonlinear transistor behaviour by analogue computer is treated in Ref. [2.18]. A computer program for calculating the elements of the charge-controlled model from the hybrid-π parameters is presented in Ref. [2.35]. A 12-element equivalent circuit suitable for computer-aided design is investigated in Ref. [2.32].

Figure 2.28 shows a nonlinear equivalent circuit of the JFET. Here the nonlinear behaviour is simulated by the dependence of both the gate capacity and the drain current on the gate voltage.

3. HIGH-FREQUENCY PROPERTIES OF LINEAR INTEGRATED CIRCUITS

3.1. Structure of monolithic integrated circuits

In monolithic integrated circuits, all circuit elements (transistors, diodes, resistors) are formed on a single chip, and the elements are connected by metallization covering the surface of the chip. Components should be located so that they require the simplest connecting network, but elements should be separated and insulated. At monolithic integrated circuit the following high-frequency problems should be resolved.

In monolithic integrated circuits, obtaining a DC level shift requires use of *pnp* transistors in addition to *npn* transistors. However, in the *lateral pnp* technologically easily-made transistors, the base width is too large and, therefore, the cut-off frequency is too low [3.11], [3.16]. *Vertical pnp* transistors can be made with much higher cut-off frequencies, but require a complicated technological procedure.

The connecting lead to the collector of the transistor contained in a monolithic integrated circuit has to be placed on the chip surface instead of the bottom of the substrate (Fig. 3.2). This introduces an increased series collector resistance and parasitic capacitance. The stray capacitances of the resistors formed by diffusion are also substantial (Fig. 3.5). Low-capacitance resistors may be formed by polysilicon stripes or NiCr metal-film resistors [3.17].

Monolithic integrated circuits have several drawbacks. For instance, capacitances and inductances required for high-frequency compensation and adjustment are not available. One method for producing compensating capacitance is the use of MOS capacitance [3.15], [3.17]. On the other hand, the decrease of inductances due to short leads is a substantial advantage.

Another shortcoming is the fact that transmission-line elements are not applicable, so these must be applied externally using hybrid (in most cases thin-film) circuits connected to the monolithic circuit [11.11]. Finally, the application of transistors with high cut-off frequencies demands highly sophisticated technological processes — this renders the use of production runs difficult. Despite these shortcomings, monolithic technology has substantial economical and technical merits.

Figure 3.1 shows an *npn* transistor and a resistor formed on a single chip. It can be seen that the basic difference between the integrated transistor and discrete transistor lies in the arrangement of the collector lead, situated at the top. This structure can be realized by several technological solutions, summarized in the following [3.4].

3.1. Monolithic integrated circuits

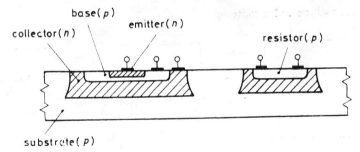

Fig. 3.1. Structure of monolithic *npn* transistor and resistor

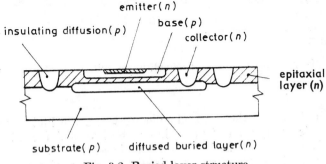

Fig. 3.2. Buried layer structure

(a) *The triple-diffusion process* is out-of-date, here all three layers are diffused from the top into the *p*-type material. This has the shortcoming that the base width is difficult to control because of the impurity profiles thus formed, and further, the collector characteristics are not favourable.

(b) *The single-epitaxial process* is carried out by growing an *n*-type collector layer of high resistance epitaxially onto the *p*-type substrate. Before the diffusion processes which produce the base and emitter layers, so-called insulation diffusion is carried out for insulating the transistors and resistors from each other. This method has the advantage of a low parasitic capacitance between the substrate and collector layers.

(c) *The double-epitaxial process* is distinguished from the previous one; in that two layers are grown epitaxially onto the *p*-type substrate. First, a low-resistance *n*-type layer is grown, and this is followed by a high-resistance *n*-type layer into which the emitter and base layers are built by the normal diffusion process. This process has the great advantage of a substantially smaller series collector resistance $r_{cc'}$ which is important for high-frequency applications.

(d) *The buried layer structure* is shown in Fig. 3.2. First, an *n*-type layer is diffused into the *p*-type substrate, and on this layer a high-resistance *n*-layer is grown. The diffused *n*-type layer is thus buried and serves only as a connection with low resistance between the active collector region

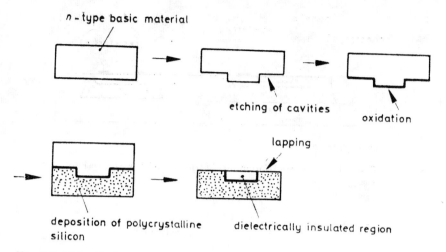

Fig. 3.3. Main fabrication steps of monolithic circuits with dielectric insulation

under the emitter, and the diffused collector layer. For high-power applications, the series collector resistance has to be low, and for this reason the horizontal dimensions of the transistor are chosen to be low.

(e) In the previous structures, the circuit elements within a single chip are separated by reverse-biased *pn* junctions, introducing hereby undesirable cross-couplings via the junction capacities. In integrated circuits which apply *dielectric insulation*, these couplings are much smaller because the separation results from an insulating oxide layer instead of a biased *pn* junction [3.1]. The main steps in the fabrication of this structure are illustrated in Fig. 3.3. First, cavities are etched from the n-type wafer. This is followed by growing an oxide layer on the surface, onto which polycrystalline silicon is deposited. The oxide layer serves for isolation, and the polycrystalline layer is used for mechanical strength. By lapping the original n-type layer, a dielectrically insulated n region is formed, into which the transistor or other type of circuit element is built by conventional methods.

(f) *The beam-lead structure* has the most favourable high-frequency properties [3.2], because of the dielectric air between the circuit elements. Onto the conventionally fabricated integrated circuit, a further thick metallization pattern is deposited, capable of holding the circuit elements (transistors, resistors) as shown in Fig. 3.4. Since the circuit elements are fastened to the metallization the bottom of the wafer is eliminated by etching, and only the islands comprising the active elements are retained. This is a rather complicated and expensive process, but is still in widespread use for circuits intended for extremely high frequencies and for transmitting fast pulses. According to measurement results published in Ref. [3.10], two operational amplifiers using the same circuit have been compared, one having a conventional structure and the other a beam-lead structure. Above 50 MHz, the

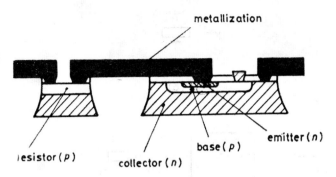

Fig. 3.4. Beam-lead monolithic structure

beam-lead structure showed favourable properties regarding input and output impedances and power gain.

Using the *PIN-insulation* method [3.9], all circuit elements are surrounded by PIN diodes; this has the effect of separating the elements. According to Section 1.2, the reverse-biased PIN diode behaves like an open circuit, and thus offers good separation even at higher frequencies. However, the three layers to be fabricated demand a complicated technological procedure, so the method is not widely used.

3.2. High-frequency properties of monolithic integrated circuits

The high-frequency properties of monolithic integrated circuits are determined by the properties of the components applied and by the circuit configuration. Most important are the transistors, but equally significant are resistors which should be frequency-independent in a wide frequency range. In monolithic circuits, capacitances as circuit elements are not preferred, so amplifiers are DC coupled in most cases. Should a capacitance be anyway necessary, it is arranged using a reverse-biased pn junction or a MOS capacitance, but the capacity values attainable are very limited.

Increasing the number of transistors applied in the amplifier will lower the cut-off frequency. This demands a high degree of feedback which, in turn, leads to instability. Thus, high-frequency monolithic amplifiers generally contain fewer transistors than low-frequency operational amplifiers.

The high-frequency behaviour of integrated circuits is decisively influenced by type of the transistor applied. Technological aspects of transistors are dealt with in Chapter 2.

However, transistors in monolithic circuits have poorer high-frequency properties than those of discrete transistors, as a result of parasitic elements and series resistances. This does not mean that a circuit using discrete transistors and other discrete elements necessarily has better high-frequency properties than its monolithic counterpart. This is because a monolithic ar-

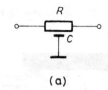

(a)

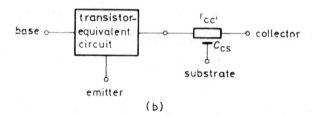

(b)

Fig. 3.5. Equivalent circuits: (a) resistor produced in a monolithic circuit, (b) transistor with the series collector resistance

rangement has many advantages, such as short connecting leads and better shielding owing to the smaller dimensions. The development trend is for integration (also justified by economical aspects), though instead of monolithic circuits, hybrid circuits using both discrete and monolithic elements are preferred, especially for circuits having high output power.

Resistors in monolithic structures have an undesirable parasitic parallel capacitance (Fig. 3.5(a)). An upper cut-off frequency may be given for such a resistor, with a value of

$$\omega_R = 1/RC. \qquad (3.2.1)$$

At higher frequencies, the resistor will act as a lossy transmission line, as the parasitic capacitance is distributed over the whole surface. A given resistor should therefore have minimal dimensions: this relates to external resistors such as collector load resistors, and also to the collector series resistance $r_{cc'}$ within the transistor. Therefore, the equivalent circuit of transistors applied in monolithic circuits is modified according to Fig. 3.5 (b).

The high-frequency behaviour of monolithic integrated circuits is characterized by the frequency response of the gain (Fig. 3.6) and by the switching times (rise time, delay time, storage time, fall time). This latter is especially important in the case of pulse amplifiers. The circuit is uniquely characterized by the frequency response shown in Fig. 3.6 for given loads and compensation, and may also be used for pulse amplifier design purposes within the linear range of operation. The figure shows a two-breakpoint Bode diagram with logarithmic scales; between the breakpoints or cut-off frequency points, straight-line approximations are possible. Breakpoints at higher frequencies are well above the operating frequency range and may thus be neglected.

3.1. Monolithic integrated circuits

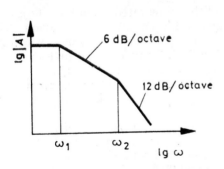

Fig. 3.6. Bode diagram of monolithic integrated amplifier

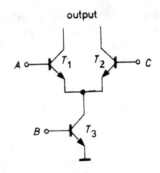

Fig. 3.7. Typical circuit of high-frequency monolithic integrated amplifiers

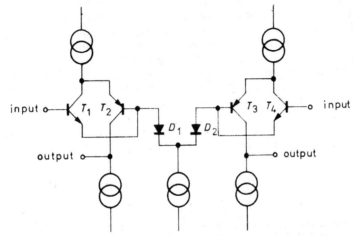

Fig. 3.8. High-frequency monolithic amplifier using complementer transistors

The positions of the breakpoints are important from the viewpoint of the usable feedback. Feedback is limited by instability introduced by increasing phase-shifts at higher frequencies. For good stability, the breakpoint frequencies ω_1 and ω_2 should be separated by at least two octaves. This question is dealt with in more detail in Section 11.2.

Figure 3.7 shows a typical circuit configuration used in monolithic circuits [3.5], [3.6]. The differential amplifier can be connected in many ways and is suitable for high-frequency application. Using point A as input with the other terminals grounded produces a common-collector–common-base two-stage amplifier. Grounding point B yields the common-emitter–common-base amplifier which is preferable at high frequencies because of low internal feedback. Using point C additionally as a driving terminal, the circuit lends itself to mixing, modulation, etc. Parameters of this circuit are dealt with in forthcoming sections.

Reference [3.11] deals with a monolithic integrated circuit made up of complementer transistors, shown in Fig. 3.8. The circuit has two advantages. The very low collector–base voltage applied permits a small base width of the npn transistors, resulting in a high cut-off frequency and high current-gain ($\beta_0 \approx 1000$). Further, the transistors of lower cut-off frequencies operate in common-base connection, and thus have less effect on the frequency response of the gain. This differential amplifier circuit allows a unity-gain frequency of $f_u = 100$–200 MHz.

In some cases, only transistors and diodes are produced in monolithic form, and external discrete passive components are applied [3.13]. This method is used primarily in extremely wideband amplifiers and in circuits utilizing hybrid technology. Reference [3.14] describes a high-frequency monolithic transistor pair which has a transition frequency of $f_T = 3$ GHz. The forward voltage difference is $\Delta V_{IN} = 2$ mV, and the noise figure with optimum generator resistance is $F = 1.6$ dB at $f = 60$ MHz.

Reference [3.26] presents a monolithic operational amplifier built up of n channel MOS transistors, indicating that MOS technique will also be used in the field of linear circuits. Bipolar and JFET's built up simultaneously on a chip are used for linear circuit in Ref. [3.25].

Finally the hybrid integrated circuit must be mentioned made by thin or thick-film technology, where transistor chips or monolithic amplifiers can also be placed on the ceramic substrate. This technique reduces the harmful interaction between active elements, permits easy arrangement of reactive elements, and has better cooling. This type allows the realization of extremely wideband and high-power amplifiers [3.3], [3.7].

3.3. Integrated-circuit delay lines

Figure 3.9 (a) shows the circuit of the bipolar bucket-brigade delay-line element. The transistors of these elements are made alternately on and off by pulses Φ_1 and Φ_2, thus transferring the charge of the input capacitor C_i successively to capacitors C_1, C_2 etc. in the manner of a bucket brigade. For a chain comprising N elements and for a repetition time T_Φ between pulses Φ_1 and Φ_2, the overall delay time will be $\tau = T_\Phi N/2$.

The time diagram showing the operation of the first element is shown in Fig. 3.9 (b). For convenience, instead of transferring the charge of C_i to C_1, C_1 is discharged into C_i, and similarly C_2 into C_1, etc. The charge thus proceeds from right to left, whereas the information proceeds from left to right. Operation phases are:

(1) Φ_2 is positive, C_1 discharges via conducting T_2 into $V_1 = V_\Phi$.

(2) Φ_1 goes into positive, voltage step V_Φ is transferred via C_i and in the first instant, $V_1 = 2V_\Phi$. However, discharge occurs by conducting T_1 into C_i; C_1 charges to the value V_Φ when T_1 becomes off as $V_{EB} = 0$. Expressing the balance of charges, we obtain

$$\Delta Q = (V_\Phi - V_1) C_i = (2V_\Phi - V_1') C_1,$$

and for the case $C_i = C_1$ the voltage will be $V_1' = V_\Phi + V_i$.

3.3. Delay lines

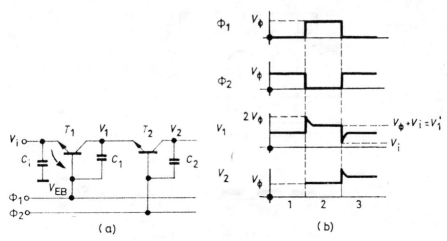

Fig. 3.9. (a) Circuit diagram of a bucket-brigade delay-line element,
(b) timing waveforms

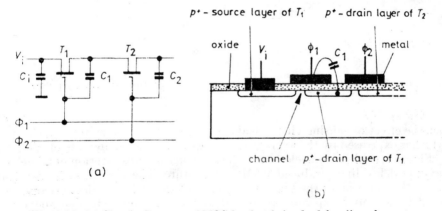

Fig. 3.10. (a) Circuit diagram of MOS bucket-brigade delay-line element,
(b) cross-section of the delay line structure

(3) During the fall time of Φ_1, V_Φ is subtracted from V'_1. This means that capacitor C_1 has the same V_1 voltage as capacitor C_i at the beginning of phase (2): the information is thus actually transferred. When transistor T_2 becomes on, a sequence similar to phase (2) takes place, with the difference that C_2 is now discharged and C_1 is charged, thus transferring the information to C_2.

At the input of the delay line, a sampling circuit samples the analog signal to be delayed with a frequency $f > 2f_M$, where f_M is the highest frequency component of the analog signal. At the output of the delay line, the analog signal is recovered by a low-pass filter. Increasing the repetition time T_Φ

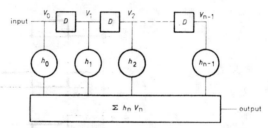

Fig. 3.11. Block diagram of transversal filter

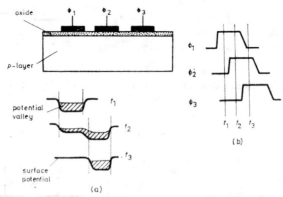

Fig. 3.12. (a) Cross-section of CCD delay-line structure, (b) timing pulses

has the effect of causing charge leakage in the storage capacitors C_1, C_2, etc., which is expressed by the average decrease of signal level (in units of mV/ms).

Figure 3.10 shows one element and the structure cross-section of a bucket-brigade delay line produced in MOS structure. The operation is identical with the operation of the bipolar version, but the leakage is less because element C_1 is a MOS capacitor produced by metallization extended above the diffused p^+ region (denoted by Φ_1).

Main application fields of delay lines are colour TV systems, time-division multiplex transmission systems, image cameras, and transversal filters. In time division multiplex speech transmission systems, the signals pertaining to different voice channels are slowly filled into the bucket-chains, and are read out in quick sequence to drive the wideband channel.

In image cameras, part of the p^+n element in the MOS structure is not covered by metallization and is exposed to incoming light. After charging every storage capacitor of the bucket-chain to voltage V_Φ, discharge takes place as the result of light radiation. Following this, the information carrying the picture content is quickly read-out.

Figure 3.11 shows the block diagram of a *transversal filter*. Samples taken from the input are delayed by elements D. Input voltage V_0 and delayed voltages V_1, V_2, ... V_{n-1} are applied to the adding block via attenuators characterized by weighting factors $h_0, h_1, \ldots h_{n-1}$. Sampling frequency may be chosen between 25 Hz and 10 MHz. Different filter characteristics may be produced by suitable choice of weighting factors.

Delay lines can also be devised by applying the CCD (*Charge Coupled Device*) principle as shown in Fig. 3.12. The structure is essentially made up of MOS capacitors, and there is no diffusion region between gates. The operation is based on the fact that below the closely spaced gates, the charge may be transferred horizontally. At time instant t_1 pulse Φ_1 produces a potential valley, in which electrons are contained. Introducing a further potential valley below the neighbouring gate by pulse Φ_2 at time instant t_2 will have the effect of transferring part of the charge below the neighbouring gate. After completion of Φ_2 at time instant t_3, the whole charge will be transferred. If the charge loss is low enough, an analog delay-line element is formed. Applying a series of these elements results in a complete delay line. A third phase Φ_3 is also needed for the operation, in order to guarantee left-to-right charge propagation.

4. CHARACTERIZATION OF SOLID-STATE DEVICES BY FOUR-POLE PARAMETERS

4.1. Four-pole equivalent circuit; four-pole parameters

Solid-state devices such as transistors and integrated circuits may be characterized by linear four-pole parameters in addition to the so-called physical-equivalent circuit. If the AC voltage levels appearing on the device are small, this device may be regarded as a linear four-pole, and thus the rules of the linear four-pole theory are applicable. The signal level permitted is determined by the requirement that the voltage swing should cover a linear part of the characteristic. In practice, this requirement is not always met, and a linear four-pole equivalent circuit is used, in spite of the fact that the AC voltage exceeds the permitted limit, though the nonlinearity error thus introduced is offset by the relatively simple method of presentation. However, this method has only limited use for the description of explicitly large-signal behaviour such as encountered in oscillators or power amplifiers.

In the following, solid-state devices will be characterized by methods applied in four-pole theory. Physical processes within the solid-state device will not be considered, and the device will be regarded as a black-box with unknown content, having two ports on which two voltages and two currents are measured.

For high-frequency applications, most frequent use is made of the y-parameter equivalent circuit in which admittances are used to relate two primary parameters (input and output currents) and two secondary parameters (input and output voltages).

For both low-frequency and high-frequency applications, the h-parameter equivalent circuit is equally applicable, and is primarily suited for the description of bipolar transistors. One impedance, one admittance and two ratios are used for describing the equivalent circuit. The most frequently used three equivalent circuits, together with a set of equations, are presented in Table 4.1.

For very high frequencies, the scattering-parameters (s parameters) are applied [4.3], [4.4]. These are measured by the reflection coefficients or standing-wave ratios appearing on transmission lines, and allow more accurate measurements than y parameters. This results from the fact that when measuring the output admittance y_{22}, a short-circuit has to be provided at the input side ($V_1 = 0$) which may be difficult to achieve at high frequencies. There is no such problem with s parameters which are defined and measured having the input and output sides terminated by the characteristic

4.1. Equivalent circuit

Table 4.1. Four-pole equivalent circuits and equation sets defining four-pole parameters

type	equivalent circuit	defining set of equations
y-equivalent circuit		$i_1 = y_{11}V_1 + y_{12}V_2$ $i_2 = y_{21}V_1 + y_{22}V_2$
z-equivalent circuit		$V_1 = z_{11}i_1 + z_{12}i_2$ $V_2 = z_{21}i_1 + z_{22}i_2$
h-equivalent circuit		$V_1 = h_{11}i_1 + h_{12}V_2$ $i_2 = h_{21}i_1 + h_{22}V_2$

impedance Z_0 (Fig. 4.1). The test item is easily accessible as the transmission line length is insignificant.

In Fig. 4.1 (b) the scattering parameters of the four-pole are shown using the signal flow graph, where e_{i1} and e_{r1} are the incident and reflected voltages at the input port, respectively, furthermore e_{i2} and e_{r2} are the voltages at the output port. The defining equations for the four-pole are:

$$e_{r1} = s_{11}e_{i1} + s_{12}e_{i2}; \qquad (4.1.1)$$
$$e_{r2} = s_{21}e_{i1} + s_{22}e_{i2},$$

In these equations s parameters are coefficients which are defined as follows.
The input-port scattering parameter s_{11} is given by

$$s_{11} = e_{r1}/e_{i1} \,|\, e_{i2} = 0, \qquad (4.1.2)$$

4. Four-pole parameters

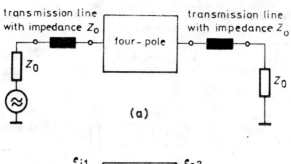

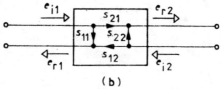

Fig. 4.1. Interpretation of scattering parameters: (a) four-pole terminated by transmission lines, (b) the signal flow graph

Table 4.2. Relations between different four-pole parameters

matrix	used parameters					
	h		z		y	
y-matrix	y_{11}	y_{12}	$\dfrac{z_{22}}{\Delta z}$	$-\dfrac{z_{12}}{\Delta z}$	$\dfrac{1}{h_{11}}$	$-\dfrac{h_{12}}{h_{11}}$
	y_{21}	y_{22}	$-\dfrac{z_{21}}{\Delta z}$	$\dfrac{z_{11}}{\Delta z}$	$\dfrac{h_{21}}{h_{11}}$	$\dfrac{\Delta h}{h_{11}}$
z-matrix	$\dfrac{y_{22}}{\Delta y}$	$-\dfrac{y_{12}}{\Delta y}$	z_{11}	z_{12}	$\dfrac{\Delta h}{h_{22}}$	$\dfrac{h_{12}}{h_{22}}$
	$-\dfrac{y_{21}}{\Delta y}$	$\dfrac{y_{11}}{\Delta y}$	z_{21}	z_{22}	$-\dfrac{h_{21}}{h_{22}}$	$\dfrac{1}{h_{22}}$
h-matrix	$\dfrac{1}{y_{11}}$	$-\dfrac{y_{12}}{y_{11}}$	$\dfrac{\Delta z}{z_{22}}$	$\dfrac{z_{12}}{z_{22}}$	h_{11}	h_{12}
	$\dfrac{y_{21}}{y_{11}}$	$\dfrac{\Delta y}{y_{11}}$	$-\dfrac{z_{21}}{z_{22}}$	$\dfrac{1}{z_{22}}$	h_{21}	h_{22}
Δy	$y_{11}y_{22}-y_{12}y_{21}$		$\dfrac{1}{\Delta z}$		$\dfrac{h_{22}}{h_{11}}$	
Δz	$\dfrac{1}{\Delta y}$		$z_{11}z_{22}-z_{12}z_{21}$		$\dfrac{h_{11}}{h_{22}}$	
Δh	$\dfrac{y_{22}}{y_{11}}$		$\dfrac{z_{11}}{z_{22}}$		$h_{11}h_{22}-h_{12}h_{21}$	

Table 4.3. Relations for calculating scattering parameters

$s_{11} = \dfrac{(z_{11} - 1)(z_{22} + 1) - z_{12}z_{21}}{(z_{11} + 1)(z_{22} + 1) - z_{12}z_{21}}$	$z_{11} = \dfrac{(1 + s_{11})(1 - s_{22}) + s_{12}s_{21}}{(1 - s_{11})(1 - s_{22}) - s_{12}s_{21}}$
$s_{12} = \dfrac{2z_{12}}{(z_{11} + 1)(z_{22} + 1) - z_{12}z_{21}}$	$z_{12} = \dfrac{2s_{12}}{(1 - s_{11})(1 - s_{22}) - s_{12}s_{21}}$
$s_{21} = \dfrac{2z_{21}}{(z_{11} + 1)(z_{22} + 1) - z_{12}z_{21}}$	$z_{21} = \dfrac{2s_{21}}{(1 - s_{11})(1 - s_{22}) - s_{12}s_{21}}$
$s_{22} = \dfrac{(z_{11} + 1)(z_{22} - 1) - z_{12}z_{21}}{(z_{11} + 1)(z_{22} + 1) - z_{12}z_{21}}$	$z_{22} = \dfrac{(1 + s_{22})(1 - s_{11}) + s_{12}s_{21}}{(1 - s_{11})(1 - s_{22}) - s_{12}s_{21}}$
$s_{11} = \dfrac{(1 - y_{11})(1 + y_{22}) + y_{12}y_{21}}{(1 + y_{11})(1 + y_{22}) - y_{12}y_{21}}$	$y_{11} = \dfrac{(1 + s_{22})(1 - s_{11}) + s_{12}s_{21}}{(1 + s_{11})(1 + s_{22}) - s_{12}s_{21}}$
$s_{12} = \dfrac{-2y_{12}}{(1 + y_{11})(1 + y_{22}) - y_{12}y_{21}}$	$y_{12} = \dfrac{-2s_{12}}{(1 + s_{11})(1 + s_{22}) - s_{12}s_{21}}$
$s_{21} = \dfrac{-2y_{21}}{(1 + y_{11})(1 + y_{22}) - y_{12}y_{21}}$	$y_{21} = \dfrac{-2s_{21}}{(1 + s_{11})(1 + s_{22}) - s_{12}s_{21}}$
$s_{22} = \dfrac{(1 + y_{11})(1 - y_{22}) + y_{21}y_{12}}{(1 + y_{11})(1 + y_{22}) - y_{12}y_{21}}$	$y_{22} = \dfrac{(1 + s_{11})(1 - s_{22}) + s_{12}s_{21}}{(1 + s_{22})(1 + s_{11}) - s_{12}s_{21}}$
$s_{11} = \dfrac{(h_{11} - 1)(h_{22} + 1) - h_{12}h_{21}}{(h_{11} + 1)(h_{22} + 1) - h_{12}h_{21}}$	$h_{11} = \dfrac{(1 + s_{11})(1 + s_{22}) - s_{12}s_{21}}{(1 - s_{11})(1 + s_{22}) + s_{12}s_{21}}$
$s_{12} = \dfrac{2h_{12}}{(h_{11} + 1)(h_{22} + 1) - h_{12}h_{21}}$	$h_{12} = \dfrac{2s_{12}}{(1 - s_{11})(1 + s_{22}) + s_{12}s_{21}}$
$s_{21} = \dfrac{-2h_{21}}{(h_{11} + 1)(h_{22} + 1) - h_{12}h_{21}}$	$h_{21} = \dfrac{-2s_{21}}{(1 - s_{11})(1 + s_{22}) + s_{12}s_{21}}$
$s_{22} = \dfrac{(1 + h_{11})(1 - h_{22}) + h_{12}h_{21}}{(h_{11} + 1)(h_{22} + 1) - h_{12}h_{21}}$	$h_{22} = \dfrac{(1 - s_{22})(1 - s_{11}) - s_{12}s_{21}}{(1 - s_{11})(1 + s_{22}) + s_{12}s_{21}}$

in the case of matched output. Similarly, the output-port scattering parameter s_{22} is given by

$$s_{22} = e_{r2}/e_{i2} \,|\, e_{i1} = 0, \qquad (4.1.3)$$

in the case of matched input. From the two transfer parameters, the gain is expressed by s_{21} and the reverse transmission or loss is expressed by s_{12}:

$$s_{21} = e_{r2}/e_{i1} \,|\, e_{i2} = 0; \qquad (4.1.4)$$

$$s_{12} = e_{r1}/e_{i2} \,|\, e_{i1} = 0. \qquad (4.1.5)$$

Any of the four sets of parameters given in the foregoing may be used for characterizing solid-state devices, and if necessary, a conversion from one system to another is possible using Tables 4.2 and 4.3. This may be useful if the analysis of an investigated circuit can be simplified by using a set of parameters other than that given in the data sheet. In this case, it is expedient to convert to the more convenient four-pole parameters using the relations given in the tables. However, it should be noted that the corresponding y, z and h parameters of Table 4.3 should be normalized to the characteristic impedance Z_0, in question, for instance,

$$z_{11\text{norm}} = z_{11}/Z_0. \tag{4.1.6}$$

4.2. Properties of terminated four-poles

In the theory of linear networks, relations for terminated four-poles are given. The most frequently used relations are the input and output impedance of the terminated four-pole, the current gain, the voltage gain and the power gain. The quantities above are expressed by the parameters y, z and h in Table 4.4. The following notations are used in the relations: Z_g is the generator impedance, Z_L is the output load impedance, R_L is the real part of the latter (Fig. 4.2).

Table 4.4. Common relations for terminated four-poles. Z_g is generator impedance, Z_L is load impedance, $R_L = \text{Re}(Z_L)$, and * denotes the conjugated value

parameter	input impedance	voltage gain $A_v = \dfrac{V_2}{V_1}$	current gain $A_i = \dfrac{i_2}{i_1}$	output impedance	power gain $G_p = \dfrac{P_{\text{out}}}{P_{\text{in}}}$
y	$\dfrac{1+y_{22}Z_L}{y_{11}+\Delta y Z_L}$	$-\dfrac{y_{21}Z_L}{1+y_{22}Z_L}$	$\dfrac{y_{21}}{y_{11}+\Delta y Z_L}$	$\dfrac{1+y_{11}Z_g}{y_{22}+\Delta y Z_g}$	$\dfrac{R_L |y_{21}|^2}{\text{Re}[(1+y_{22}Z_L)(y_{11}+\Delta y Z_L)^*]}$
z	$\dfrac{\Delta z + z_{11}Z_L}{z_{22}+Z_L}$	$\dfrac{z_{21}Z_L}{\Delta z + z_{11}Z_L}$	$-\dfrac{z_{21}}{z_{22}+Z_L}$	$\dfrac{\Delta z + z_{22}Z_g}{z_{11}+Z_g}$	$\dfrac{R_L |z_{21}|^2}{\text{Re}[(\Delta z + z_{11}Z_L)(z_{22}+Z_L)^*]}$
h	$\dfrac{h_{11}+\Delta h Z_L}{1+h_{22}Z_L}$	$-\dfrac{h_{21}Z_L}{h_{11}+\Delta h Z_L}$	$\dfrac{h_{21}}{1+h_{22}Z_L}$	$\dfrac{h_{11}+Z_g}{\Delta h + h_{22}Z_g}$	$\dfrac{R_L |h_{21}|^2}{\text{Re}[(h_{11}+\Delta h Z_L)(1+h_{22}Z_L)^*]}$

In the last column of the table, we show the power gain, which is the ratio of the power dissipated in the useful load (P_{out}) to the power used by the transistor (P_{in}). The relation expressed in terms of y parameters may be put into the following modified form:

$$G_p = \frac{P_{\text{out}}}{P_{\text{in}}} = \frac{g_L |y_{21}|^2}{|y_{22}+y_L|^2 \text{Re}(y_{\text{in}})}. \tag{4.2.1}$$

Here the input admittance is given by

$$y_{\text{in}} = y_{11} - \frac{y_{12}y_{21}}{y_{22}+y_L}; \tag{4.2.2}$$

further the load admittance is $y_L = 1/Z_L$ whose real part is $g_L = \text{Re}(y_L)$.

4.3. Stability

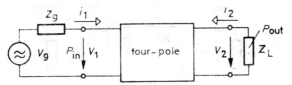

Fig. 4.2. Circuit for calculating the parameters of the terminated four-pole

In order to determine the power gain as expressed by (4.2.1), the input power P_{in} has to be known, calling for the measurement of the *input current*. Since this presents problems, it is expedient to use the *transducer gain* instead of the relation (4.2.1). This is expressed as the ratio of output power P_{out} to the maximum available generator power $P_{g,max}$:

$$G_T = \frac{P_{out}}{P_{g,max}}, \qquad (4.2.3)$$

where $P_{g,max} = V_g^2/4R_g$. Substituting into eqn. (4.2.3) the values of the y parameters, we have

$$G_T = \frac{4 g_g g_L |y_{21}|^2}{|(y_g + y_{11})(y_{22} + y_L) - y_{12} y_{21}|^2}, \qquad (4.2.4)$$

where $y_g = 1/Z_g$ is the generator admittance, having real part $g_g = \text{Re}(y_g)$.

4.3. Stability of active four-poles

Active four-poles between given terminations may become unstable, offering spontaneous oscillations independently of the driving signal. The condition for oscillation depends on both the active four-pole itself and the terminations. One group of four-poles is stable even under extreme termination conditions, and there is no oscillation with terminations having any real positive component. These are called *unconditionally stable* four-poles.

Another group of four-poles operates normally with given terminations, but the four-pole may become unstable for a certain range of terminations. These four-poles are not unconditionally stable and are called *potentially unstable* four-poles. In these cases, the termination ranges which yield stable operation should be determined by the designer.

Let us characterize the four-pole by y parameters, and let the terminations be the generator admittance y_g and the load admittance y_L (Fig. 4.3). The terminating admittances and the y_{11} and y_{22} parameters, respectively, may be combined. Expressing the set of equations which define the y parameters applying to the four-pole thus constructed, we have

$$0 = (y_{11} + y_g) V_1 + y_{12} V_2;$$
$$0 = y_{21} V_1 + (y_{22} + y_L) V_2, \qquad (4.3.1)$$

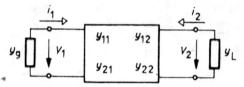

Fig. 4.3. Terminated four-pole characterized by four-pole parameters

(see Fig. 4.3). According to these equations, the value of external currents flowing into the overall four-pole is zero.

If there are finite internal V_1 and V_2 voltages without external driving currents, these may originate only from the four-pole itself, i.e. the four-pole oscillates. The condition for oscillation is given by the above set of equations for non-zero values of V_1 and V_2, according to the following relation:

$$(y_{11} + y_g)(y_{22} + y_L) - y_{12} y_{21} = 0 . \tag{4.3.2}$$

This complex equation may be transformed into two scalar equations. Writing from these equations

$$\begin{aligned} \operatorname{Im}(y_{11} + y_g) &= b_{11} + b_g ; \\ \operatorname{Im}(y_{22} + y_L) &= b_{22} + b_L , \end{aligned} \tag{4.3.3}$$

we have two quadratic equations for these imaginary parts, and thus the sum of the two imaginary terms may be easily expressed. The condition for oscillation is a positive discriminant value in the quadratic-solving formula (the value under the root sign). In this case we get real scalar values of the imaginary part, expressing the termination values which will produce oscillations.

If we have a negative discriminant value in the quadratic-solving formula, no real root will exist, implying that there will be no oscillation with any termination value. Thus for a four-pole operating under general terminating conditions, the criterion for stability is a negative discriminant. Mathematically, this means that the so-called Stern factor (k) should exceed unity:

$$k = \frac{2(g_{11} + g_g)(g_{22} + g_L)}{|y_{12} y_{21}| + \operatorname{Re}(y_{12} y_{21})} \geq 1, \tag{4.3.4}$$

where $g_{11} = \operatorname{Re}(y_{11})$ and $g_{22} = \operatorname{Re}(y_{22})$.
The criterion for unconditional stability is a higher-than-unity Stern factor even in the worst case of $g_g = g_L = 0$, i.e.

$$k_{us} = \frac{2 g_{11} g_{22}}{|y_{12} y_{21}| + \operatorname{Re}(y_{12} y_{21})} = \frac{1}{\delta(1 + \cos \varphi)} \geq 1, \tag{4.3.5}$$

where the following notations have been introduced:

$$\delta = |y_{12} y_{21}|/2 g_{11} g_{22}, \tag{4.3.6}$$

$$\varphi = \varphi_{12} + \varphi_{21}.$$

In calculations of the unconditional stability, another quantity is also used. This is called the Linvill factor, C, whose reciprocal

$$K = 1/C = \frac{2 g_{11} g_{22} - \operatorname{Re}(y_{12} y_{21})}{|y_{12} y_{21}|} \geq 1. \tag{4.3.7}$$

The stability factors expressed by eqns. (4.3.5) and (4.3.7) are related by:

$$K = 1/\delta - \cos \varphi = k_{us}(1 + \cos \varphi) - \cos \varphi. \tag{4.3.8}$$

Let us return to eqn. (4.3.4) which is a function of the terminating conductances g_g and g_L. In a high-frequency amplifier, these terminations may depend on both adjustments and gain control, and may be functions of frequency within the amplifier pass band. The actual stability of high-frequency amplifier is characterized by the so-called minimum stability factor k_{min} which is the minimum value of eqn. (4.3.4), determined from the minimum g_g and g_L values.

4.4. Power transfer of unconditionally stable four-poles

The most obvious case of unconditional stability is the case of zero reverse transmission, i.e. $y_{12} = 0$. The factor according to eqn. (4.3.5) is then $k_{us} = \infty$, i.e. the four-pole is unilateral. The maximum value of the power transfer according to eqn. (4.2.4) is obtainable through a conjugated match of the input side to the generator and of the output side to the load, i.e. under the conditions $g_{11} = g_g$, $b_{11} = -b_g$, $g_{22} = g_L$, $b_{22} = -b_L$. Under these conditions, the transducer gain is called the *maximum available gain* given by the following equation:

$$MAG = \frac{|y_{21}|^2}{4 g_{11} g_{22}}. \tag{4.4.1}$$

For practical, active four-poles, $y_{12} \neq 0$, so the above considerations do not apply. If, however, the four-pole is unconditionally stable ($k_{us} \geq 1$), in spite of the reverse transmission, then conjugated matching can be applied according to the following equations:

$$y_g = y_{in}^*;$$
$$y_L = y_{out}^*, \tag{4.4.2}$$

where $y_{in}(y_L)$ and $y_{out}(y_g)$ are the input and output admittances, respectively, which depend on the terminations. Solving eqn. (4.4.2), we have the matched terminations yielding the MAG.

$$\operatorname{Re}(y_g) = g_g = g_{11}\sqrt{(1 - \delta \cos \varphi)^2 - \delta^2};$$
$$\operatorname{Im}(y_g) = b_g = -b_{11} + g_{11} \delta \sin \varphi;$$
$$\operatorname{Re}(y_L) = g_L = g_{22}\sqrt{(1 - \delta \cos \varphi)^2 - \delta^2};$$
$$\operatorname{Im}(y_L) = b_L = -b_{22} + g_{22} \delta \sin \varphi.$$
(4.4.3)

Here the notations according to (4.3.6) have been used. The transducer gain thus computed is given by

$$G_{T,\max} = \frac{|y_{21}|^2}{4 g_{11} g_{22}} \frac{2}{1 - \delta \cos \varphi + \sqrt{(1 - \delta \cos \varphi)^2 - \delta^2}}. \quad (4.4.4)$$

Note that the first factor of this expression is the MAG value according to (4.4.1), which may be increased or decreased by the second factor, depending on the nature of the reverse transmission, i.e. the value of the phase angle φ.

Let us rearrange eqn. (4.4.4) by introducing the K factor according to eqn. (4.3.7), expressing the unconditional stability of the four-pole:

$$G_{T,\max} = \left|\frac{y_{21}}{y_{12}}\right| \left(K - \sqrt{K^2 - 1}\right). \quad (4.4.5)$$

By using this well known expression, the maximum transducer gain of the unconditionally stable four-pole can be readily determined. Expression (4.4.5) is meaningless for $K < 1$, and in the limiting case of $K = 1$,

$$G_{T,\max}(K = 1) = \left|\frac{y_{21}}{y_{12}}\right|. \quad (4.4.6)$$

4.5. Power transfer of potentially unstable four-poles

In the case of potentially unstable four-poles, the conjugated matching according to eqn. (4.4.2) is not applicable, so either the reverse transmission has to be decreased (by neutralization) or mismatch has to be used to increase the stability. The neutralization method will be dealt with in a later section, so only the method of mismatch will be treated here.

The principle of the *mismatch method* is that the Stern factor according to (4.3.4) is increased for obtaining $k > 1$ by applying suitably high terminating conductances g_g and g_L. Practical k values are in the range 2—4. The terminating conductance values are realized by the suitable choice of turns ratio applied in the transformers, filters etc. Mismatch may be applied simultaneously at both ends, but principally, a suitably chosen mismatch at only the input or only the output is sufficient.

The transducer gain, G_T, according to (4.2.4) is decreased as a consequence of mismatch. Therefore, the problem is to determine those values of conductances g_g and g_L which, under mismatch conditions, result in a k factor of suitable value and simultaneously a maximum value of transducer gain G_T. This problem may be solved generally using a computer. However, good approximation is achieved by a simplified method, whereby we assume the tuning out of the input and output reactances, i.e.

$$b_g + b_{11} = 0;$$
$$b_L + b_{22} = 0, \qquad (4.5.1)$$

and also the identity of the mismatches at the input and output, i.e.

$$g_L/g_{22} = g_g/g_{11}. \qquad (4.5.2)$$

This method will be dealt with in connection with specific design problems.

Since the transducer gain is decreased as a consequence of mismatch, the maximum usable gain, defined by

$$MUG = \gamma \left| \frac{y_{21}}{y_{12}} \right|, \qquad (4.5.3)$$

in accordance with eqn. (4.4.5), may be introduced. The proportionality constant in the above expression in most practical cases can be approximated by $\gamma \simeq 0.4$.

Another design procedure is described in Ref. [4.5] where the added power, $P_{out} - P_{in}$ is related to the input power, defined as the *maximally efficient gain*.

4.6. Connection of four-poles

For the design of multi-stage amplifiers, the overall parameters which are valid for connected four-poles are required. These will be given in the following.

The parallel–parallel connection of four-poles is shown in Fig. 4.4, where both inputs and outputs are parallel. The y matrix of the overall four-pole is given by the

$$y_R = \begin{vmatrix} y'_{11} + y''_{11} & y'_{12} + y''_{12} \\ y'_{21} + y''_{21} & y'_{22} + y''_{22} \end{vmatrix}. \qquad (4.6.1)$$

The series–series connection is shown in Fig. 4.5. The corresponding circuit elements of the equivalent circuit are connected in series, and the z parameters are added. The z matrix of the overall four-pole is given by

$$z_R = \begin{vmatrix} z'_{11} + z''_{11} & z'_{12} + z''_{12} \\ z'_{21} + z''_{21} & z'_{22} + z''_{22} \end{vmatrix}. \qquad (4.6.2)$$

4. Four-pole parameters

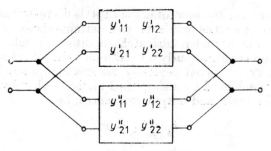

Fig. 4.4. Parallel–parallel connection of four-poles

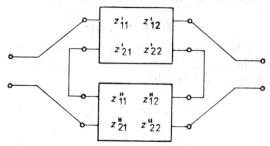

Fig. 4.5. Series–series connection of four-poles

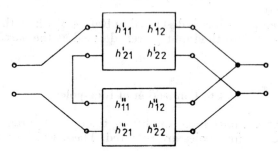

Fig. 4.6. Series–parallel connection of four-poles

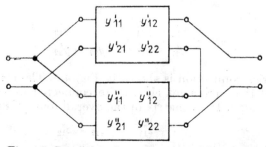

Fig. 4.7. Parallel–series connection of four-poles

4.6. Connection of four-poles

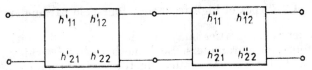

Fig. 4.8. Cascade connection of four-poles

The series–parallel connection is shown by Fig. 4.6. Input terminals are series-connected and output terminals are parallel-connected. The h matrix of the overall four-pole is given by

$$h_R = \begin{vmatrix} h'_{11} + h''_{11} & h'_{12} + h''_{12} \\ h'_{21} + h''_{21} & h'_{22} + h''_{22} \end{vmatrix}. \tag{4.6.3}$$

The parallel–series connection is shown by Fig. 4.7. Input terminals are parallel-connected and output terminals are series-connected. The y parameters of the overall four-pole are given by

$$y_{11R} = y'_{11} + y''_{11} + \frac{(y'_{12} - y''_{12})(y'_{21} - y''_{21})}{y'_{22} + y''_{22}};$$

$$y_{12R} = \frac{y'_{12} y''_{22} + y''_{12} y'_{22}}{y'_{22} + y''_{22}};$$

$$y_{21R} = \frac{y'_{21} y''_{22} + y''_{21} y'_{22}}{y'_{22} + y''_{22}}; \tag{4.6.4}$$

$$y_{22R} = \frac{y'_{22} y''_{22}}{y'_{22} + y''_{22}}.$$

Figure 4.8 shows the *cascade connection* of four-poles. The h parameters of the overall four-pole are given by

$$h_{11R} = \frac{h'_{11} + \Delta h' \, h''_{11}}{h'_{22} h''_{11} + 1};$$

$$h_{12R} = \frac{h'_{12} h''_{12}}{h'_{22} h''_{11} + 1};$$

$$h_{21R} = \frac{h'_{21} h''_{21}}{h'_{22} h''_{11} + 1}; \tag{4.6.5.}$$

$$h_{22R} = \frac{h'_{22} \Delta h'' + h''_{22}}{h'_{22} h''_{11} + 1}.$$

4.7. Types of high-frequency four-pole parameters

High-frequency properties may be characterized by either equivalent circuits or four-pole parameters. Both methods are in widespread use, the former primarily for discrete semiconductor devices such as high-frequency diodes and transistors. For integrated circuits, the equivalent network is too complicated to be of practical use.

The advantage of the equivalent circuit is its ability to describe also the frequency dependence. On the other hand, the four-pole parameters apply for a single frequency or single working point, so root-loci are used for characterization. In these cases, the parameters in question are plotted in either rectangular or polar coordinate systems as functions of frequency, for various working points. Intermediate working points may be determined by interpolation.

For *bipolar transistors*, the four-pole parameters are usually given only for one of the three basic connections, in most cases for the common-emitter or common-base connection. The transfer to another connection is facilitated by Tables 4.5 and 4.6, in which the transfer formulae of admittance and h parameters are presented for the three basic connections. Accurate formulae are given, but substantial simplifications can be attained sacrificing only a little accuracy.

It may occur that the four-pole transistor parameters have to be determined from the equivalent circuit. This presents no special problems but it should be borne in mind that accuracy of the complex high-frequency parameters thus determined will be much worse than the directly given or measured values.

Let us first investigate the calculation of the four-pole parameters of bipolar transistors based on the equivalent circuit of Fig. 2.15(b). The high-frequency h parameters for common-emitter configuration can be given as follows:

$$\begin{aligned} h_{11e} &= r_b + \beta r_e \; ; \\ h_{12e} &= \beta r_e Y_c \; ; \\ h_{21e} &= \beta \; ; \\ h_{22e} &= \beta Y_c \, . \end{aligned} \qquad (4.7.1)$$

In the above expressions, the following simplified notations have been introduced:

$$\beta = \beta(\omega) = \frac{\beta_0}{1 + j\omega/\omega_\beta} \; ; \qquad (4.7.2)$$

$$Y_c = j\omega C_c \, . \qquad (4.7.3)$$

Further, for clarity, the base resistance has been denoted $r_b = r_{b'b}$, and the dynamic resistance of the emitter diode r_e.

The calculated h parameters are expressed by relatively simple expressions, being a considerable simplification for further circuit-design calculations.

4.7. Types of the parameters

Table 4.5. Relations between different admittance parameters

circuit	y_b	y_e	y_c
Common-base circuit	y_{11b} y_{12b} y_{21b} y_{22b}	$(y_{11e}+y_{12e}+y_{21e}+y_{22e})$ $-(y_{12e}+y_{22e})$ $-(y_{21e}+y_{22e})$ y_{22e}	y_{22c} $-(y_{21c}+y_{22c})$ $-(y_{12c}+y_{22c})$ $(y_{11c}+y_{12c}+y_{21c}+y_{22c})$
Common-emitter circuit	$(y_{11b}+y_{12b}+y_{21b}+y_{22b})$ $-(y_{12b}+y_{22b})$ $-(y_{21b}+y_{22b})$ y_{22b}	y_{11e} y_{12e} y_{21e} y_{22e}	y_{11c} $-(y_{11c}+y_{12c})$ $-(y_{11c}+y_{21c})$ $(y_{11c}+y_{12c}+y_{21c}+y_{22c})$
Common-collector circuit	$(y_{11b}+y_{12b}+y_{21b}+y_{22b})$ $-(y_{11b}+y_{12b})$ $-(y_{11b}+y_{21b})$ y_{11b}	$(y_{11e}+y_{12e}+y_{21e}+y_{22e})$ $-(y_{11e}+y_{12e})$ $-(y_{11e}+y_{21e})$ y_{11e}	y_{11c} y_{12c} y_{21c} y_{22c}

Table 4.6. Relations between different h parameters

circuit	h_b	h_e	h_c
Common-base circuit	h_{11b} h_{12b} h_{21b} h_{22b}	$\dfrac{h_{11e}}{1+\Delta h_e + h_{21e} - h_{12e}}$ $\dfrac{\Delta h_e - h_{12e}}{1+\Delta h_e + h_{21e} - h_{12e}}$ $\dfrac{-(\Delta h_e + h_{21e})}{1+\Delta h_e + h_{21e} - h_{12e}}$ $\dfrac{h_{22e}}{1+\Delta h_e + h_{21e} - h_{12e}}$	$\dfrac{h_{11c}}{\Delta h_c}$ $\dfrac{h_{21c}+\Delta h_c}{\Delta h_c}$ $\dfrac{h_{12c}-\Delta h_c}{\Delta h_c}$ $\dfrac{h_{22c}}{\Delta h_c}$
Common-emitter circuit	$\dfrac{h_{11b}}{1+\Delta h_b + h_{21b} - h_{12b}}$ $\dfrac{\Delta h_b - h_{12b}}{1+\Delta h_b + h_{21b} - h_{12b}}$ $\dfrac{-(h_{21b}+\Delta h_b)}{1+\Delta h_b + h_{21b} - h_{12b}}$ $\dfrac{h_{22b}}{1+\Delta h_b + h_{21b} - h_{12b}}$	h_{11e} h_{12e} h_{21e} h_{22e}	h_{11c} $1-h_{12c}$ $-(1+h_{21c})$ h_{22c}
Common-collector circuit	$\dfrac{h_{11b}}{1+\Delta h_b + h_{21b} - h_{12b}}$ $\dfrac{1+h_{12b}}{1+\Delta h_b + h_{21b} - h_{12b}}$ $\dfrac{h_{12b}-1}{1+\Delta h_b + h_{21b} - h_{12b}}$ $\dfrac{h_{22b}}{1+\Delta h_b + h_{21b} - h_{12b}}$	h_{11e} $1-h_{12e}$ $-(1+h_{21e})$ h_{22e}	h_{11c} h_{12c} h_{21c} h_{22c}

Table 4.7. Relations for calculating common-emitter and common-base admittance parameters from the hybrid-π equivalent circuit

circuit	y parameters
common-emitter circuit	$y_{11e} = \dfrac{y_{b'e} + y_{b'c}}{1 + r_{bb'}(y_{b'e} + y_{b'c})}$ $y_{12e} = \dfrac{-y_{b'c}}{1 + r_{bb'}(y_{b'e} + y_{b'c})}$ $y_{21e} = \dfrac{y_m - y_{b'c}}{1 + r_{bb'}(y_{b'e} + y_{b'c})}$ $y_{22e} = \dfrac{y_{ce} + y_{b'c} + r_{bb'}[y_{b'e}(y_{ce} + y_{b'c}) + y_{b'c}(y_m + y_{ce})]}{1 + r_{bb'}(y_{b'e} + y_{b'e})}$ $\Delta y_e = \dfrac{y_{b'e}(y_{ce} + y_{b'c}) + y_{b'c}(y_m + y_{ce})}{1 + r_{bb'}(y_{b'e} + y_{b'e})}$
common-base circuit	$y_{11b} = \dfrac{y_m + y_{b'e} + r_{bb'}[y_{b'e}(y_{ce} + y_{b'c}) + y_{b'c}(y_m + y_{ce})]}{1 + r_{bb'}(y_{b'e} + y_{b'c})}$ $y_{12b} = \dfrac{-y_m - r_{bb'}[y_{b'e}(y_{ce} + y_{b'c}) + y_{b'c}(y_m + y_{ce})]}{1 + r_{bb'}(y_{b'e} + y_{b'c})}$ $y_{21b} = \dfrac{-y_m - y_{ce} - r_{bb'}[y_{b'e}(y_{ce} + y_{b'c}) + y_{b'c}(y_m + y_{ce})]}{1 + r_{bb'}(y_{b'e} + y_{b'c})}$ $y_{22b} = \dfrac{y_{b'c} + y_{ce} + r_{bb'}[y_{b'e}(y_{ce} + y_{b'c}) + y_{b'c}(y_m + y_{ce})]}{1 + r_{bb'}(y_{b'e} + y_{b'c})}$ $\Delta y_b = \dfrac{y_{b'e}(y_{ce} + y_{b'c}) + y_{b'c}(y_m + y_{ce})}{1 + r_{bb'}(y_{b'e} + y_{b'c})}$

The design of tuned amplifiers is based usually on y parameters. The determination of y parameters from the equivalent circuit is a more complicated task, especially when taking into account that the operating frequency of tuned amplifiers is generally higher. Therefore, the neglection of many parameters, justified at lower frequencies, is no longer applicable.

The y parameters of common-emitter and common-base configurations, based on the hybrid-π equivalent circuit shown in Fig. 4.9, are given by Table 4.7.

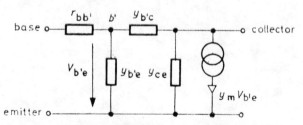

Fig. 4.9. Hybrid-π equivalent circuit of bipolar transistors

4.7. Types of the parameters

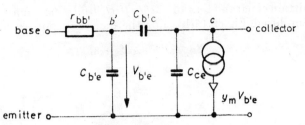

Fig.4.10. Simplified high-frequency hybrid-π equivalent circuit of bipolar transistors

At high frequencies where the condition $\omega \gg \omega_\beta$ is fulfilled, the real term in the admittance $y_{b'e}$ may be neglected. Similarly, the real part of admittance y_{ce} may also be neglected, and the circuit element $y_{b'c}$ may be considered as a pure capacitance (see Fig. 4.10):

$$y_{b'e} = j\omega C_{b'e};$$
$$y_{ce} = j\omega C_{ce}; \qquad (4.7.4)$$
$$y_{b'c} = j\omega C_{b'c}.$$

Calculating the y parameters of the common-emitter connection yields the scalar components as given in Table 4.8.

Table 4.8. Scalar components of the y parameters

$y_{11e} = g_{11e} + j\omega C_{11e}$	$g_{11e} = 1/r_b(1+s^2)$ $C_{11e} = C_{b'e}s^2/(1+s^2)$
$y_{12e} = g_{12e} + j\omega C_{12e}$	$-g_{12e} = C_{b'c}/r_b C_{b'e}(1+s^2)$ $-C_{12e} = C_{b'c}s^2/(1+s^2)$
$y_{21e} = \|y_{21e}\| e^{j\varphi_{21e}}$	$\|y_{21e}\| = g_m s/(1+s^2)(1+\omega^2/\omega_m^2)$ $-\varphi_{21e} = \arctan(\omega/\omega_m) + \arctan(1/s)$
$y_{22e} = g_{22e} + j\omega C_{22e}$	$g_{22e} = \dfrac{g_m \omega r_b C_{b'c}(\omega/\omega_m + 1/s)}{(1+\omega^2/\omega_m^2)(1+1/s^2)}$ $C_{22e} = C_{b'c}\left[1 + \dfrac{g_m r_b(1-\omega/\omega_m s)}{(1+\omega^2/\omega_m^2)(1+1/s^2)}\right] + C_{ce}$
notations	$s = 1/\omega r_b C_{b'e} \qquad \omega > \omega_\beta$ $y_m = \dfrac{g_m}{1+j\omega/\omega_m}$

The admittance parameters of *field-effect transistors* may be calculated from the simplified Fig. 2.18(b). For the common-source connection,

$$y_{11} = j\omega(C_{gs} + C_{gd});$$
$$y_{12} = -j\omega C_{gd};$$
$$y_{21} = g_m - j\omega C_{gd};$$
$$y_{22} = j\omega(C_{gd} + C_{ds}).$$

(4.7.5)

As with the equivalent circuit, the above parameters can only be applied with sufficient accuracy in the frequency range $\omega < \omega_m$.

The computer generation of the small-signal equivalent circuit, based on scattering parameters, is treated in [2.18]. The determination of the hybrid-π equivalent circuit by computer calculation, based on the nonlinear charge-controlled model, is given in [2.35]. Further computer simulations are presented in Refs. [2.28] and [2.32].

5. PROPERTIES OF WIDEBAND AMPLIFIERS

5.1. Analysis by the pole-zero method

Amplifiers are basically characterized by the transfer function. Let us investigate the transfer-function properties in the complex frequency range $p = \sigma + j\omega$. The transfer function is given by

$$A(p) = \frac{v_{\text{out}}(p)}{v_{\text{in}}(p)}, \tag{5.1.1}$$

where v_{out} and v_{in} are Laplace transforms of the output and input time functions, respectively.

The input-to-output amplitude ratio in the frequency range is expressed by the amplitude characteristic

$$a(\omega) = |A(p)|_{p=j\omega}, \tag{5.1.2}$$

which is the most important characteristic of all amplifiers. In certain cases, the delay characteristic may be of equal importance:

$$\tau(\omega) = \frac{d\varphi(A)}{d\omega}\bigg|_{p=j\omega}, \tag{5.1.3}$$

where $\varphi(A)$ is the phase of the transfer function. Additional to these characteristics, the unit-step response-time function may also be important because of the determination of rise time and possible overshoot. The Laplace transform used to shift from the time domain to the complex frequency domain may be solved in most cases by the commonly used set of formulae.

Expressing the transfer function by using either the equivalent circuit or the four-pole parameters, we have a ratio function with real parameters:

$$A(p) = A_0 \frac{1 + a_1 p + a_2 p^2 + \ldots + a_m p^m}{1 + b_1 p + b_2 p^2 + \ldots + b_n p^n}, \tag{5.1.4}$$

where A_0 is the zero-frequency transmission, and the denominator has a higher degree than the numerator ($n > m$). The poles p_{pk} and zeros p_{oj} of the transfer function are determined by changing over to the factored form expression:

$$A(p) = A_0' \frac{(p - p_{01})(p - p_{02}) \ldots (p - p_{0j})}{(p - p_{p1})(p - p_{p2}) \ldots (p - p_{pk})}. \tag{5.1.5}$$

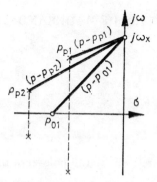

Fig. 5.1. Pole-zero pattern of circuits

Showing these poles and zeros, which may be real or conjugated complex numbers, on the complex-frequency plane, we arrive at the so-called pole-zero pattern of the circuit in question (Fig. 5.1). The transfer-function absolute value at the frequency p is given by the product and ratio, respectively, of the pole vector and zero vector lengths, and the transfer-function phase is given by the sum and difference, respectively, of the vector phases.

Several important conclusions may be deduced from the pole-zero pattern, e.g., regarding stability, characteristic shape, step-function response etc. It is for this reason that the pole-zero pattern is widely used for performing circuit analysis. In the following, however, the treatment of the transfer functions is based on the equations applying to the ω frequency domain.

5.2. Transfer functions and frequency properties

Rewriting the transfer function as given by eqn. (5.1.4) for the ω domain, we have

$$\frac{v_{out}(\omega)}{v_{in}(\omega)} = A(\omega) = A_0 \frac{1 + ja_1\omega - a_2\omega^2 - ja_3\omega^3 + a_4\omega^4 - \ldots}{1 + jb_1\omega - b_2\omega^2 - jb_3\omega^3 + \ldots}. \quad (5.2.1)$$

where A_0 denotes the zero-frequency transmission (low-frequency transmission); a and b are real coefficients, and the denominator always has a higher degree than the numerator.

Investigating the frequency dependence of the transfer function, the amplitude and phase responses have to be determined. The amplitude-frequency response may be written in the following form:

$$a(\omega) = |A(\omega)| = A_0 \frac{1 + c_2\omega^2 + c_4\omega^4 + c_6\omega^6 + \ldots}{1 + d_2\omega^2 + d_4\omega^4 + d_6\omega^6 + \ldots}, \quad (5.2.2)$$

where, for all c and d coefficients, the following relations hold:

$$c = c(a_1 \ldots a_m);$$
$$d = d(b_1 \ldots b_n).$$

5.2. Transfer functions

It can be seen that the amplitude response is an even function. Let us assume that the following condition holds for the amplitude response:

$$\left.\frac{d^k a(\omega)}{d\omega^k}\right|_{\omega=0} = 0, \quad (5.2.3)$$

i.e. all derivatives of the function disappear at zero frequency. In this case we have the so-called maximally flat amplitude response which means that the Taylor series of the function is tangential to the low-frequency A_0 value with the highest order. Naturally, the relation (5.2.3) should only hold for even degree numbers since the function is even. Equating these derivatives to zero yields a set of equations for the a and b coefficients. Solving this set will give the coefficients of the maximally flat amplitude response.

As an example, let us consider the transfer function

$$A(\omega) = A_0 \frac{1 + ja_1 \omega}{1 + jb_1 \omega - b_2 \omega^2}, \quad (5.2.4)$$

which has a quadratic denominator and linear numerator. For practical reasons, let us take the absolute value squared, and writing the equation

$$\frac{da^2}{d\omega^2} = 0,$$

we arrive at the following characteristic equation:

$$a_1^2 = b_1^2 - 2b_2. \quad (5.2.5)$$

Equation (5.2.5) may be applied to calculate the 3-dB cut-off frequency of the maximally flat response:

$$\omega_H = \sqrt{\frac{a_1^2 + \sqrt{a_1^4 + 4b_2^2}}{2b_2^2}}. \quad (5.2.6)$$

For the case of $a_1 = 0$, the expression may be simplified to $\omega_H = 1/\sqrt{b_2}$.

A similar reasoning may be applied for the phase response of the transfer function. By taking the derivatives of the delay frequency response

$$\tau(\omega) = \frac{d\varphi(\omega)}{d\omega}, \quad (5.2.7)$$

which may be calculated from the phase response, and equating these to zero at zero frequency as before, we have

$$\left.\frac{d^k \tau(\omega)}{d\omega^k}\right|_{\omega=0} = 0, \quad (5.2.8)$$

which represents a maximally flat delay response.

An important characteristic of the transfer function is the cut-off frequency ω_H pertaining to the 3-dB attenuation of the amplitude response, and possibly also the response peak which may appear. The pulse response is characterized by the rise time between 10% and 90% levels (t_r), and by the overshoot which may appear on the output response function $\varrho(t)$ in the case of unit-step input drive.

In the following, the parameters mentioned will be discussed for the more important elementary functions. For simplicity, the low-frequency transmission A_0 will be taken as unity.

5.3. Transfer function with first-degree denominator

The transfer function is given by

$$A(\omega) = \frac{1}{1 + j\omega b_1} = \frac{1}{1 + j\omega/\omega_0}, \qquad (5.3.1)$$

where the notation $b_1 = 1/\omega_0$ has been introduced. The amplitude response is given by

$$a(\omega) = 1/\sqrt{1 + (\omega/\omega_0)^2}, \qquad (5.3.2)$$

yielding the 3-dB cut-off frequency of $\omega_H = \omega_0$. The response is monotonous, having no extreme value. The phase characteristic is given by

$$\varphi(\omega) = \arctan(-\omega/\omega_0), \qquad (5.3.3)$$

and the response to the unit step is

$$\varrho(t) = 1 - \exp(-\omega_0 t), \qquad (5.3.4)$$

which is shown in Fig. 5.2 by the curve $n = 1$. The rise time is $t_r = 2.2/\omega_0$. The output response is seen to have no overshoot.

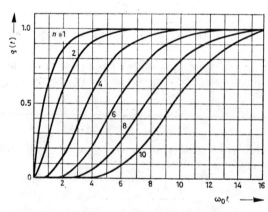

Fig. 5.2. Transient response of n identical transfer functions, each having first degree denominator, for various degree numbers

5.4. Transfer function made up of n identical first-degree denominators

Using eqn. (5.3.1), the transfer function is given by

$$A(\omega) = 1/(1 + j\omega/\omega_0)^n. \qquad (5.4.1)$$

The amplitude response is given by

$$a(\omega) = 1/[1 + (\omega/\omega_0)^2]^{n/2}, \qquad (5.4.2)$$

from which the 3-dB cut-off frequency is $\omega_H = \omega_0 \sqrt{2^{1/n} - 1}$. The phase response is given by

$$\varphi(\omega) = n \arctan(-\omega/\omega_0), \qquad (5.4.3)$$

and the response to the unit-step drive is expressed by

$$\varrho(t) = 1 - \sum_{k=0}^{k-1} [(\omega_0 t)^k \exp(-\omega_0 t)/k!], \qquad (5.4.4)$$

which is plotted for different n values in Fig. 5.2. The approximate value of the rise time is

$$t_r \simeq 2.2 \sqrt{n}/\omega_0. \qquad (5.4.5)$$

It is interesting to note that while n is increasing, the cut-off frequency is decreased and the rise time is increased, the product of these two quantities staying approximately constant, i.e. $\omega_H t_r \approx 2$.

5.5. Transfer function made up of n first-degree denominators having different cut-off frequencies

A drawback of the characteristic discussed in Section 5.4 is that the cut-off frequency ω_H is substantially decreased with increasing n. However, by proper choice of the different cut-off frequencies, both the amplitude response and the phase response can be shaped to be maximally flat.

The functions yielding maximally flat amplitude responses up to the value $n = 5$ are shown in Table 5.1. Here the notation $\omega/\omega_0 = \Omega$ has been applied to denote the relative frequency. The Butterworth-type amplitude response in this case is given by

$$a(\omega) = 1/\sqrt{1 + (\omega/\omega_0)^{2n}}, \qquad (5.5.1)$$

which has a monotonous shape (see Fig. 5.3), and the 3-dB cut-off frequency is given by $\omega_H = \omega_0$, independent of n. Rise time and overshoot for different n values are presented in Table 5.2. The overshoot is seen to be substantial, and in some cases when this may not be permitted, the coefficients may be chosen according to the maximally flat delay response as shown in Table 5.3. The 3-dB cut-off frequencies, the rise times and overshoot values per-

Table 5.1. Transfer functions of maximally flat Butterworth amplitude characteristics for $n = 1$–5

$$A = \frac{1}{1+j\Omega} \qquad (n = 1)$$

$$A = \frac{1}{1+j\,1.41\Omega - \Omega^2} \qquad (n = 2)$$

$$A = \frac{1}{1+j\,2\Omega - 2\Omega^2 - j\Omega^3} \qquad (n = 3)$$

$$A = \frac{1}{1+j\,2.61\Omega - 3.41\Omega^2 - j\,2.61\Omega^3 + \Omega^4} \qquad (n = 4)$$

$$A = \frac{1}{1+j\,3.24\Omega - 5.24\Omega^2 - j\,5.24\Omega^3 + 3.24\Omega^4 + j\Omega^5} \qquad (n = 5)$$

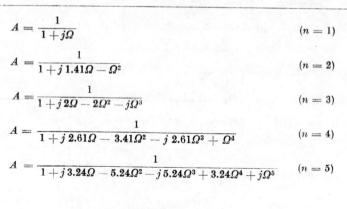

Fig. 5.3. Maximally flat amplitude characteristics for various degree numbers

taining to different degree characteristics are summarized in Table 5.4. The overshoot is seen to be consistently extremely high. Finally, it should be noted that the product $\omega_H t_r$ is again approximately constant, independent of the degree value.

Table 5.2. Normalized rise time t_r and overshoot percentage of transfer functions given in Table 5.1, for different degree numbers

n	$t_r \omega_0$	overshoot [%]
1	2.2	0
2	2.15	4.3
3	2.29	8.15
4	2.43	10.9
5	2.56	12.8

Table 5.3. Transfer functions pertaining to maximally flat delay time characteristics, for $n = 1$–5

$$A = \frac{1}{1 + j\Omega} \qquad (n = 1)$$

$$A = \frac{1}{1 + j\,1.73\Omega - \Omega^2} \qquad (n = 2)$$

$$A = \frac{1}{1 + j\,2.47\Omega - 2.4\Omega^2 - j\Omega^3} \qquad (n = 3)$$

$$A = \frac{1}{1 + j\,3.2\Omega - 4.39\Omega^2 - j\,3.12\Omega^3 - \Omega^4} \qquad (n = 4)$$

$$A = \frac{1}{1 + j\,3.94\Omega - 6.89\Omega^2 - j\,6.78\Omega^3 - 3.81\Omega^4 + j\Omega^5} \qquad (n = 5)$$

Table 5.4. Normalized values of the 3-dB cut-off frequency ω_H, rise time t_r and percentage values of overshoot for the transfer functions given in Table 5.3, for different degree numbers

n	ω_H/ω_0	$t_r\omega_0$	overshoot [%]
1	1	2.2	0
2	0.786	2.73	0.43
3	0.7	3.0	0.7
4	0.65	3.3	0.83
5	0.6	3.5	0.8

5.6. Transfer function with second-degree denominator

The two special cases of transfer functions having second-degree denominators are given by the relations pertaining to $n = 2$ in Section 5.5. Investigating the transfer properties generally, the transfer function is written in the following form:

$$A(\omega) = \frac{1}{1 + jb_1\omega - b_2\omega^2} = \frac{1}{1 + j2\zeta\omega/\omega_0 - (\omega/\omega_0)^2}, \qquad (5.6.1)$$

where the following notations have been introduced:

$$b_2 = 1/\omega_0^2, \quad b_1 = 2\zeta/\omega_0,$$

where ζ is called damping factor. The amplitude and phase characteristics are calculated as follows:

$$a(\omega) = 1/\sqrt{[1 - (\omega/\omega_0)^2]^2 + (2\zeta\omega/\omega_0)^2}\,; \qquad (5.6.2)$$

$$\varphi(\omega) = \arctan\left[\frac{-2\zeta\omega/\omega_0}{1 - (\omega/\omega_0)^2}\right]. \qquad (5.6.3)$$

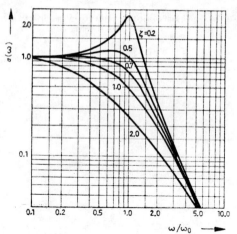

Fig. 5.4. Amplitude characteristic of transfer function with second-degree denominatot for various damping factors

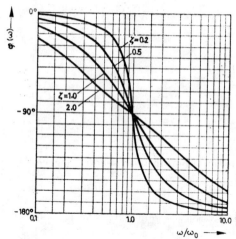

Fig. 5.5. Phase characteristic of transfer function with second-degree denominator for various damping factors

These are shown in Figs. 5.4 and 5.5, and the unit-step response is plotted in Fig. 5.6. It can be seen from Fig. 5.4 that a response peak is formed for values $\zeta < 0.707$, and the overshoot is substantially increased.

5.7. Practical expressions for transfer functions of higher degree

There are a few practical relations which have sound application, even for transfer functions which are mathematically complicated. One of these relations, as stated earlier, is given by the approximately constant product of

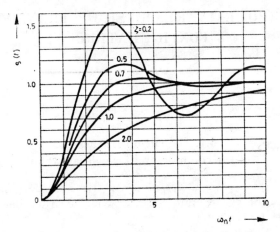

Fig. 5.6. Transient response of transfer function with second-degree denominator fo various damping factors

the rise time and the cut-off frequency:

$$t_r \omega_H \approx 2.2. \tag{5.7.1}$$

This relation is sometimes written in the form $B \simeq 0.34/t_r$ where B is the required bandwidth. The overshoot in the unit-step response is related to the amplitude-response peak and to the slope of the high-frequency part. It has been seen in the cases investigated that the increase of any of these parameters will increase the overshoot. Finally, a practical statement which applies for low-degree transfer functions: in contrast with the zero rise-time unit step, the overshoot in response to the input step function having a finite rise time of

$$t_1 = 1/2\omega_H, \tag{5.7.2}$$

may generally be neglected.

6. FREQUENCY RESPONSE OF TRANSISTOR CONFIGURATIONS

6.1. The cut-off frequency of the common-emitter configuration

To calculate the frequency response of transistor amplifiers the frequency response of the common-emitter configuration has to be known. The configuration will be investigated for resistive generator resistance R_g, and for resistive load resistance R_L, as shown in Fig. 6.1. The problem is to calculate the voltage gain V_2/V_g as a function of frequency.

The calculation will be carried out according to the equivalent circuit shown in Fig. 6.2 (a) where the following simplified notations have been used: r_b (base resistance), r_e (dynamic emitter resistance), and C_c (collector-base capacitance).

The ratio V_2/V_g may be expressed from the equations which are written for the circuit to be investigated, and the ω-dependent factor will give the voltage frequency response.

The method applied for the determination of the frequency response is to eliminate the feedback capacitance C_c, thus separating the input and output part of the circuit. The circuit is thereby divided into two parts, connected by the internal transconductance alone.

The feedback capacitance may only be disregarded provided the transfer properties of the circuit are not altered. This means that the effect of this capacitance must somehow be taken into account in both the input and output parts.

In the input part, the effect of C_c may be accounted for taking the so-called Miller admittance between points b–e, which has the following value:

$$Y_{Mi} = -A_v^0 C_c, \qquad (6.1.1)$$

where A_v^0 is the internal voltage gain from point b' to the collector, and is assumed to be much higher than unity: $|A_v^0| \gg 1$. For the case of a resistive load resistance R_L, this voltage gain will be real and the Miller admittance purely imaginary, i.e. capacitive:

$$\frac{Y_{Mi}}{j\omega} = C_{Mi} = \frac{R_L}{r_e} C_c = (1-\psi) \frac{1}{\omega_T r_e}, \qquad (6.1.2)$$

where $\psi = 1 + \omega_T C_c R_L$.

The factor ψ, which will be frequently used in the following and which takes into account the load-resistance dependent feedback rarely differs much from unity.

6.1. Common-emitter configuration

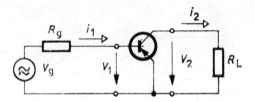

Fig. 6.1. Circuit for calculating frequency response of common-emitter configuration

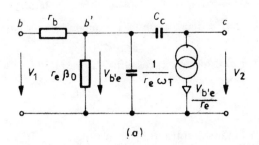

(a)

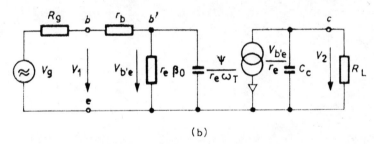

(b)

Fig. 6.2. Common-emitter configuration with (a) application of equivalent circuit and (b) elimination of feedback capacitance

The modified input circuit is shown in Fig. 6.2 (b). Capacitance C_0 is a function of resistance R_L, so it should be emphasized that ignoring of the feedback is only valid for a specific R_L value:

$$C_0 = \frac{\psi}{\omega_T r_e}. \tag{6.1.3}$$

In the output part, the effect of C_c is accounted for by placing a capacitance C_c in parallel with the output. This may be understood directly from the fact that for high voltage gain A_v^0, the other end of the capacitance C_c, as seen from the collector point, is approximately at earth potential.

The divided equivalent circuit, valid only for a specific load-resistance value R_L, is presented in Fig. 6.2 (b). Calculating the ratio V_2/V_g according

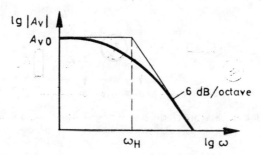

Fig. 6.3. Bode diagram of common-emitter circuit

to this circuit, the following expression is derived:

$$\frac{V_2}{V_g} = A_v = A_{v0} \frac{1}{1 + j\omega/\omega_H}, \qquad (6.1.4)$$

where the low-frequency voltage gain is given by

$$A_{v0} = -\frac{\beta_0 R_L}{\beta_0 r_e + r_b + R_g} = -\frac{\beta_0 R_L}{r_{1E} + R_g}, \qquad (6.1.5)$$

and the 3-dB cut-off frequency is given by

$$\omega_H = \frac{\omega_\beta}{\psi} \frac{\beta_0 r_e + r_b + R_g}{r_b + R_g} = \frac{\omega_\beta}{\psi} \frac{r_{1E} + R_g}{r_b + R_g}, \qquad (6.1.6)$$

where $r_{1E} = r_b + \beta_0 r_e$ is the common-emitter circuit low-frequency input resistance. Equation (6.1.4) is approximate, the exact relation is more complicated and the denominator contains several factors. However, the remaining cut-off frequencies (e.g., the cut-off frequency of the output circuit) are much higher than ω_H given in (6.1.6), so these factors may be neglected.

The frequency dependence is thus described by expression (6.1.4) with a single cut-off frequency ω_H, shown in Fig. 6.3. The cut-off frequency ω_H may be regarded as the reciprocal value of the input RC time constant; $\omega_H = 1/R_0 C_0$, where C_0 is the capacitance according to eqn. (6.1.3), and R_0 is the resistance comprised of the input elements which have real values: $R_0 = (r_b + R_g) \| \beta_0 r_e$. It has been assumed here that $\beta_0 \gg 1$ and $R_L \ll 1/\omega_H C_c$.

From the general expression given by (6.1.4), the voltage gain which applies to the voltage controlled case is given by the substitution $R_g = 0$, which is the transistor voltage gain. After simplification, the expression (6.1.5) will give the low-frequency voltage gain:

$$A_{v0}(R_g = 0) = \frac{-\beta_0 R_L}{\beta_0 r_e + r_b}. \qquad (6.1.7)$$

6.1. Common-emitter configuration

The cut-off frequency according to eqn. (6.1.6) will be denoted by subscript v, corresponding to the voltage-controlled case:

$$\omega_{\mathrm{H}v}(R_g = 0) = \frac{\omega_\beta}{\psi} \frac{\beta_0 r_e + r_b}{r_b} = \frac{\omega_\beta}{\psi} \frac{r_{1\mathrm{E}}}{r_b}. \tag{6.1.8}$$

Comparing this expression with (6.1.6), which applies for the general generator resistance R_g, it may be seen that decreasing the generator resistance will substantially increase the cut-off frequency, limited only by the value of r_b. For the case of $R_g = 0$ and $\beta_0 r_e \gg r_b$, the cut-off frequency will be

$$\omega_{\mathrm{H}v} = 1/r_b C_{b'e} = \omega_T r_e/r_b. \tag{6.1.9}$$

This is the highest cut-off frequency which is attainable for the common-emitter configuration. Increasing the generator resistance decreases the cut-off frequency, and the lowest value is reached at infinite generator resistance.

However, there is no voltage gain with infinite generator resistance, and current gain must, therefore, be considered. Substituting $R_g = \infty$, the current gain is given by

$$\frac{i_2}{i_g} = A_i(R_g = \infty) = -\frac{\beta_0}{1 + j\omega/\omega_{\mathrm{H}i}}, \tag{6.1.10}$$

where the 3-dB cut-off frequency is given by

$$\omega_{\mathrm{H}i} = \omega_\beta/\psi, \tag{6.1.11}$$

where the subscript i stands for current drive.

The following can be stated for the cut-off frequencies. The minimum obtainable cut-off frequency is given by (6.1.11), applying to current drive and a relatively high valued load resistance, R_L. Decreasing the generator resistance will increase the cut-off frequency, and the maximum realizable value, given by (6.1.8), is reached with voltage drive. The cut-off frequency lies generally between these limits, according to the generator resistance used.

A figure of merit of wideband amplifiers is the gain-bandwidth product GB. Let us calculate the GB product for this configuration, so that we can later evaluate different methods for increasing the bandwidth. For a general value of generator resistance R_g, if follows from (6.1.5) and (6.1.6) that

$$GB_v = |A_{v0}| \omega_H = \frac{\omega_T}{\psi} \frac{R_L}{r_b + R_g}, \tag{6.1.12}$$

where GB_v stands for the gain-bandwidth product expressed by the voltage gain. Similarly, the gain-bandwidth product expressed by the current gain is given by

$$GB_i = \beta_0 \omega_{\mathrm{H}i} = \omega_T/\psi. \tag{6.1.13}$$

7 Kovács: High-Frequency

6.2. Input and output impedances of the common-emitter configuration

The input impedance of the common-emitter configuration presented in Fig. 6.4, follows directly from the simplified equivalent circuit shown in Fig. 6.2 (b). In contrast with the transistor equivalent circuit shown in Fig. 6.2 (a), the input capacitance is increased by the factor ψ as a result of feedback. The input impedance Z_{in}, may be evaluated from this figure.

The frequency function of the input impedance on the complex plane is plotted in Fig. 6.5. At infinite frequency, capacitance C_0 represents a short-circuit, so the impedance is determined by the base resistance.

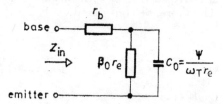

Fig. 6.4. Input impedance of common-emitter circuit

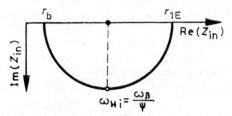

Fig. 6.5. Plot of the common-emitter circuit input impedance on the complex plane

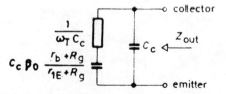

Fig. 6.6. Output impedance of the common-emitter circuit

The output impedance of the common-emitter configuration is presented in Fig. 6.6. Instead of a mathematical derivation, only qualitative analysis will be given. The series RC elements which are parallel to the output terminals are due to the feedback caused by C_c. For a high generator resistance R_g, this capacity is approximately equal to $\beta_0 C_c$. At low frequencies, the series resistance may be neglected, making the output impedance capacitive and with a value of $(1 + \beta_0)C_c$.

The role of the series resistance increases with higher frequencies, and at very high frequencies, the series capacitance may be regarded as a short-circuit, causing the output conductance $\omega_T C_c$ to appear at the output. It should be noted that in the circuit of Fig. 6.6, the effects of conductances g'_{bc} and g_{ce} have been neglected.

6.3. Frequency dependence of the common-base and common-collector configurations

In the majority of wideband amplifiers, the common-emitter configuration is applied. However, common-base configurations are also in use. The common-collector (emitter-follower) configuration is mainly used as an impedance converter, and is applied in wideband amplifiers for this purpose. In the following, the basic properties of these two configurations will be investigated briefly.

The input impedance of the *common-base* transistor is shown in Fig. 6.7. At low frequencies, the input resistance is given by the well known expression $r_{1B} = r_e + r_b/\beta_0$. At very high frequencies the importance of the capacity $1/r_e \omega_\alpha$, which is in parallel with resistance r_e, will be more pronounced. For a practical case the elements related to the base resistance r_b may be neglected. In wideband amplifiers, the upper cut-off frequency is usually well below ω_α, so it is permitted to take the input resistance of the common-base configuration to be equal to r_e.

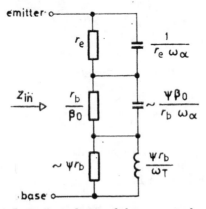

Fig. 6.7. Input impedance of the common-base circuit

Owing to the low input impedance of the common-base configuration, this circuit is driven by a current generator in most practical multi-stage wideband amplifiers. Writing the current gain for the circuit of Fig. 6.8, the following approximate cut-off frequency is derived:

$$\omega_H = \frac{\omega_a}{1 + \omega_a R_L C_c}. \qquad (6.3.1)$$

The cut-off frequency is decreased by the non-zero value of load resistance R_L, similarly to the factor ψ known from the common-emitter configuration. For zero load resistance, we arrive back at the cut-off frequency of the short-circuit current-gain factor.

The output impedance of the common-base stage usually exceeds the input resistance of the next stage. The approximation of the output impedance for $R_g = \infty$ is shown by Fig. 6.9 where the low-frequency output resistance of the common-base circuit is denoted by r_c.

The *emitter-follower* is frequently used for its high input impedance. The approximation of the input impedance is shown in Fig. 6.10. At low frequencies, the input resistance is given by $r_{1C} = r_b + (R_L + r_e)\beta_0$ which may become very high with high load resistances. The frequency dependence of the input impedance is primarily determined by the shunt capacitance. It can be seen that the characteristic frequency of the RC network shown is the well known quantity ω_β/ψ, which is the cut-off frequency of current gain in the common-emitter configuration. The effect of the series-connected parallel RL circuit is usually negligible.

According to the circuit shown in Fig. 6.11, the voltage gain of the emitter-follower is given by the ratio V_2/V_g which may be expressed approximately as:

$$\frac{V_2}{V_g} = A_{v0} \frac{1}{1 + j\omega/\omega_H}, \qquad (6.3.2)$$

where the low-frequency voltage gain is given by

$$A_{v0} = \frac{R_L}{R_L + r_e + \beta_0(r_b + R_g)}, \qquad (6.3.3)$$

and the 3-dB cut-off frequency is given by

$$\omega_H \simeq \omega_a \frac{R_L + r_e + (r_b + R_g)\beta_0^{-1}}{R_L + (1 + \varphi_0)(r_b + R_g)}. \qquad (6.3.4)$$

Note that in the case of infinite generator resistance $R_g = \infty$, the cut-off frequency as given by (6.3.4) is $\omega_H = \omega_{Hi}$, i.e. it is equal to the cut-off frequency of the current gain in the common-emitter configuration.

The cut-off frequency is increased by decreasing the generator resistance. With zero generator resistance the cut-off frequency of the voltage gain is

6.3. Common-base and common-collector configurations

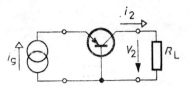

Fig. 6.8. Circuit for calculating the frequency response of the common-base circuit

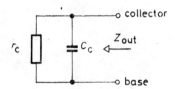

Fig. 6.9. Output impedance of the common-base circuit

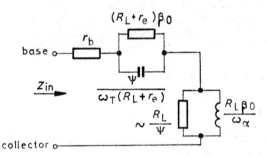

Fig. 6.10. Input impedance of common-collector circuit

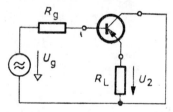

Fig. 6.11. Circuit for calculating the frequency response of the common-collector circuit

determined by the load resistance R_L. For zero load resistance,

$$\omega_H(R_g = R_L = 0) = \omega_\beta r_{1E}/r_b, \qquad (6.3.5)$$

which is seen to equal expression (6.1.8) applying for the common-emitter configuration, in the limiting case of $R_L = 0$. This means that for voltage drive and for low-valued load resistances, the voltage-gain cut-off frequen-

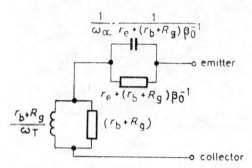

Fig. 6.12. Output impedance of the common-collector circuit

cies for the common-emitter and common-collector configurations are identical. This is explained by the facts that with increasing R_L, the cut-off frequency of the common-emitter configuration decreases, and that of the common-collector configuration increases, tending to ω_α according to (6.3.4):

$$\omega_H(R_g = 0,\ R_L = \infty) = \omega_a. \tag{6.3.6}$$

The output impedance of the common-collector configuration for any generator resistance R_g is shown by Fig. 6.12. At low frequencies, the output resistance is

$$r_{out} = r_e + (r_b + R_g)/\beta_0, \tag{6.3.7}$$

which is a rather low value for low generator resistances. For $R_g = 0$, we have the input resistance of the common-base configuration: $r_{out} = r_{1B} = r_e + r_b/\beta_0$. The frequency dependence of the output impedance is governed by the shunt capacity and the inductance.

6.4. Field-effect transistor configurations

Figure 6.13 shows the equivalent circuit of the common-source JFET amplifier. In contrast with previous considerations, the capacitive loading has also been taken into account. As the transconductance g_m of field-effect transistors is much lower than that of bipolar transistors for the case of high emitter currents, a high load resistance is needed for sufficiently high gain. This results in a higher shunting effect of the parallel-connected capacitances C_L, at high frequencies. The voltage gain is calculated to be approximately [2.15]

$$A(\omega) = \frac{V_2}{V_{gs}} \approx \frac{A_0}{1 + j\omega R'_L C'_L} = \frac{A_0}{1 + j\omega/\omega_H}, \tag{6.4.1}$$

where the low-frequency gain is given by $A_0 = -R'_L g_m$ and the 3-dB cut-off frequency by $\omega_H = 1/R'_L C'_L$. The circuit elements defining the cut-off

6.4. Field-effect transistor configurations

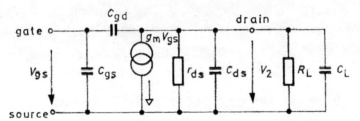

Fig. 6.13. Equivalent circuit of the common-source JFET amplifier

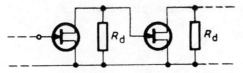

Fig. 6.14. Cascade of common-source stages

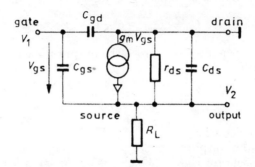

Fig. 6.15. Circuit for calculating the frequency response of source-follower circuit

frequency are the following:

$$R'_L = R_L \| r_{ds};$$
$$C'_L = C_L + C_{ds} + C_{gd}. \qquad (6.4.2)$$

It has been assumed during the calculations that $|A_0| \gg 1$.

Figure 6.14 shows a cascade connection of identical stages. The cut-off frequency of the stages is given by

$$\omega_H = 1/R'_d C_{equ}, \qquad (6.4.3)$$

where $R'_d = R_d \| r_{ds}$ and the equivalent load capacitance is given by

$$C_{equ} = C_{gs} + C_{ds} + C_{gd}(1 - A_0), \qquad (6.4.4)$$

where the last term is the Miller capacity originating from feedback.

Figure 6.15 shows a source-follower configuration. The calculation of the voltage gain [6.4], based on the equivalent circuit, has the following result:

$$A_v = \frac{V_2}{V_1} = A_0 \frac{1 + j\omega/\omega_1}{1 + j\omega/\omega_2}. \quad (6.4.5)$$

The low-frequency gain, assuming $r_{ds} \ll R_L$, is given by

$$A_0 = \frac{g_m R_L}{1 + g_m R_L}, \quad (6.4.6)$$

and the cut-off frequencies are given by

$$\omega_1 = g_m/C_{gs}; \quad (6.4.7)$$

$$\omega_2 = \frac{1 + g_m R_L}{R_L(C_{gs} + C_{ds})}. \quad (6.4.8)$$

The input impedance is capacitive, increased by the Miller capacity:

$$C_{in} = C_{gd} + C_{gs}(1 - A_0). \quad (6.4.9)$$

In the case of low generator resistance the output impedance is given by the following expression:

$$Z_{out} = \frac{1}{g_m(1 + j\omega\, C_{gs}/g_m)}. \quad (6.4.10)$$

At low frequencies when $\omega C_{gs} \ll g_m$, this is simplified to give the value $R_{out} = 1/g_m$ for the output impedance, as anticipated.

Figure 6.16, based on Ref. [6.3], shows a wideband source-follower configuration with high input impedance. The transconductance of the field-effect transistor applied is $g_m \approx 12$ mA/V. Resistor R_2, capacitor C_1, and di-

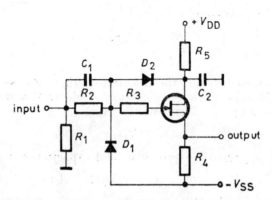

Fig. 6.16. Source-follower circuit with high input impedance

odes D_1 and D_2 serve for overload protection, R_4 and R_5 are included for bias adjustment. The real part of the input impedance is given by R_1. This connection is commonly used for the input stage of oscilloscope wideband vertical amplifiers.

In Ref. [6.2], further source-follower configurations which decrease the input capacitance are given. Reference [6.7] contains an amplifier on account of very low input current (10^{-14} A), 0.4 pF input capacity and 4 MHz bandwidth, and a temperature drift of approximately 100 μV/°C.

The common-gate configuration of field-effect transistors is seldom applied in wideband amplifiers, so this will not be dealt with here.

7. FREQUENCY RESPONSE OF MULTI-STAGE WIDEBAND AMPLIFIERS

7.1. Frequency response of a two-stage common-emitter amplifier

Let us investigate the frequency response of the two-stage common-emitter amplifier shown in Fig. 7.1. The amplifier is driven by a generator having a resistance R_g, and feeds a load resistance R_L. Resistance R is connected in parallel between the two transistors, primarily for DC connection purposes, but is also used frequently for achieving a suitable frequency response.

In circuit analysis an approximation is needed in the factor ψ_1 of the first stage, because of the frequency-dependent real part of the load impedance:

$$R_{L1}(\omega) = R \parallel \mathrm{Re}(Z_{\mathrm{in}2}), \tag{7.1.1}$$

where R_{L1} is the load resistance of the first stage, and $Z_{\mathrm{in}2}$ is the input impedance of the second stage. This may be approximated by taking the mean value between the frequency response maximum and minimum values, always taking into account the parallel resistance R. The real part of impedance $Z_{\mathrm{in}2}$ has a maximum value of r_{1E} (at low frequencies) and a minimum value of r_b, so that

$$R_{L1} \approx (R \parallel r_{1E} + R \parallel r_b)/2. \tag{7.1.2}$$

This average value of the resistance is accounted for by the factor ψ_1 of the first transistor which is given by $\psi_1 = 1 + \omega_T C_c R_{L1}$. Writing the equations for the circuit of Fig. 7.1, and noting the approximation of R_{L1}, the voltage gain is expressed by

$$\frac{V_2}{V_g} = A_{v0} \frac{1}{(1 + j\omega/\omega_{H1})(1 + j\omega/\omega_{H2})}, \tag{7.1.3}$$

where A_{v0} is the low-frequency voltage gain of the two-stage amplifier:

$$A_{v0} = \frac{\beta_0^2 R R_L}{(r_{1E} + R_g)(r_{1E} + R)}. \tag{7.1.4}$$

The two cut-off frequencies are

$$\omega_{H1} = \frac{\omega_\beta}{\psi_1} \frac{r_{1E} + R_g}{r_b + R_g}; \tag{7.1.5}$$

7.2. Common-emitter–common-base cascade connection

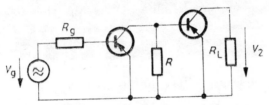

Fig. 7.1. Two-stage common-emitter amplifier

$$\omega_{H2} = \frac{\omega_\beta}{\psi} \frac{r_{1E} + R}{r_b + R}. \qquad (7.1.6)$$

Remembering that the actual generator resistance of the second stage is the resistance R, we see that the cut-off frequencies according to eqn. (6.1.6) are given in both expressions.

Let us investigate the conditions applying in the limit cases $R = \infty$ and $R_g = \infty$. As the resistance R_{L1} only has a secondary effect on ψ, let us assume that $\psi_1 = \psi$. In this case, $\omega_{H1} \approx \omega_{H2} = \omega_\beta/\psi$, i.e. the current-gain cut-off frequency is given by (6.1.11).

Without making the approximation $\psi_1 = \psi$, the value R_{L1} calculated from (7.1.2) and pertaining to $R = \infty$ should be taken into account:

$$R_{L1} \cong (r_{1E} + r_b)/2. \qquad (7.1.7)$$

7.2. Frequency response of a common-emitter — common-base cascade connection

In wideband amplifiers, the common-emitter–common-base cascade connection shown in Fig. 7.2 is frequently applied [7.3]. The effect of shunting by the resistance R between the two transistors is generally negligible when in parallel to the input impedance to the common-base stage, and this will not show up in further calculations because of the current-generator drive. For similar reasons, feedback capacity C_c may be neglected in the first stage, as the first-stage voltage gain is very low.

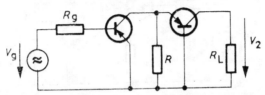

Fig. 7.2. Common-emitter–common-base two-stage amplifier

The following expression is obtained for the first-stage voltage gain:

$$\frac{V_2}{V_g} = A_v = A_{v0} \frac{1}{(1+j\omega/\omega_{H1})(1+j\omega/\omega_{H2})}, \quad (7.2.1)$$

where the low-frequency voltage gain is given by

$$A_{v0} = -\frac{\beta_0 R_L}{r_{1E} + R_g}. \quad (7.2.2)$$

The two cut-off frequencies are given by the following expressions:

$$\omega_{H1} = \omega_\beta \frac{r_{1E} + R_g}{r_b + R_g}, \quad (7.2.3)$$

which corresponds to the cut-off frequency (6.1.6) of the common-emitter configuration ($\psi = 1$ because of the loading), and

$$\omega_{H2} = \frac{\omega_\alpha}{1 + \omega_\alpha C_c R_L}. \quad (7.2.4)$$

It has again been assumed that the parameters of the two transistors are identical.

The gain reduction is determined primarily by ω_{H1}; ω_{H2} is substantially higher and thus has a smaller effect, especially with a high generator resistance R_g.

7.3. Frequency response of common-collector — common-emitter cascade connection

A common-collector configuration results in high input impedance, but having no voltage gain, it is usually followed by a common-emitter stage. The frequency response of the two-stage amplifier thus formed is very favourable, and so it is frequently used as wideband amplifier (Fig. 7.3).

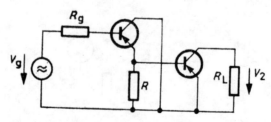

Fig. 7.3. Common-collector – common-emitter two-stage amplifier

7.3. Common-collector–common-emitter connection

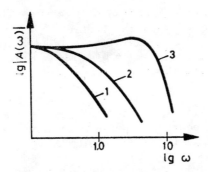

Fig. 7.4. Two-stage amplifier frequency responses. (1) Common-emitter – common-emitter stages, (2) common-emitter – common-base stages, (3) common-collector – common-emitter stages

The calculation has been omitted, but the end result for the voltage gain of this connection is given by

$$\frac{V_2}{V_g} = A_v = A_{v0} \frac{1 + j\omega/\omega_\alpha}{1 + j\omega(1/\omega_{H1} + 1/\omega_{H2}) - \omega^2/\omega_{H1}\omega_{H2}}, \qquad (7.3.1)$$

where the low-frequency voltage gain is given by

$$A_{v0} = \beta_0 R_L / r_{1E}, \qquad (7.3.2)$$

and the cut-off frequencies are the following:

$$\omega_{H1} = \frac{\omega_\beta}{\psi} \frac{r_{1E}}{r_b + r_e + R_g/\beta_0}; \qquad (7.3.3)$$

$$\omega_{H2} = \omega_\alpha \frac{r_{1E} R}{r_{1E} R + (1 + \varphi_0)(r_{1E} + R)(r_b + R_g)}; \qquad (7.3.4)$$

$$\omega_{H3} = \omega_\alpha \frac{r_b + r_e + R_g/\beta_0}{r_b + (1 + \varphi_0)(r_b + R_g)}. \qquad (7.3.5)$$

It was assumed for the calculation that the two transistors have identical parameters. Further, the generator resistance is low enough that $R_g \ll \beta_0 r_{1E}$, and the shunt resistance between the two transistors is much higher than the base resistance.

As seen from eqn. (7.3.1), the frequency response is not always monotonous but may show a voltage peak. This is due to the capacitive input impedance of the common-emitter stage and to the inductive output impedance of the common-collector stage.

Normalized frequency responses of cascade amplifiers utilizing identical transistors are shown in Fig. 7.4. The cut-off frequency is lowest for response

1, applicable to the common-emitter configuration. However, this configuration has the highest power gain.

The common-emitter–common-base combination has higher bandwidth but lower power gain as shown by response *2*. Response *3* applies for the common-collector–common-emitter combination showing the widest bandwidth with a small voltage peak.

7.4. Frequency response of a common-emitter three-stage configuration

A common-emitter three-stage amplifier with usual loadings is shown in Fig. 7.5. The voltage gain can be expressed in the following form:

$$\frac{V_2}{V_g} = A_{v0} \frac{1}{(1+j\omega/\omega_{H1})(1+j\omega/\omega_{H2})(1+j\omega/\omega_{H3})}. \quad (7.4.1)$$

The calculation has been omitted, and the end result alone is presented. The low-frequency voltage gain is given by

$$A_{v0} = -\beta_0^3 \frac{R_1 R_2 R_L}{(r_{1E}+R_g)(r_{1E}+R_1)(r_{1E}+R_2)}. \quad (7.4.2)$$

The cut-off frequencies are the following:

$$\omega_{H1} = \frac{\omega_\beta}{\psi_1} \frac{r_{1E}+R_g}{r_b+R_g}; \quad (7.4.3)$$

$$\omega_{H2} = \frac{\omega_\beta}{\psi_2} \frac{r_{1E}+R_1}{r_b+R_1}; \quad (7.4.4)$$

$$\omega_{H3} = \frac{\omega_\beta}{\psi} \frac{r_{1E}+R_2}{r_b+R_2}. \quad (7.4.5)$$

The factors ψ in the cut-off frequency expressions depend on the load resistances of the stages as follows:

$$\psi_1 = 1 + \omega_T C_c R_{L1}; \quad (7.4.6)$$

$$R_{L1} = (R_1 r_{1E} + R_1 r_b)/2, \quad (7.4.7)$$

and further,

$$\psi_2 = 1 + \omega_T C_c R_{L2}; \quad (7.4.8)$$

$$R_{L2} = (R_2 r_{1E} + R_2 r_b)/2, \quad (7.4.9)$$

where r_{1E} is the low-frequency input resistance of the common-emitter stage.

For infinite inter-stage resistances, i.e. $R_1 = R_2 = \infty$, the above expressions for cut-off frequencies ω_{H2} and ω_{H3} will simplify. For current-

7.5. Monolithic integrated multi-stage amplifiers

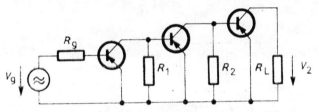

Fig. 7.5. Three-stage common-emitter amplifier

generator drive ($R_g = \infty$), and assuming that load resistance R_L is nearly equal to r_{1E}, we have

$$\frac{i_{out}}{i_{in}} = A_{i0} \frac{1}{(1 + j\omega/\omega_{H0})^3}, \quad (7.4.10)$$

where the low-frequency gain is $A_{i0} = -\beta_0^3$ and the cut-off frequency is given by

$$\omega_{H0} \simeq \frac{\omega_\beta}{1 + \omega_T C_c r_{1E}}. \quad (7.4.11)$$

From this expression, the frequency response of a cascade connection of any number of common-emitter stages can be determined, assuming of course that the inter-stage shunt resistances have negligible effect (these are needed for DC connection). The current gain of an n-stage amplifier is thus given by

$$A_i \simeq \beta_0^n \frac{1}{(1 + j\omega/\omega_{H0})^n}. \quad (7.4.12)$$

The finite values of inter-stage shunt resistances have the effect of decreasing the amplification, but simultaneously, the bandwidth is increased. A compromise has thus to be reached when choosing the shunt resistances, depending on the bandwidth and gain requirements of a specific case.

The gain-bandwidth product of the amplifier is dependent on the inter-stage shunt resistances, and has a maximum value at shunt resistance R_{opt} which is given by

$$R_{opt} = \sqrt{r_b/\omega_T C_c} \ll r_{1E}. \quad (7.4.13)$$

In Ref. [7.4], the overall h parameters of the circuit made up of n identical stages are calculated.

7.5. Monolithic integrated multi-stage wideband amplifiers

In monolithic integrated multi-stage transistor amplifiers, both transistors and resistors are produced by the same technology on a single chip. As explained in Section 3.2, these amplifiers are primarily DC coupled in

7. Multi-stage wideband amplifiers

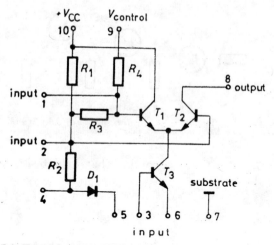

Fig. 7.6. Typical circuit of high-frequency monolithic amplifier

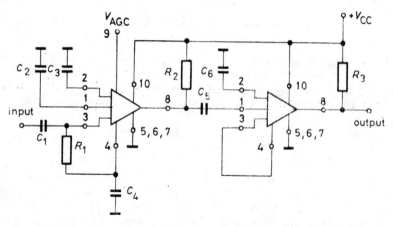

Fig. 7.7. Two-stage wideband amplifier consisting of monolithic blocks

order to avoid the need for coupling capacitors which would be difficult to arrange.

Monolithic integrated amplifiers are made accessible at a relatively high number of internal points in order that the circuit properties may be modified by compensations, short-circuits, loadings and feedback connections. In this section, we shall only deal with monolithic amplifiers without feedback. The most general form of monolithic amplifier is the differential amplifier (also called "long-tail pair") shown in Fig. 7.6.

In one of the applications, voltage drive is via input 3, and the remaining input points are grounded for AC. The two-stage amplifier thus formed is a common-emitter–common-base cascade connection having a high-fre-

7.5. Monolithic integrated multi-stage amplifiers

quency response according to Section 7.2. High-frequency properties are governed by the common-emitter stage as the cut-off frequency of the common-base stage is much higher. The reverse transmission is extremely low which is especially advantageous for tuned amplifiers (i.e. at high load impedances). We shall return to this in Section 17.1.

In another application, voltage drive at input 1 is used, and the remaining points are again grounded for AC. A common-collector–common-base two-stage amplifier is thus formed, transistor T_3 acting merely as a resistor. This connection, although it does find some application, has less significance for wideband amplifiers.

In Fig. 7.7, a two-stage video amplifier consisting of monolithic blocks according to Fig. 7.6 is shown. The first block is connected as a common-emitter–common-base amplifier, and the second block as a common-collector–common-base amplifier. Changing the control voltage V_{AGC} at the control input of the first block, amplification can be adjusted in a wide range, without substantial modification of the frequency response. With resistances $R_1 = 50$ ohm, $R_2 = 1.5$ kohm, $R_3 = 3.3$ kohm, and with blocking and coupling capacitors $C = 1$ μF, the following gain values are obtained:

$$A = 55 \text{ dB at } V_{AGC} = 0 \text{ V},$$

$$A = 25 \text{ dB at } V_{AGC} = 4.5 \text{ V}.$$

Bandwidth is within the range $B = 6.4 - 8.2$ MHz, depending on gain control.

8. COMPENSATED WIDEBAND AMPLIFIERS

8.1. Common-emitter configuration with output parallel compensation

The low bandwidth of the common-emitter configuration may be increased considerably by the use of compensating inductances, giving a load impedance increasing with frequency (Fig. 8.1). The voltage across inductance L_2 increases with frequency as does the cut-off frequency. This effect is greatly influenced by the loading of the circuit following the stage. We shall assume in the following that the input impedance of the next stage does not load the series RL circuit. The load impedance of the common-emitter stage across which voltage V_2 appears is then given by

$$Z_L = R_L + j\omega L_2 = R_L(1 + j\omega/\omega_2), \qquad (8.1.1)$$

where $\omega_2 = R_L/L_2$ is the characteristic frequency of the load. Relating the output signal to the input current results in the transfer impedance given by $Z_T = V_2/i_1$. In the following, this transfer is investigated using the h parameters of the common-emitter configuration, giving the equivalent circuit shown in Fig. 8.2. Here the complex impedance Z_L is connected in parallel to the $1/h_{22e}$ output impedance of the transistor.

The transfer impedance is the product of the current gain and the load

$$Z_T = A_i Z_L = \frac{h_{21e}}{1 + h_{22e} Z_L} Z_L. \qquad (8.1.2)$$

Substituting the values of the h parameters from (4.4.1) results in the following expression:

$$Z_T = \beta_0 R_L \frac{1 + j\omega/\omega_2}{1 + j\psi\omega/\omega_\beta - \omega^2/\omega_c^2}, \qquad (8.1.3)$$

where ω_c, the resonance frequency characterizing the output circuit, is given by

$$\omega_c = 1/\sqrt{L_2 \beta_0 C_c}, \qquad (8.1.4)$$

as the output capacitance is β_0 times the capacitance C_c for the case of current-generator drive, and $\psi = 1 + \omega_T C_c R_L$.

The frequency responses with various inductances L_2, are shown in Fig. 8.3. Gradual increase in inductance results in a flattening of the response, and

8.1. Output parallel compensation

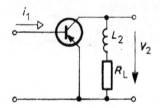

Fig. 8.1. Common-emitter amplifier with output parallel compensation

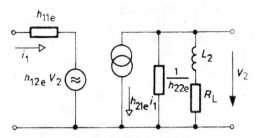

Fig. 8.2. Equivalent circuit of amplifier with output parallel compensation

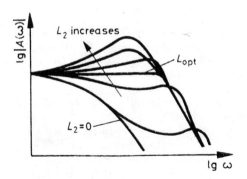

Fig. 8.3. Frequency responses of common-emitter amplifier having output parallel compensation, for various inductances

a certain inductance L_{opt} yields a maximally flat response. A further increase in inductance will result in a high-frequency peak. L_{opt} may be calculated from the relation

$$\frac{d|a(\omega)|^2}{d\omega^2} = 0, \qquad (8.1.5)$$

where $a(\omega)$ is the relative response. Calculating from this relation the characteristic equation, and replacing the coefficients using expression (8.1.3),

8. Compensated wideband amplifiers

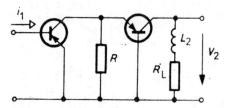

Fig. 8.4. Two-stage amplifier with output parallel compensation

the optimum inductance is determined:

$$L_{\text{opt}} = \frac{R_L}{\omega_\beta} (\sqrt{2\delta^2 + 2\delta + 1} - \delta), \tag{8.1.6}$$

where δ is defined as the following quantity:

$$\delta = \psi - 1 = \omega_T C_c R_L. \tag{8.1.7}$$

If $\delta \ll 1$, eqn. (8.1.6) is simplified to yield $L_{\text{opt}} = R_L/\omega_\beta$, or expressed in another form, $\omega_2 = \omega_\beta$. The 3-dB cut-off frequency ω_H for the case $\omega_c > \omega_\beta$ is given by

$$\omega_H \simeq \omega_c^2/\omega_\beta. \tag{8.1.8}$$

This compensation method is extremely effective, in spite of its simple form. With suitable design, an increase in cut-off frequency of 4–5 times can be achieved. The condition for this is to avoid the loading of the input impedance of the next stage which would have a decreasing effect on the cut-off frequency.

The design and measurement results of a video amplifier with inductive compensation is treated in Ref. [8.3].

Parallel compensation can be also applied in multi-stage amplifiers, connecting the compensating inductance in series with, say, the collector resistance of the last-stage transistor. The frequency response of the multi-stage amplifier may thus be arbitrarily shaped by choosing suitable inductances.

A frequently used type of wideband amplifier is shown in Fig. 8.4. The first stage of the two-stage amplifier is a common-emitter configuration; this gives the current gain and determines the cut-off frequency of the whole amplifier. The second stage is a common-base configuration, and the compensating inductance is in the collector circuit of this stage. Correct choice of this inductance results in an extremely wide frequency range as the cut-off of the common-base configuration is not significant.

8.2. Common-emitter configuration with input parallel compensation

A parallel-connected compensating inductance at the input side results in a substantial bandwidth increase (Fig. 8.5). The current gain expression for this circuit is given by

$$A_i = \frac{i_2}{i_1} = A_{i0} \frac{1 + ja_1\omega}{1 + jb_1\omega - b_2\omega^2}, \qquad (8.2.1)$$

where A_{i0} is the low-frequency current gain, and the coefficients are functions of the transistor parameters and loadings (and thus of the compensat-

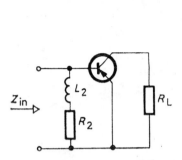

Fig. 8.5. Common-emitter amplifier with input parallel compensation

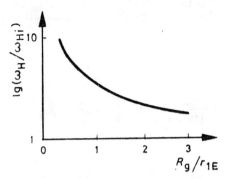

Fig. 8.6. Relative increase of cut-off frequency as a function of generator resistance with optimum compensation

ing elements). By expressing the condition (8.1.5) for the maximally flat amplitude response, we obtain the following relation, giving the optimum value of the compensating inductance:

$$L_{opt} \simeq \frac{R_g}{\omega_{Hi}} \left[\sqrt{\frac{R_g^2}{R_s^2} + \frac{(R_g + r_b)^2}{R_s r_{1E}}} - \frac{R_g}{R_s} \right], \qquad (8.2.2)$$

where the notation (6.1.11) and the notation $R_s = 2R_g + r_{1E}$ have been used. For $R_g \approx r_{1E}$, the fairly good approximation of $L_{opt} = 0.3 R_g/\omega_{Hi}$ is valid. It is seen that for a given cut-off frequency, the optimum inductance resulting in a maximally flat response depends on resistance R_g. Since the bandwidth is highly dependent on R_g, this can best be determined from the desired bandwidth. In Fig. 8.6, the relative cut-off frequency is plotted as a function of the ratio R_g/r_{1E} for maximally flat response, i.e. optimum compensation.

8.3. Common-emitter configuration with input series compensation

The stage bandwidth may be increased by placing a compensating inductance L_1 in series with the base, assuming a sufficiently low generator resistance R_g. The transfer function for the circuit shown in Fig. 8.7 is given by

$$A_v = \frac{V_2}{V_g} = A_{v0} \frac{1}{1 + jb_1\omega - b_2\omega^2}, \qquad (8.3.1)$$

where A_{v0} is the low-frequency voltage gain, and coefficients b_1 and b_2 are functions of the compensating inductance. Expressing the condition for the maximally flat response, we have

$$L_{opt} = \frac{r_{1E}}{\omega_{Hi}} \left[1 - \sqrt{1 - \left(\frac{R_g + r_b}{r_{1E}}\right)^2} \right], \qquad (8.3.2)$$

where the condition $R_g + r_b < r_{1E}$ has to be satisfied. In this case, the 3-dB cut-off frequency is given by

$$\omega_H = \omega_{Hi} \sqrt{\frac{1 + R_g/r_{1E}}{1 - \sqrt{1 - (R_g + r_b)^2/r_{1E^2}}}}. \qquad (8.3.3)$$

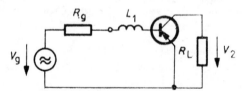

Fig. 8.7. Common-emitter amplifier with series compensation

It can be seen from the expressions above that the series compensation is effective for low R_g generator resistances, and this cannot be always, arranged.

8.4. Common-emitter configuration with combined compensation

The parallel compensation treated in Section 8.1 is not effective if impedance Z_L is loaded by the next stage. The loading effect is decreased by the series inductance L_1 which practically represents a series compensation (Fig. 8.8). However, the circuit is compensated in a combined way as both series

8.4. Combined compensation

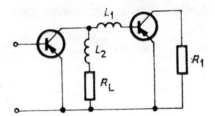

Fig. 8.8. Two-stage amplifier with simultaneous series and parallel compensation

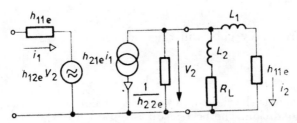

Fig. 8.9. Equivalent circuit of amplifier with series and parallel compensation

and parallel compensations are used. The parallel impedance is given by $Z_L = R_L + j\omega L_2$.

In the following, the transfer function should be determined, and the optimum inductances L_1 and L_2 should be determined from the transfer function. However, the calculation of the exact transfer function presents serious difficulties because of the number of frequency-dependent elements. The transfer function thus determined would not be useful for design purposes, so the numerical investigation of the connection will be made using another method. This is less exact but simple and usable.

Figure 8.9 shows the h-parameter equivalent circuit of the amplifier of Fig. 8.8. It can be seen that the impedance Z_L may be included in the transistor parameter h_{22}. A new four-pole is thus arrived at, with all parameters unchanged except the parameter h_{22} which is given by

$$H_{22} = h_{22e} + Y_1, \tag{8.4.1}$$

where Y_1 is the admittance of the series circuit:

$$Y_1 = 1/(R_L + j\omega L_2) = 1/Z_L. \tag{8.4.2}$$

The load impedance of the amplifier is $Z_2 = h_{11e} + j\omega L_1$. The current-gain factor is given by the well known relation

$$A_1 = \frac{i_2}{i_1} = \frac{h_{21e}}{1 + H_{22} Z_2}. \tag{8.4.3}$$

The design is initiated by setting the current gain at the upper cut-off frequency ω_H equal to the low-frequency gain:

$$A_{l0} = A_l(\omega_H). \tag{8.4.4}$$

At low frequencies, the parameters simplify to

$$\begin{aligned} h_{21e}(\omega = 0) &= \beta_0; \\ h_{22e}(\omega = 0) &\simeq 0; \\ \mathrm{Re}(1/Y_1) &= R_L; \\ h_{11e}(\omega = 0) &= r_{1E}. \end{aligned} \tag{8.4.5}$$

The low-frequency gain will thus be

$$A_{l0} = -\frac{\beta_0}{1 + r_{1E}/R_L}. \tag{8.4.6}$$

The next step is the right-hand side expansion of eqn. (8.4.4) which gives the design condition. The best result is obtained if the reactance of inductance L_1 is just equal to the capacitive reactance of h_{11e} at frequency $\omega = \omega_H$:

$$-\omega_H L_1 = \mathrm{Im}(h_{11e}). \tag{8.4.7}$$

This forms a series circuit at the output which will increase the current i_2. With this condition,

$$A_l(\omega_H) = \frac{h_{21e}(\omega_H)}{1 + H_{22}(\omega_H)\,\mathrm{Re}[h_{11e}(\omega_H)]}, \tag{8.4.8}$$

where

$$H_{22}(\omega_H) = h_{22e}(\omega_H) + 1/(R_L + j\omega_H L_2). \tag{8.4.9}$$

The design is carried out in several steps. In the first step, resistance R_L is calculated from eqn. (8.4.6) knowing A_{l0}.

For the second step, inductance L_1 is determined from eqn. (8.4.7) by taking into account the highest frequency ω_H to be transmitted. The remaining task is the calculation of parameters h_{21e}, h_{22e}, Y_1 and h_{11e} at the frequency ω_H. L_2 may be determined from eqn. (8.4.8), from the calculated values and from Y_1. This last task requires most mathematical operations during the whole design.

Another form of combined compensation, the π compensation, is shown in Fig. 8.10. The frequency response can be varied within wide limits by the choice of the three inductances. The analysis of this circuit is rather complicated, and will not be presented here.

In Fig. 8.10, a four-pole is inserted between the two stages, and the suitable response of this four-pole compensates the cut-off of the transistors.

The number of the four-pole elements can be increased further, thus producing complicated filter networks for achieving extremely wideband responses. Reactive and lossy filters are treated by the network theory in

8.5. Frequency-independent input impedances

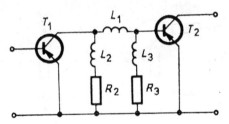

Fig. 8.10. Two-stage amplifier with compensation, using an inductive π network

detail, and the transmission of a large variety of filter networks is plotted or tabulated. However, these relations are valid for resistive terminations of the filter network, and this is not true for transistor input and output impedances. In the case of complex terminations, the reactive terms may be included in the filter network if the equivalent circuit of the termination is known; the problem is thus simplified and the relations may be applied. The method is not applicable, however, if the equivalent circuit of the termination (e.g. the transistor output as generator) is not known accurately enough. As an example this may result from the modification of the filter network connected to the base because of reverse transmission parameters. Circuit design may then become very inaccurate, with substantial deviations from the theoretical results.

In spite of these facts, reactive four-poles for matching and compensation purposes are frequently applied to give substantial extension to the transmission bandwidth. These methods will be dealt with in a forthcoming chapter in which extremely wideband amplifiers will be treated.

8.5. Stages with frequency-independent input impedances

The design of wideband amplifiers often requires the production of a constant, frequency-independent input impedance within the transmission bandwidth. However, this constant input impedance can only be approximated in most cases. One of the methods utilizes negative feedback for arbitrarily increasing (or decreasing) the transistor input impedance and making possible the use of a series (or parallel) resistance at the input, which produces a frequency-independent input impedance. The condition for this is a substantially higher (or lower) additional resistance compared with the actual input impedance of the transistor stage. The method is shown in Fig. 8.11, where the transistor input impedance, decreased by negative feedback, is denoted by Z_{in1}, and the impedance increased by feedback is denoted by Z_{in2}.

The input impedance can also be made constant without negative feedback, by making use of compensation.

Series compensation is shown in Fig. 8.12(a), and parallel compensation in Fig. 8.12(b). By correct choice of compensating elements, the input impedance

8. Compensated wideband amplifiers

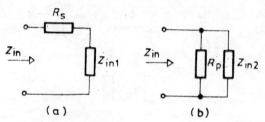

Fig. 8.11. Design of frequency-independent input resistances by an additional resistance. (a) Series resistance, (b) parallel resistance

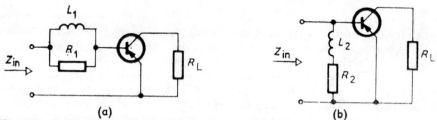

Fig. 8.12. Design of frequency-independent input resistance by (a) series compensation, (b) parallel compensation

of the common-emitter stage is constant up to about five times the cut-off frequency f_α. Using notation as in the figures, the element values for series compensation are

$$R_1 = \beta_0 r_e; \tag{8.5.1}$$

$$L_1 = r_{1E}/\omega_{Hi}, \tag{8.5.2}$$

and those for parallel compensation are

$$R_2 = r_b(1 + r_b/r_{1E}); \tag{8.5.3}$$

$$L_2 = r_b^2 \psi / r_e \omega_T. \tag{8.5.4}$$

The frequency-independent character of the input impedance may be increased by a series resistor in the first case, and a shunt resistor in the second case, as in the method shown in Fig. 8.11.

9. SINGLE-STAGE FEEDBACK AMPLIFIERS

9.1. General characteristics of feedback amplifiers

Negative feedback is frequently used to increase the bandwidth of high-frequency amplifiers. This method modifies not only the bandwidth but also all other characteristics of the circuit.

The principle of a feedback amplifier is shown in Fig. 9.1, where $A(\omega)$ is the transfer function of the amplifier without feedback, and $\beta^*(\omega)$ is the transfer function of the feedback network (the asterisk is used to distinguish from the current-gain factor of the common-emitter stage). Two voltages are shown as input and output parameters, but input and output currents can also be used in certain cases. Using notations as in the figure, the gain with feedback is given by the well known expression:

$$A_f(\omega) = \frac{A(\omega)}{1 - A(\omega)\beta^*(\omega)}. \qquad (9.1.1)$$

The product in this expression is the loop gain:

$$T(\omega) = A(\omega)\beta^*(\omega), \qquad (9.1.2)$$

and the denominator is the feedback factor:

$$F(\omega) = 1 - A(\omega)\beta^*(\omega). \qquad (9.1.3)$$

For high-loop gains, an approximation for eqn. (9.1.1) can be used, and the transmission in these cases is determined only by the transfer function of the feedback network alone. The advantages of this method are well known. However, the increase in loop gain is limited; this question will be dealt with later when the stability problem is investigated.

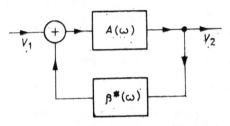

Fig. 9.1. General block diagram of feedback amplifier

The four basic feedback types are summarized in Table 9.1 which also gives the impedance modifications resulting from feedback. It is seen that series interconnections will increase the impedance at both the input and output side, and parallel interconnections will decrease the impedance. The impedance change in all cases is equal to the feedback factor F.

Table 9.1. Input and output impedances of the four basic feedback four-pole types

type of feedback	figure	input impedance	output impedance
parallel–parallel	4.4	Z_i/F	Z_0/F
series–series	4.5	FZ_i	FZ_0
series–parallel	4.6	FZ_i	Z_0/F
parallel–series	4.7	Z_i/F	FZ_0

The amplifier bandwidth may be increased substantially by using frequency-independent feedback which decreases the gain. For single-stage amplifiers, the change of the feedback factor will introduce a change in gain such that the gain bandwidth product remains approximately constant. However, this is not the exact case because of the loading effect of the feedback network, but the decrease in gain-bandwidth product is rarely very significant. Using a frequency-dependent feedback network, the gain-bandwidth product can even be increased. This may be accomplished using a reactive circuit element to decrease the feedback factor at the upper end of the frequency range, thus increasing the gain.

For *two-stage* amplifiers with maximally flat responses, the gain-bandwidth product of the feedback amplifier will equal, to good approximation, the geometric mean of the single-stage gain-bandwidth products, assuming frequency-independent feedback. Thus the gain-bandwidth product remains approximately constant with identical stages. The situation is worse with stages having different bandwidth products, but the overall product can still be increased by frequency-dependent feedback.

For *three-stage* amplifiers, it is still true that the gain-bandwidth product of the feedback amplifier equals the geometric mean of the single stage products. However, stability requirements impose severe restrictions on the frequency response of the loop gain, and such a circuit is not practically realizable because of stray parameters. This is even more relevant to feedback networks spanning more than three stages. They can seldom be used in practice at high frequencies.

Finally, the effect of feedback on circuit nonlinearity should be mentioned. The nonlinearity percentage value is given by

$$\mathrm{NL}(\%) = \frac{A(\omega)_{\max} - A(\omega)_{\min}}{A(\omega)_{\max}}, \tag{9.1.4}$$

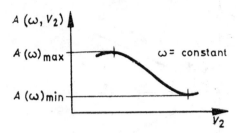

Fig. 9.2. Gain as a function of signal level

where $A(\omega)_{max}$ and $A(\omega)_{min}$ are the maximum and minimum transfer gain values measured at different output voltage levels V_2, at a fixed frequency (Fig. 9.2). It can be shown that the decrease in nonlinearity is proportional to the feedback factor, assuming a linear (passive) feedback network:

$$\mathrm{NL}_f \approx \frac{\mathrm{NL}}{1 - A(\omega)\beta^*(\omega)}. \qquad (9.1.5)$$

If the feedback factor is strongly level-dependent, then the maximum value, i.e. the worst case, must to be taken into account in eqn. (9.1.5).

The linearizing effect of feedback is only substantial at high-loop gains, so conditions become worse above the amplifier cut-off frequency.

9.2. Stability of feedback amplifiers

If the loop gain according to eqn. (9.1.2) becomes unity, then the denominator in eqn. (9.1.1) giving the transmission of the feedback amplifier is zero, and the amplifier will oscillate. Plotting the loop gain in the complex plane we obtain the so-called Nyquist diagram (Fig. 9.3). According to eqn. (9.1.1), the plot may not encircle the real point $+1$. In Fig. 9.3 (a), the frequency response of a stable amplifier is shown, while the frequency response of Fig. 9.3 (c) represents an unstable amplifier. Figure 9.3 (b) is characteris-

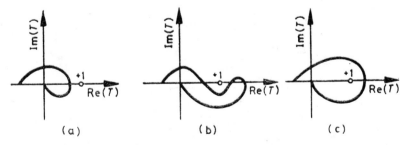

Fig. 9.3. Types of Nyquist diagram. (a) Stable amplifier, (b) potentially unstable amplifier, (c) unstable amplifier

9. Single-stage feedback amplifiers

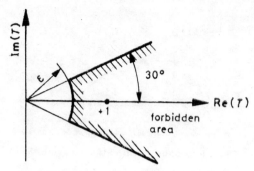

Fig. 9.4. Forbidden loop-gain area on the complex plane

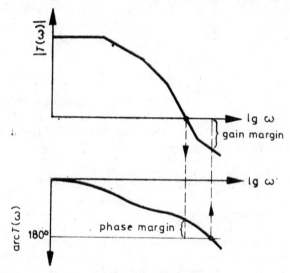

Fig. 9.5. Relative stability criteria on the Bode diagram

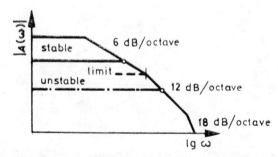

Fig. 9.6. Estimate of stability from the Bode diagram

tic for a potentially unstable amplifier, when the decrease in loop gain may shift the response curve such that it must pass through the point $+1$.

The *relative stability* or stability margin is characteristic for the behaviour of the response curve in the vicinity of point $+1$. This will also determine whether the feedback may be increased without causing instability. The shaded area in Fig. 9.4, through which the response curve is not allowed to pass, may be characterized by two numbers, the distance from the origin ε and the phase angle. In practice, a minimum phase margin $\varphi = 30°$ and a minimum distance from the origin

$$\varepsilon[\text{dB}] \approx 1/3n, \qquad (9.2.1)$$

are needed, where n is the number of amplifier stages. For a two-stage amplifier this means a 15-dB gain margin.

The relative stability can be determined not just in the complex plane, but also from the frequency response of the loop-gain absolute value and phase shift. Figure 9.5 shows the Bode diagram of the loop gain. Subtracting the phase shift at $|T(\omega)| = 1$ from 180° we obtain the phase margin. The attenuation at the frequency where the loop-gain phase shift is 180° is called the gain margin.

A quick estimate of stability is possible from the Bode diagram of the open-loop gain (Fig. 9.6). The slope of the response $|A(\omega)|$ is increased at each cut-off frequency point. Feedback has the effect of decreasing the low-frequency gain, and thus the response of the feedback amplifier is a straight line which is shifted downwards as the feedback is increased, and merges with the open-loop response at high frequencies. At the intersection of the two curves, the absolute value of the loop gain $T(\omega)$ is unity, so the phase shift at this frequency can be ascertained from the slope of the $A(\omega)$ characteristic. In the range where the slope of the absolute value characteristic is not more than 6 dB/octave, the phase shift is less than 90°. Thus the following statement holds as a rough estimate: if the intersection of the open-loop and feedback responses is in the range where the slope of the open-loop response is 6 dB/octave then the amplifier is stable. If the intersection lies within the 12-dB/octave section where the phase shift may reach 180° then the amplifier may become unstable. Thus, according to this commonly used rule, the stability limit is the cut-off frequency point separating the 6-dB/octave and 12-dB/octave sections.

9.3. Amplifier with emitter feedback

The circuit diagram of the amplifier is shown in Fig. 9.7. Feedback, as is well known, is caused by the resistance R_E. The transistor is driven by a generator of resistance R_g; load resistance is R_L. The feedback efficiency is determined by the generator resistance; the feedback is ineffective for current drive.

This feedback amplifier may be analyzed in several ways. In one method, the transistor four-pole defined by the h parameters and the shunting four-

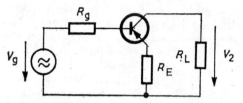

Fig. 9.7. Amplifier with emitter feedback

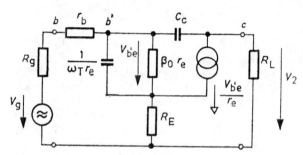

Fig. 9.8. Equivalent circuit of emitter feedback amplifier

pole made up of resistance R_E are formally contracted, using the series–series connection rule of linear network theory.

Another procedure utilizes the equivalent circuit method, according to Ref. [9.1]; Fig. 9.8 shows the equivalent circuit, the feedback resitance R_E and the terminations. Omitting the calculations, the voltage gain is calculated to be

$$A_v = \frac{V_2}{V_g} = A_{v0} \frac{1}{1 + j\omega/\omega_H}, \qquad (9.3.1)$$

where the low-frequency voltage gain is

$$A_{v0} = -\frac{\beta_0 R_L}{\beta_0(r_e + R_E) + r_b + R_g} \approx \frac{R_L}{R_E}, \qquad (9.3.2)$$

and the cut-off frequency is

$$\omega_H = \frac{\omega_\beta}{\psi} \frac{r_b + R_g + \beta_0(r_e + R_E)}{r_b + R_g + R_E} \approx \frac{\omega_T}{\psi}, \qquad (9.3.3)$$

where a high valued resistance R_E has been assumed for both approximations.

It is seen from these results that because of this feedback resistance R_E, the low-frequency gain is decreased and the cut-off frequency is increased

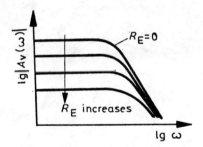

Fig. 9.9. Frequency responses for various emitter resistances

since in eqn. (9.3.3), the numerator increases more rapidly with R_E than does the denominator.

The effectiveness of feedback is expressed by the gain-bandwidth product:

$$GB = |A_{v0}|\omega_H = \frac{\omega_T}{\psi}\frac{R_L}{r_b + R_g + R_E}. \qquad (9.3.4)$$

Comparing this with the value given by (6.1.12) for the case without feedback, it is seen that feedback has the effect of decreasing the gain-bandwidth product by the factor

$$\eta = 1 + R_E/(r_b + R_g). \qquad (9.3.5)$$

The frequency response of the voltage gain is shown in Fig. 9.9. The response curves have single cut-off frequencies, and are shifted downwards with increasing resistance R_E.

9.4. Emitter-feedback amplifier with compensation

The bandwidth of an emitter-feedback amplifier may be increased by shunting the emitter resistance R_E by a suitably chosen capacitance C_E (Fig. 9.10). This capacitance lowers the voltage feedback near the high-frequency band limit and thus raises the amplification. In the following, an optimum C_E value yielding a maximally flat response is to be determined.

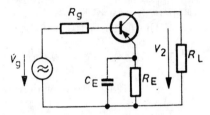

Fig. 9.10. Emitter-feedback amplifier with compensation

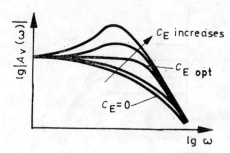

Fig. 9.11. Frequency responses for various compensating capacities

The circuit may be analyzed by substituting a complex impedance Z_E for R_E in the relations derived for resistive feedback:

$$Z_E = R_E \parallel 1/j\omega C_E = R_E/(1 + j\omega/\omega_E), \qquad (9.4.1)$$

where $\omega_E = 1/R_E C_E$ is the characteristic frequency of the RC circuit.

After substitution, the voltage gain is expressed in the following form:

$$A_v = A_{v0} \frac{1 + j a_1 \omega}{1 + j b_1 \omega - b_2 \omega^2}, \qquad (9.4.2)$$

where the low-frequency voltage gain A_{v0} is given by (9.3.2). Coefficients a_1, b_1 and b_2 are functions of capacitance C_E. The frequency responses for various C_E values are shown in Fig. 9.11. The bandwidth is increased by increasing the compensating capacitance, and a maximally flat response is achieved at the value $C_{E\,opt}$. This can be calculated from eqn. (5.2.5). For simplicity, the frequency defined by the equation

$$\omega_{opt} = 1/R_E C_{E\,opt}, \qquad (9.4.3)$$

will be calculated.

From the characteristic eqn. (5.2.5), the feedback time constant yielding the maximally flat response and expressed in angular frequency is given by

$$\omega_{opt} = \omega_H \frac{\eta(1 - q^2)}{\eta q - 1 + \sqrt{1 - 2\eta q + \eta^2}}. \qquad (9.4.4)$$

In the above expression, ω_H is the cut-off frequency of the feedback amplifier without compensation ($C_E = 0$) given by relation (9.3.3); η has the value given by (9.3.5), and the factor q is given by:

$$q = (r_{1E} + R_g)/(r_{1E} + R_g + \beta_0 R_E). \qquad (9.4.5)$$

Assuming strong feedback, $\eta \approx 1$ and $q \ll 1$. Expression (9.4.4) then simplifies to the form $\omega_{opt} = 2.42 \omega_H$ from which the optimum emitter capacitance can be calculated:

$$C_{E\,opt} = 0.41/R_E \omega_H. \qquad (9.4.6)$$

9.4. Emitter-feedback amplifier with compensation

The emitter capacity is thus dependent on the feedback value.

In the optimum case, i.e. with maximally flat response, the bandwidth will be 1.7 times the uncompensated value:

$$\omega_{H\,opt} = 1.7\omega_H. \qquad (9.4.7)$$

This results, simultaneously, in a similar increase in gain-bandwidth product. Using C_E values higher than the optimum value will produce response peaks as seen in Fig. 9.11 which can be used to advantage in certain cases, e.g., for equalizing the frequency roll-off in other stages.

The frequency response can be modified further and the bandwidth increased using a *series resonant circuit* instead of compensating capacity C_E, as shown in the circuit diagram of Fig. 9.12.

The end result is given, omitting the calculation. The method used is similar to that above, but an impedance Z_E equivalent to the series resonant circuit is substituted:

$$Z_E = R_E \parallel (j\omega L_E + 1/j\omega C_E). \qquad (9.4.8)$$

The inductance yielding a maximally flat response is:

$$L_{E\,opt} = \frac{R_E^2 C_{E\,opt}}{2\eta[\eta + (\eta q - 1)\omega_H/\omega_{opt}]}. \qquad (9.4.9)$$

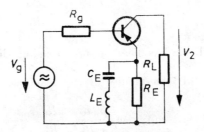

Fig. 9.12. Emitter-feedback amplifier with compensation using a series resonant circuit

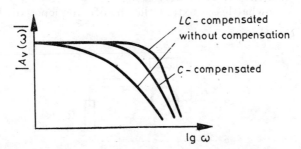

Fig. 9.13. Frequency responses with different compensations

Assuming conditions $\eta \approx 1$ and $q \ll 1$ again, this expression simplifies to

$$L_{E\,opt} \approx 0.8\, R_E^2\, C_{E\,opt}. \tag{9.4.10}$$

Figure 9.13 shows the frequency responses obtained by the two different compensation methods. The best result is obtained with the compensation using the series resonant circuit according to Fig. 9.12. For both compensations, the maximally flat responses are plotted, and the response of the uncompensated amplifier is also shown for comparison.

9.5. Single-stage amplifier with current feedback

The circuit diagram of the single-stage amplifier with current feedback is shown by Fig. 9.14. The feedback resistance R_F has the effect of stabilizing the current gain of the stage and decreasing the input and output resistances.

In practice, the decreased input resistance is much lower than the resistance of the driving generator, and therefore the current gain of this circuit will be analyzed:

$$\frac{i_2}{i_g} = A_{10}\frac{1}{1 + j\omega/\omega_H}. \tag{9.5.1}$$

The low-frequency current gain is given by

$$A_{10} = \frac{-\beta_0 R_F}{\beta_0 R_L + R_F} \approx -\frac{R_F}{R_L}, \tag{9.5.2}$$

where a strong feedback has been assumed using the approximation for which $\beta_0 R_L \gg R_F$.

Furthermore, the 3-dB frequency of the response with a single cut-off frequency is given by

$$\omega_H = \omega_\beta \frac{\beta_0 R_L + R_F}{\psi R_F + R_L}. \tag{9.5.3}$$

In the two expressions above, it was assumed that $R_F \gg r_b$. The feedback resistance R_E has the effect of decreasing the low-frequency gain and increasing to a somewhat lesser extent the cut-off frequency of the stage. The

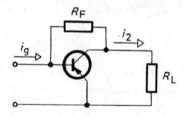

Fig. 9.14. Single-stage amplifier with current feedback

gain-bandwidth product is

$$GB = |A_{i0}|\omega_H = \frac{\omega_T}{\nu\psi},\qquad(9.5.4)$$

where the factor expressing the decrease of the gain-bandwidth product due to feedback is given by

$$\nu = 1 + R_L/\psi R_F.\qquad(9.5.5)$$

This factor also expresses how much the increase in cut-off frequency is below the expected. The frequency response shapes with different R_F values are similar to those in Fig. 9.9. The responses are shifted downwards with decreasing R_F resistances, while the bandwidth is increased.

In Ref. [9.7], the analysis of a single-stage voltage- and current-feedback amplifier is given for extremely high frequencies, using scattering parameters.

9.6. Single-stage compensated amplifier with current feedback

The current feedback may also be compensated by a complex impedance Z_F, similarly to the voltage feedback. If this impedance is inductive, then the feedback decreases at high frequencies, thus increasing the amplification. The circuit diagram is shown in Fig. 9.15. The feedback impedance is given by

$$Z_F = R_F + j\omega L_F = R_F(1 + j\omega/\omega_F),\qquad(9.6.1)$$

where ω_F is the characteristic frequency of the feedback. The transfer function may be written as:

$$A_i = A_{i0}\frac{1 + ja_1\omega}{1 + jb_1\omega - b_2\omega^2},\qquad(9.6.2)$$

where A_{i0} is again the low-frequency gain as given in (9.5.2). Coefficients a_1, b_1 and b_2 are functions of feedback impedance.

The optimum frequency may be calculated from eqn. (5.2.5) pertaining to the maximally flat response:

$$\omega_{opt} = 2.42\,\omega_H,\qquad(9.6.3)$$

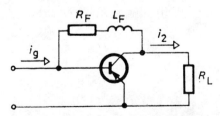

Fig. 9.15. Compensated amplifier with current feedback

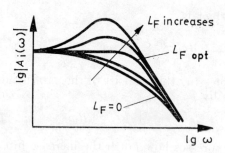

Fig. 9.16. Frequency responses for various compensation inductances

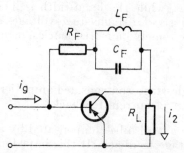

Fig. 9.17. Current-feedback amplifier, using compensation with parallel resonant circuit

where ω_H is the cut-off frequency without compensation as given in (9.5.3). The optimum inductance may be calculated from this expression, taking into account (9.6.1):

$$L_{F\,opt} = 0.41\, R_F/\omega_H. \tag{9.6.4}$$

The feedback time constant is calculated to be the same as for the voltage feedback in Section 9.4. The two feedbacks are dual versions, the relations are similar, and the frequency responses are identical. Figure 9.16 shows responses for various L_F values. A maximally flat response is given by inductance $L_{F\,opt}$, and higher inductances will introduce response peaks. In the optimum case, the bandwidth increase is 1.7 times according to (9.4.7).

The high-frequency response may be improved further by including a parallel resonant circuit in impedance Z_F (Fig. 9.17):

$$Z_F = R_F + j\omega L_F \,\|\, 1/j\omega C_F. \tag{9.6.5}$$

The improvement is approximately the same as with the resonant-circuit compensation shown in Fig. 9.13.

10. MULTI-STAGE WIDEBAND AMPLIFIERS WITH FEEDBACK

10.1. Two-stage amplifier with current feedback

Two-stage amplifiers with feedback spanning both stages have much better properties than single-stage feedback amplifiers [10.1], [10.4], [10.5], [10.8], [10.9]. Best results are obtained with the two-stage feedback amplifier shown in Fig. 10.1, and so the design method of this amplifier will be dealt with in detail.

Feedback is obtained using elements Z_E and Z_F. Calculation is simplified by assuming a modified circuit in which the second transistor is used in a common-collector configuration. The load resistance R_L usually has a low value and does not interfere with this modification. As a first calculation step, the current gain of the two-stage amplifier without feedback is determined; this can be written in the following form:

$$A_i = A_{i0} \frac{1}{1 + jb_1\omega - b_2\omega^2}, \qquad (10.1.1)$$

where A_{i0} is the low-frequency current gain, and coefficients b_1 and b_2 are functions of the feedback elements.

Current i_f is fed back to the input as a result of the feedback elements. The feedback factor is given by the ratio of current i_f and current i_2:

$$\frac{i_f}{i_2} = f = -\frac{Z_E}{Z_E + Z_F}. \qquad (10.1.2)$$

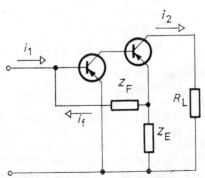

Fig. 10.1. Two-stage amplifier with current feedback

10. Multi-stage wideband amplifiers with feedback

The current gain of the feedback amplifier:

$$\frac{i_2}{i_1} = A_{if} = \frac{A_i}{1 - A_i f}. \qquad (10.1.3)$$

Let us first investigate the case where both Z_F and Z_E are real valued resistances. Equation (10.1.2) then simplifies to the following form:

$$f = -\frac{R_E}{R_E + R_F} = -f_0. \qquad (10.1.4)$$

Substituting this into eqn. (10.1.3), we have

$$A_{if} = A_{of} \frac{1}{1 + j2\zeta\omega/\omega_0 - \omega^2/\omega_0^2}. \qquad (10.1.5)$$

In this expression, the low-frequency current gain is given by

$$A_{of} = \frac{A_{i0}}{1 + A_{i0} f_0} = \frac{\beta_0 R_F}{\beta_0 R_E + r_e}, \qquad (10.1.6)$$

and the coefficients of the frequency-dependent part are calculated as follows:

$$\zeta = (r_e + 2 R_q \beta_0^{-1} + \omega_T C_c R_s^2)/\sqrt{R_E R_q \psi_p}; \qquad (10.1.7)$$

$$\omega_0 = \omega_T \sqrt{R_E/R_q \psi_p}. \qquad (10.1.8)$$

The following notations have been used in these two equations:

$$R_q = r_b + R_E + R_F;$$
$$R_s^2 = (R_F + r_b) R_E + r_e(R_E + R_F); \qquad (10.1.9)$$
$$R_p^2 = r_b (R_E + R_F);$$
$$\psi_p = 1 + \omega_T C_c R_p.$$

It has been assumed in the calculation that

$$R_E \gg R_q/\beta_0 \quad \text{and} \quad R_E \gg r_e/\beta_0.$$

For eqn. (10.1.5), the characteristic equation giving the maximally flat response may be applied. According to Chapter 5, $\zeta_{opt} = 0.7$ pertains to the maximally flat response. If the factor ζ is higher than the optimum value, then the initial slope is much higher, and thus cut-off occurs earlier. If, on the other hand, ζ is lower than the optimum value, then a peak is obtained at the band limit which may be extremely high, depending on the value of ζ.

It would be rather complicated to adjust the optimum ζ value by feedback resistance R_F, since it is difficult to express resistance R_F through eqn. (10.1.7), and the low-frequency gain required also presents problems.

This means that the factor ζ will have a certain value as a result of the given feedback resistance R_F. However, there is evidently a possibility of obtaining a maximally flat response by compensation while leaving the low-frequency gain unchanged. This may be accomplished by shunting iether R_E or R_F by a capacity which will change the response at the upper band limit.

If the ζ value obtained is lower than optimum ($\zeta < 0.7$), then the shunting of resistance R_F will give better results. In this case, feedback element Z_F is complex:

$$Z_F = R_F \parallel 1/j\omega C_F = R_F/(1 + j\omega R_F C_F). \qquad (10.1.10)$$

The feedback factor (10.1.2) will also be complex:

$$f = -\frac{R_E}{R_E + Z_F} = -f_0(1 + j\omega R_F C_F). \qquad (10.1.11)$$

Substituting this into eqn. (10.1.3) and writing the characteristic equation for the coefficients of the obtained response, the compensating capacity which results in a maximally flat response may be calculated:

$$C_{F\,opt} = \frac{\sqrt{2} - 2\zeta}{\omega_0 R_F f_0 A_{0f}}. \qquad (10.1.12)$$

In this equation, ω_0 and ζ are response coefficients which apply to the uncompensated amplifier; A_{0f} is the low-frequency current gain, and f_0 is the low-frequency feedback factor. It is seen that, from this expression, zero shunting capacity is obtained for $\zeta = \zeta_{opt} = 0.7$, and the expression has no meaning for higher ζ values.

If we have a resistive feedback impedance R_F, and $\zeta > 0.7$, then the emitter resistance R_E may be shunted in order to approximate the maximally flat response. Omitting the calculations, the optimum shunt capacity is given by

$$C_{E\,opt} \approx 0.4/\omega_0 R_E. \qquad (10.1.13)$$

The 3-dB cut-off frequency for the maximally flat response is $\omega_H = \omega_0$.

Response curves are plotted in Figs. 10.2 and 10.3. Figure 10.2 shows the uncompensated responses for various R_F values. Decreasing R_F values will also decrease the value of ζ, so the response is shifted downwards and bandwidth is increased. Bandwidth increase is approximately proportional to the decrease in low-frequency gain. For a given resistance R_F, ζ will reach the optimum value and the response will be maximally flat. A further increase in feedback will introduce a peak in the response.

Figure 10.3 (a) shows responses for amplifiers having small feedback ($\zeta > 0.7$). In this case, the shunting capacity C_E of resistance R_E will increase the bandwidth. Optimum capacity yields a maximally flat response, and above-optimum capacity results in a response peak.

Figure 10.3 (b) shows responses for amplifiers having large feedback ($\zeta < 0.7$). The shunting of resistance R_F has the effect of decreasing the

10. Multi-stage wideband amplifiers with feedback

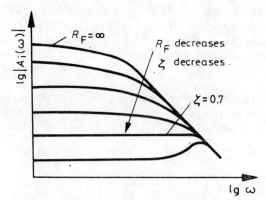

Fig. 10.2. Frequency responses of two-stage uncompensated current-feedback amplifier for various feedback factors

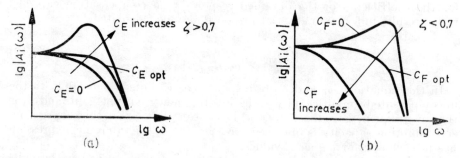

Fig. 10.3. Frequency response of two-stage current-feedback amplifier (a) for low feedback factors and various capacitances C_E, (b) for high feedback factors and various capacitances C_F

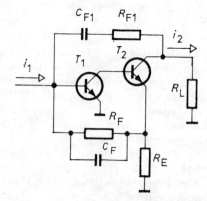

Fig. 10.4. Circuit diagram of two-stage current-feedback amplifier with positive feedback

10.1. Two-stage amplifier

peak, and with a specific $C_{F\,opt}$ capacity, the response will again be maximally flat.

Applying additional elements R_{F1} and C_{F1} which introduce *positive* feedback according to Fig. 10.4, the two-stage current-feedback amplifier will show a sharp response peak near the upper band limit which can be utilized for selective (narrow band) amplification. The current gain for such an amplifier is given approximately by

$$\frac{i_2}{i_1} = A_{if}^* \approx \left(1 + \frac{R_F}{R_E}\right) \frac{1 - j\,\omega_{F1}/\omega}{(1 - j\,\omega_{F1}/\omega)(1 + j\omega/\omega_F) - R_F\,R_L/R_E\,R_{F1}}, \quad (10.1.14)$$

where $\omega_F = 1/R_F C_F$ and $\omega_{F1} = 1/R_{F1} C_{F1}$; the "resonant frequency" of the response peak is

$$\omega_R = \sqrt{\omega_F\,\omega_{F1}}, \quad (10.1.15)$$

and the bandwidth is

$$B = \omega_{F1} - \omega_F(R_F\,R_L/R_E\,R_{F1} - 1), \quad (10.1.16)$$

which can be made very small by correct choice of feedback resistances. A feature of this connection is the possibility of adjusting the bandwidth with the load resistance R_L independently of the resonant frequency ω_R. This circuit produced in a Q factor of 50 at $f_R = 5$ MHz with reasonable stability [10.19]. However, stability is critical for this connection, temperature changes may cause a substantial shift in transistor parameters and thus in the frequency response.

10.2. Multi-stage feedback amplifiers

The two-stage voltage-feedback circuit shown in Fig. 10.5 is the dual circuit of the two-stage current-feedback amplifier investigated in the preceding section. The voltage gain is stabilized, and determined by the resistance ratio R_{F1}/R_{E1} for large feedback. The bandwidth is increased

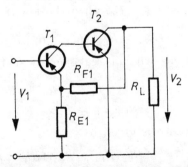

Fig. 10.5. Two-stage voltage-feedback amplifier

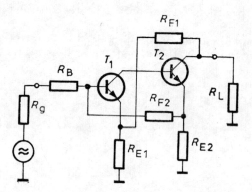

Fig. 10.6. Circuit diagram of two-stage amplifier with double feedback

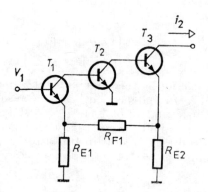

Fig. 10.7. Three-stage overall feedback amplifier

approximately in proportion to the decrease in low-frequency gain. The amplifier is especially useful for applications requiring high input resistance.

Figure 10.6 shows a two-stage amplifier with double feedback [10.20]. Resistance R_B exists to satisfy the requirement $R_{in} \approx R_{out}$, important in line amplifiers. If $R_g = R_L = R_0$, then the low-frequency voltage gain is given by

$$A_{v0} = (R_{F1} + R_{E1})(R_{in} - R_B)/R_{E1} R_{in}, \qquad (10.2.1)$$

where the input resistance which is real over a wide frequency range is given by

$$R_{in} \approx R_B + \frac{R_0 R_{E1}(R_{F2} + R_{E2})}{R_{E2}(R_{F1} + R_{E1} + R_0) + R_{E1} R_0}, \qquad (10.2.2)$$

and the output resistance is given by

$$R_{out} \approx \frac{R_{E2}(R_{F1} + R_{E1})(R_B + R_0)}{R_{E1}(R_{F2} + R_{E2} + R_B + R_0) + R_{E2}(R_B + R_0)}. \qquad (10.2.3)$$

10.3. Cascade connection of single-stage amplifiers

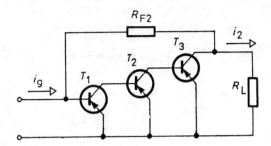

Fig. 10.8. Three-stage current-feedback amplifier

The increase in cut-off frequency due to feedback is approximately equal to the decrease in low frequency gain.

Using an additional emitter follower in the circuit of Fig. 10.5, and applying feedback from its output, produces the amplifier shown in Fig. 10.7. The transconductance of the three-stage overall feedback amplifier is constant over a wide frequency range [10.20], [11.7], and has the value of

$$g_m = \frac{i_2}{V_1} = (R_{E1} + R_{E2} + R_{F1})/R_{E1}R_{E2}. \qquad (10.2.4)$$

The monolithic realization of this amplifier is dealt with in Ref. [10.15], and takes into account the effect of the compensating capacity shunting R_{F1}.

Figure 10.8 shows a three-stage overall current-feedback amplifier. The current gain, stabilized by the feedback, has the value of R_{F2}/R_L in the case of large feedback. Phase conditions produce a tendency for the circuit to oscillate; stable operation can be attained by using phase-correction elements between stages [10.2]. Both input and output impedances are small, and thus the circuit can be used as a line amplifier. Theoretically, the feedbacks of Figs. 10.7 and 10.8 may be combined [10.20], but this method is seldom used because of the tendency for oscillations to occur.

It is not practically possible to use overall feedback over more than three transistor stages because of instability.

10.3. Cascade connection of single-stage feedback amplifiers

The cascade connection of single-stage feedback amplifiers is widely used in extremely wideband amplifiers. Bandwidth is increased and gain is decreased by the *simultaneous* application of series and parallel feedback, and the change of input and output impedances is not as high as that produced with solely series or parallel feedback. Omitting detailed calcula-

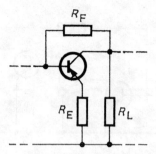

Fig. 10.9. Simultaneous application of voltage-feedback and current-feedback in cascaded-amplifier chains

tions, the final results will be presented for the circuit shown in Fig. 10.9, assuming cascade connection of identical stages. According to Ref. [11.11], the low-frequency gain is

$$A_{v0} = -\frac{R_L(R_F - R_E)}{R_E(R_F + R_L)} \simeq -\frac{R_L}{R_E}, \qquad (10.3.1)$$

and the input and output resistances are

$$R_{in} \simeq R_E \frac{R_F + R_L}{R_E + R_L}; \qquad R_{out} \simeq R_E \frac{R_F + R_g}{R_E + R_g}, \qquad (10.3.2)$$

where R_g is the generator resistance. The 3-dB cut-off frequency is given by

$$\omega_H = \omega_T/A_{v0} \psi_E \simeq \omega_T R_E/R_L \psi_E, \qquad (10.3.3)$$

where the factor ψ_E is characteristic for the loading effect of the next identical stage, and is given by

$$\psi_E = 1 + \omega_T C_c [R_L \parallel (r_e + R_E)]. \qquad (10.3.4)$$

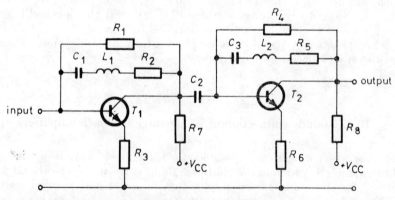

Fig. 10.10. Circuit diagram of two-stage amplifier with voltage- and current-feedback and inductive compensation

Figure 10.10 shows a two-stage wideband amplifier circuit based on Ref. [10.16]. Parallel feedback is realized by resistances R_3 and R_7, and series feedback is realized by resistances R_4 and R_8. Inductances L_1 and L_2 serve to compensate the gain reduction at the upper frequency limit. The cut-off frequency of the applied transistors is $f_T = 1.4$ GHz. The amplifier has the following data: $A_0 = 13 \pm 0.7$ dB, bandwidth = 40–860 MHz, VSWR < 2 in the whole frequency range, at both the input and output sides.

In Ref. [9.10], a single-stage compensated feedback amplifier is presented for the frequency range 0.1–1 GHz, using computer optimization.

10.4. Two-stage amplifier with stage-by-stage feedback

Two-stage amplifiers with stage-by-stage feedback and suitable compensation are widely used in wideband amplifiers. Of the various possible combinations, the arrangement shown in Fig. 10.11 has the best properties as regards input and output impedances and bandwidth.

Analysis of this connection will be presented on the following assumptions: capacitances C_c are sufficiently small ($\psi = 1$), emitter feedback resistance R_E is relatively high so that $\beta_0 R_E \gg r_b + R_g$, and the compensations meet the following requirements:

$$\omega_T R_E C_E = 1 ; \tag{10.4.1}$$

$$\omega_T L_F/R_F = (r_b + R_g)/(R_E + r_e) . \tag{10.4.2}$$

In this case, the transfer function is simplified, and the denominator is quadratic. The low-frequency voltage gain is given by

$$A_{v0} = \frac{V_2}{V_g} \approx \frac{R_F}{R_E + r_e} \approx \frac{R_F}{R_E} , \tag{10.4.3}$$

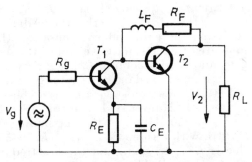

Fig. 10.11. Circuit diagram of two-stage wideband amplifier with feedback and compensation

with the assumption that $\beta_0 R_L \gg R_F$, i.e. the feedback is large. The 3-dB cut-off frequency is given by

$$\omega_H = k\, \omega_T \sqrt{\frac{R_L (R_E + r_e)}{R_F (R_g + r_b)}}, \qquad (10.4.4)$$

where the value of the proportionality constant is $k = 0.8$–1.4.

In another version of the circuit shown in Fig. 10.11, a shunting capacity C_F of resistance R_F is used for compensation instead of series inductance L_F, in order to compensate for the small response peak caused by transistor T_1. The value of C_F is usually very low and of the same order as the capacity C_c of transistor T_2.

In Ref. [10.6], the circuit of Fig. 10.11 is applied in a pulse amplifier. The principal data of the applied transistors are: $f_T = 800$ MHz, $r_b = 100$ ohm, $C_c = 1.7$ pF. Component data are $R_E = 22$ ohm, $R_F = 180$ ohm, $C_E = 10$–20 pF, variable; $L_F = 120$ nH, amplification $A_{vo} = 12$ dB, rise time $t_r \approx 2$ ns. A shunt resistance $R_1 = 50$ ohm shunting the base of transistor T_1 is used for producing a suitable input resistance.

10.5. Multi-stage amplifiers with several feedback loops

Figure 10.12 shows a cascade connection of amplifiers compensated by series feedback inductances. In Ref. [10.3], a five-stage amplifier utilizing the stages shown in Fig. 10.12 is presented.

The main data of the applied transistors are: $f_T = 400$ MHz, $r_b = 50$ ohm; $C_c = 2$ pF. The low-frequency gain between terminations $R_g = R_L = 50$ ohm is $A_0 = 50$ dB, with feedback resistances $R_F = 220$ ohm. Maximum bandwidth with $L_F = 0.5$ μH is $B = 100$ MHz. Input impedance over the whole frequency range stays within $Z_{in} = 40$–100 ohm. Output power without distortion is approximately -7 dBm.

In the frequency range above 100 MHz, accuracy of circuit analysis is substantially influenced by parasitic elements due to the transistor itself and to the external circuit. These parasitic elements introduce response fluctuations which can be compensated by the circuit shown in Fig. 10.13. The parallel resonant circuit comprising elements L_2, R_2 and C_2 and having a high damping factor has the effect of lowering the feedback at the resonant frequency, thus increasing the gain and compensating the response. Response peaks are eliminated by the shunting effect of the series resonant circuit made up of L_3, C_3 and R_3. Inductance L_4 has a series compensation effect, matching the low input impedance of the next stage to the output.

In Refs. [10.7] and [10.11], a seven-stage amplifier consisting of identical stages as shown in Fig. 10.12 is presented. The transition frequency of the applied transistors is $f_T = 850$ MHz, and the maximum power gain at $f = 500$ MHz is $G = 7$ dB, with terminations $R_g = R_L = 50$ ohm. The feedback resistance value is $R_1 = 82$ ohm, and the low-frequency gain of the complete amplifier is $A_0 = 52$ dB. The resonant frequency of the compensating circuit in the feedback path is $f_2 = 200$ MHz, $R_2 = 68$ ohm, $C_2 = 5$ pF.

10.5. Feedback loops

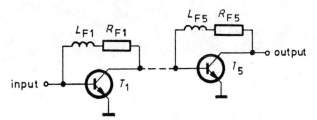

Fig. 10.12. Circuit diagram of five-stage wideband amplifier with current-feedback and compensated stages

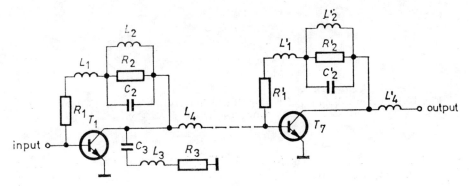

Fig. 10.13. Circuit diagram of seven-stage wideband amplifier with current-feedback and multiple compensated stages

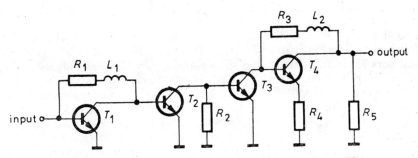

Fig. 10.14. Circuit diagram of four-stage wideband amplifier

The resonant frequency of the shunting series resonant circuit is $f_3 = 370$ MHz, $R_3 = 68$–100 ohm; $C_3 = 5$–10 pF. Response flatness in a band of $B = 500$ MHz is a maximum of ± 2 dB. Low-frequency cut-off is at $f = 100$ kHz, maximum undistorted output power is approximately 0 dBm.

The simplified circuit diagram of a four-stage wideband amplifier is shown in Fig. 10.14 (see Ref. [10.13]). The amplifier consists of two common-emitter pair of stages, one pair with feedback and the other without. The tran-

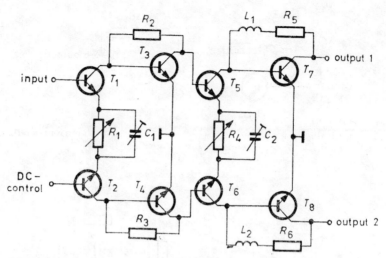

Fig. 10.15. Circuit diagram of push-pull wideband amplifier with feedback and compensation

sistor transition frequency is $f_T = 1.3$ GHz. Feedback resistance values are $R_1 = 2$ kohm, $R_3 = 200$ ohm, $R_4 = 24$ ohm. The output standing-wave ratio is improved by the parallel resistance $R_5 = 100$ ohm. Compensating inductances are $L_1 = L_2 = 0.1$ μH. In the frequency range $f = 50$–500 MHz, amplification is $A_v = 20$–23 dB, input and output impedances are $Z_0 = 50$ ohm, input and output VSWR < 2. Undistorted output power is approximately 5 dBm.

The simplified circuit diagram of a symmetrical push-pull wideband amplifier is shown in Fig. 10.15 (see Ref. [10.14]). A long-tail pair is realized by transistors T_1 and T_2 acting as a phase splitter for the input signal and as a push-pull driving stage for the two symmetrical amplifier branches. In both transistor emitter circuits, resistance $R_1/2$ provides feedback, and capacity $2C_1$ compensation. The phase splitter is followed by amplifiers with parallel feedback, and further by two stages with feedback and compensation. The response is adjusted by the emitter resistances and capacities. The push-pull output voltage is used to drive a delay line in an oscilloscope vertical amplifier. The other input of the phase splitter serves for DC-level adjustment. Transition frequency is $f_T = 1.2$ GHz, collector capacity is $C_c = 1.5$ pF. Feedback resistances are $R_1 = 150$ ohm, $R_2 = R_3 = 470$ ohm, $R_4 = 100$ ohm; $R_5 = R_6 = 470$ ohm. Inductances are $L_1 = L_2 = 0.2$ μH for compensation, and by also utilizing a low-capacity trimmer capacitor, amplification is $A_0 = 38$ dB, and bandwidth is $B = 300$ MHz.

Figure 10.16 shows the symmetrical output stage of the amplifier above. Emitter-feedback transistors T_1–T_4 are driven by symmetrical input signals. The output stage is made up from the parallel-connected transistors T_5–T_7 and T_6–T_8, respectively, both also having emitter feedback, and the output

10.5. Feedback loops

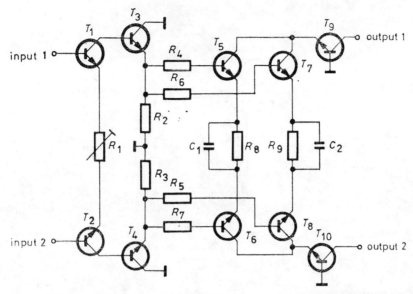

Fig. 10.16. Circuit diagram of push-pull wideband power amplifier with feedback

stages T_9 and T_{10} driven by these pairs. Transition frequency is $f_T = 1$ GHz. Component values are: $R_1 = 100$ ohm, $R_2 = R_3 = 2.7$ kohm, $R_4 = R_5 = R_6 = R_7 = 68$ ohm, $R_8 = R_9 = 82$ ohm, $C_1 = C_2 = 1$ pF; amplification is $A_0 = 26$ dB, bandwidth is $B = 300$ MHz, with an output load of $R_L = 165$ ohm. Maximum output voltage is 2×18 V.

A symmetrical wideband amplifier made up of the double feedback stages shown in Fig. 10.6 is presented in Ref. [10.20] (see Fig. 10.17). The second element of the feedback amplifier is the Darlington circuit made up of transistors $T_2 - T_3$ and $T_5 - T_6$, respectively. Balun's Tr_1 and Tr_2 serve for balancing, inductances L_1 and L_2 serve for high-frequency matching. Utilizing transistors with transition frequencies $f_T = 5$ GHz, a flat response in the frequency range 3–300 MHz is achieved with a gain of $A_v \simeq 10$ dB and a noise figure of $F = 7$ dB. Second and third order intermodulation distortions are both below $d_{IM} < -75$ dB. The main application of this circuit is in preamplifiers and power amplifiers used for CATV purposes.

A four-transistor wideband amplifier for the frequency range 30–300 MHz is presented in Ref. [12.14]. The first stage is a single transistor amplifier with current feedback and inductive compensation. This is followed by a three-transistor stage with double feedback and a Darlington circuit second element.

An eight-stage push-pull wideband amplifier is presented in Ref. [10.12]. Push-pull pairs are made up of complementer transistors according to Fig. 10.18 which shows the output stage of the amplifier. The preamplifier part has similar stages. Transistors T_1 and T_2 are common-emitter complementer stages, with L_1 compensating inductances in series with load resistance R_1.

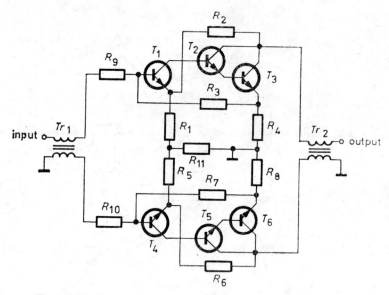

Fig. 10.17. Circuit diagram of symmetrical wideband amplifier

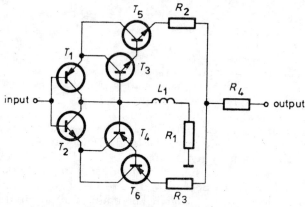

Fig. 10.18. Circuit diagram of push-pull wideband amplifier with complementer transistors

Driving transistors T_3 and T_4 and output stages T_5 and T_6 are emitter followers.

Resistances R_2, R_3, and R_4 are used to provide a specified output resistance. The push-pull arrangement results in good linearity even at higher levels. An amplification of $G = 35$ dB with a load of $R_L = 50$ ohm has been realized in a bandwidth of $B = 145$ MHz. Distortion is less than 1% at a level 13 dBm. Output impedance is $Z_{out} \simeq 50$ ohm.

10.5. Feedback loops

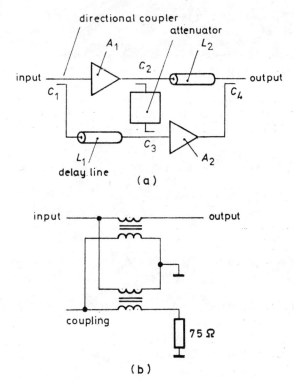

Fig. 10.19. "Error-controlled" amplifier (a) circuit principle, (b) directional coupler

A wideband pulse amplifier with variable gain is presented in Ref. [10.18]; bandwidth is not influenced by the gain control. Control and compensating elements are two JFET's.

"Error-controlled" amplifiers [12.8], [12.14] having extremely low distortion may be considered as falling into a separate group of wideband amplifiers. The circuit principle is shown in Fig. 10.19 (a). The input signal is sampled by directional coupler C_1, the sampled signal is then delayed by delay line L_1, and from this signal, the signal passing through main amplifier A_1 and the attenuator, consisting of noise and distortion, is subtracted by directional coupler C_3. Amplifier A_2 thus only amplifies noise and distortion components, and these are subtracted from the main amplifier signal delayed by L_2; subtraction is performed by directional coupler C_4. With correct subtractions for amplitude and phase, the nonlinear distortion of active elements is cancelled, resulting in extremely low intermodulation distortion.

Figure 10.19 (b) shows the circuit principle of the wideband directional coupler which has toroidal construction. According to Ref. [12.14], a thin-film implementation of the above principle resulted in $d_{IM} < -95$ dB in the frequency range 30–300 MHz.

10.6. Input impedance of feedback amplifiers

The input impedance of transistor amplifiers is increased by voltage feedback and decreased by current feedback (see Table 9.1). Strong current feedback may result in a change of the impedance type, i.e. a real or inductive impedance may result, instead of the capacitive input impedance of the common-emitter stage without feedback.

Let us first investigate the input impedance of the single-stage amplifier with emitter feedback (Fig. 10.20(a)). Resistance is increased and the paralle

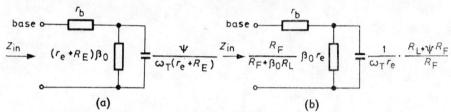

10.20. Input impedance of single-stage amplifier with (a) emitter feedback (b) current feedback

capacitance is decreased by feedback, the time constant given by these two elements remaining unchanged.

The input impedance of the current-feedback amplifier is shown in Fig. 10.20 (b). In contrast with the previous case, the RC time constant is decreased by increasing feedback, because the capacitance increase is slower than the resistance decrease. With high enough feedback, input impedance is approximately resistive and tends to the value r_b.

The input impedance change of two-stage amplifiers is similar, but numerical values are different. The input impedance of the two-stage current feedback amplifier treated in Section 10.1 is decreased by much less feedback so as to make it first real, and then inductive. Extremely low input impedances can be produced in this type of amplifier, making possible the use of series-resistance extensions providing constant, frequency-independent input resistances.

11. WIDEBAND APPLICATIONS OF FEEDBACK INTEGRATED AMPLIFIERS

11.1. Types of integrated amplifiers

Developments in solid-state technology have made it possible to construct multi-stage amplifiers on a single chip [11.4]. In the following, the main characteristics of these circuits will be surveyed, primarily from the high-frequency application aspect.

The group of linear integrated circuits with the largest variety consists of *operational amplifiers*. These are mainly produced in monolithic form, with DC coupling, differential input, very high gain and low thermal drift. The bandwidth is only a few MHz, even with high feedback. The two basic types of feedback are shown in Fig. 11.1. Figure (a) shows the current feedback for which the relation

$$A_v = V_2/V_1 = - R_2/R_1, \qquad (11.1.1)$$

holds. In Fig. (b), voltage feedback is represented, with the relation

$$A_v = (R_1 + R_2)/R_2. \qquad (11.1.2)$$

In both cases, the voltage between the inverting (−) and the noninverting (+) input terminals of the operational amplifier is negligible compared with the input voltage V_1.

Operational amplifiers are rarely applied in wideband amplifiers, especially when the required bandwidth of the amplifier is higher than 1 MHz. The values of bandwidth and the necessary compensating elements are included in the set of responses which are given for each type of feedback. The compensation problem will be dealt with later.

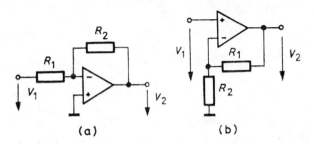

Fig. 11.1. Monolithic operational amplifiers with (a) current feedback and (b) voltage feedback

Another group of linear integrated amplifiers is formed by the *wideband* monolithic amplifiers. These are multi-stage feedback amplifiers also containing the feedback network within the integrated circuit, and requiring only external compensating capacitances. The number of stages in this amplifier is much less than in operational amplifiers, and in most cases, no differential input is provided. However, the bandwidth is much higher, in some cases exceeding 100 MHz. In the applied technology, as opposed to the epitaxial technology of operational amplifiers, dielectric isolation or beam-lead construction is used (see Section 3.1).

The *symmetrical* wideband monolithic amplifiers are mainly applied in vertical-deflection amplifiers of oscilloscopes [3.3], [3.7].

For extremely high frequency applications, *hybrid* types are mainly used. These are characterized by a resistance network deposited on a ceramic substrate which also holds transistors, monolithic amplifiers and capacitors. This technology is suited to high-frequency applications since the advantages given by the monolithic technology are retained, without the limitations of a single chip (parasitic capacitances, dissipation limits, etc.).

11.2. Compensation of monolithic integrated amplifiers

According to the basic rule given in Section 9.2, the feedback amplifier is unconditionally stable if the intersection point of the open-loop and closed-loop-responses lies within the 6-dB/octave section of the open-loop response (see Fig. 9.6). Instability may result if the intersection lies within the 12-dB/octave section. Good stability thus requires a long 6-dB/octave section, i.e. the frequency difference between the two cut-off frequencies should be high (at least two octaves).

There are two basic compensation methods. When applying the *phase-compensation* method, the open-loop response is shaped to have a suitable form, and the feedback network is frequency-independent. When applying the *closed-loop compensation*, the required phase margin is produced by using a frequency-dependent feedback network. Both methods are treated in the following.

The Bode diagram for phase compensation is shown in Fig. 11.2. The open-loop characteristic $A(\omega)$ has three cut-off frequencies and may thus be expressed in the following form:

$$A(\omega) = \frac{A_0}{(1+j\omega/\omega_1)(1+j\omega/\omega_2)(1+j\omega/\omega_3)}, \qquad (11.2.1)$$

where ω_1, ω_2 and ω_3 are the cut-off frequencies. Figure 11.3 shows part of the amplifier circuit diagram. Let us assume that frequency ω_1 is the cut-off frequency of stage T_2 which has an input termination of R_{L1} and an output termination of R_{L2}. The first factor in the denominator of eqn. (11.2.1) is thus given by this stage. Placing the series feedback network consisting of R_1 and C_1 between collector and base of this stage, we obtain the

11.2. Compensation of monolithic amplifiers

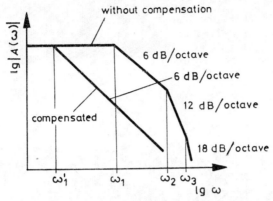

Fig. 11.2. Bode diagram of the open-loop gain with and without compensation

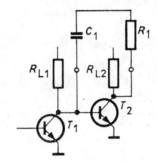

Fig. 11.3. Part of monolithic amplifier with external compensation circuit

following expression for the response of this stage:

$$A_2(\omega) = A_{20} \frac{1 + j\omega/\omega_2'}{1 + j\omega/\omega_1'}, \qquad (11.2.2)$$

where the two cut-off frequencies of interest are

$$\omega_1' = 1/R_{L1} C_1 (1 + g_m R_{L2}); \qquad (11.2.3)$$

$$\omega_2' = 1/R_1 C_1. \qquad (11.2.4)$$

In eqn. (11.2.3), g_m is the transconductance, and the factor in parentheses is the gain. Replacing the factor which has cut-off frequency ω_1 in eqn. (11.2.1), by the response (11.2.2), and assuming a correct design for which $\omega_2' = \omega_2$, we obtain a transfer function of the form

$$A'(\omega) = \frac{A_0}{(1 + j\omega/\omega_1')(1 + j\omega/\omega_3)}. \qquad (11.2.5)$$

This compensated response as shown in Fig. 11.2 has a long 6-dB/octave section and is thus suited for strong feedback. Cut-off frequency ω_3 is usually much higher than the upper cut-off and thus may be neglected.

154   11. Applications of feedback integrated amplifiers

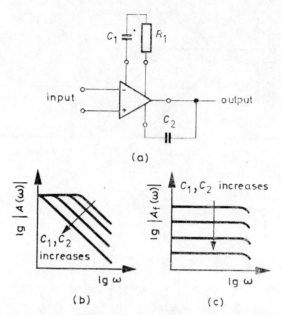

Fig. 11.4. Compensation of monolithic amplifier. (a) Circuit diagram, (b) frequency responses with open-loop gain, (c) frequency responses with closed-loop gain, for various compensating elements

From the responses specified for monolithic amplifiers, the compensating element values for different gains can be determined (Fig. 11.4). The open-loop response also gives stability information, found by noting the response slope at the intersection point with the feedback response. In some cases, the phase-frequency responses for several compensating element values are also given, in addition to the absolute values of open-loop gain response. From these two responses, the phase margin may be determined for any feedback.

It was assumed in the foregoing that the feedback factor β^* was frequency-independent. This implies that the open-loop response $A(\omega)$ and the feedback response $T(\omega)$ are identical and have the same cut-off frequencies. However, stability may be increased by choosing a frequency-dependent feedback factor β^*. This is called *closed-loop* compensation. The positive phase shift of the feedback factor having the form

$$\beta^* = \beta_0^*(1 + j\omega/\omega_1^*), \qquad (11.2.6)$$

is substracted from the negative phase angle of the feedback response $T(\omega)$, thus increasing the phase margin according to Fig. 9.5. Stability conditions are then checked by investigation of the feedback response determined according to Fig. 11.5,

$$T(\omega) = {}_iV_2/V_1. \qquad (11.2.7)$$

11.2. Compensation of monolithic amplifiers

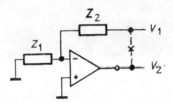

Fig. 11.5. Determination of loop gain by breaking the feedback loop

This can be accomplished either by measurement of the Nyquist diagram, or by satisfying the requirement according to which the slope of the absolute value response should not be higher than 6 dB/octave, but should be less than 12 dB/octave at the value $|T(\omega)| = 1$.

Monolithic operational amplifiers utilize *pnp* transistors because of the required DC level shift, and these are realized by lateral transistors with low cut-off frequencies. This requires a low cut-off frequency ω'_i during the phase compensation, resulting in a low bandwidth. Bandwidth may be increased by using vertical *pnp* transistors [3.22], but this demands a much more complicated technology.

Bandwidth may also be increased by shunting the lateral *pnp* transistor by a capacitor [3.16], [3.20]. This is the so-called "feedforward" method. The DC level shift can also be realized by an RC network utilizing a MOS

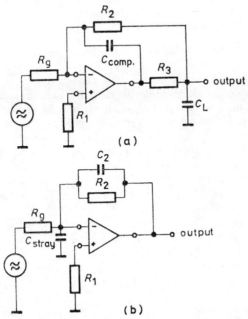

Fig. 11.6. Decreasing the effect of (a) load capacitance, and (b) input stray capacitance

capacitor for AC shunt [3.15]. With this method, the frequency for unity gain, ω_u, may be increased to 10–15 MHz.

A special operational-amplifier circuit is presented in Ref. [11.3]. The amplifier has a low-gain stage with high cut-off frequency (A_{10}, ω_1) and a high-gain stage with low cut-off frequency (A_{20}, ω_2). However, the stages are not only cascaded, but the output of the first stage is added via a summing network to the amplifier output. The overall gain is thus given by

$$A(\omega) = A_1 + A_1 A_2. \qquad (11.2.8)$$

As soon as gain A_2 falls below unity, the frequency response and the phase shift is determined by amplifier A_1. Thus the response slope may not exceed 6 dB/octave even in this frequency range, and this is advantageous for stability.

The effect of a capacitive output load may be decreased by the circuit shown in Fig. 11.6 (a). The capacitive load C_L is separated by resistance R_3 from the amplifier. High-frequency feedback via capacitor $C_{\text{comp.}}$ is directly from the amplifier output, thus the loop gain is not affected by the output phase shift.

The input stray capacitance can cause instability in amplifiers with high input impedances (e.g. FET input stages), especially if the feedback resistance is also high [3.18]. Instability may then be eliminated by a capacity C_2 in parallel with the feedback resistance (Fig. 11.6 (b)) which has a value approximately equal to the stray capacity divided by the feedback gain. Viewed from the output port, the circuit is a frequency-compensated resistive divider.

11.3. Wideband monolithic amplifiers

For monolithic amplifiers which use dielectric isolation technology, the 6-dB/octave loop-gain slope may be realized over a wide frequency range without compensation. The use of such an amplifier with 17 transistors as a voltage follower is shown schematically in Fig. 11.7. The output signal is fed back completely to the inverting input, thus producing unity gain. The amplifier is used as an impedance transformer, with a 7-MHz bandwidth, 15% overshoot and 40-ns rise time.

Figure 11.8 shows the circuit diagram of a three-stage monolithic amplifier. The feedback via resistors R_1 and R_2 corresponds to the circuit of Fig. 10.16, but the difference is given by the emitter follower T_3 following the two-stage amplifier and supplying the feedback voltage. Zener diode D_1 is used to set the working point. Since the loop gain is much higher than unity, the feedback gain is determined by the passive feedback network:

$$A = (R_1 + R_2)/R_1. \qquad (11.3.1)$$

Compensation has to be used to satisfy stability requirements: this is accomplished using external capacitance C_1 which has a low value and is placed between the base and collector of transistor T_2. With this circuit,

11.3. Wideband monolithic amplifiers

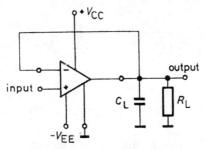

Fig. 11.7. Wideband voltage-follower circuit

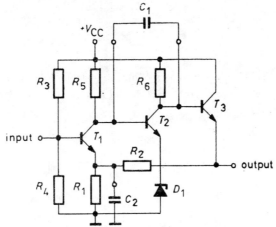

Fig. 11.8. Three-stage monolithic wideband amplifier

a gain of $A_v = 100$ may be attained with a cut-off frequency of $f_H = 100$ MHz.

The emitter of transistor T_1 is accessible, and by connecting a capacitance between this point and ground, the feedback may be decreased. If C_2 has a low value, feedback is decreased at the upper band limit and thus a response peak is formed. With large C_2, the feedback may be completely eliminated. Without feedback, the gain is approximately 60 dB, and the cut-off frequency is $f_H = 5$ MHz. Output impedance is very low, approximately $R_{out} = 1$ ohm, rendering the amplifier insensitive to loading.

The circuit of a four-stage monolithic amplifier is shown in Fig. 11.9 [11.4]. A two-stage feedback amplifier is made up of transistors T_1 and T_2; feedback is determined by resistances R_1 and R_2. Transistor T_3 is a stage with emitter feedback, which drives the emitter follower T_4. Since no more than two stages are bridged by feedback, the stability is very good. A bandwidth of 25 MHz may be attained with a gain of 30 dB.

Figure 11.10 shows a symmetrical push-pull wideband amplifier [11.6]. The four transistors with extremely high cut-off frequencies are produced on

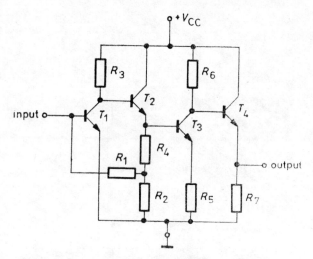

Fig. 11.9. Four-stage monolithic wideband amplifier

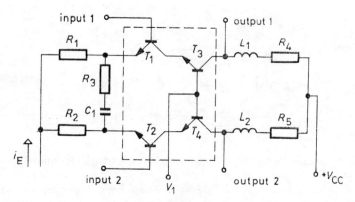

Fig. 11.10. Push-pull wideband amplifier with monolithic transistor quad

a single chip [3.13], and discrete components are used for the other circuit elements. Monolithic transistors allowed short connecting leads and low stray capacitances. The first stages are emitter-feedback amplifiers, with feedback resistances R_1 and R_2. Network $R_3 C_1$ is there for compensation near the upper band limit. The second stages are common-base amplifiers with collector inductances L_1 and L_2 also present for compensation. The transistor current is determined by current generator I_E. The compensation may be adjusted by using variable R_3, C_1 elements. With a cascade connection of several of these stages, vertical deflection oscilloscope amplifiers with a bandwidth of 250 MHz have been designed.

11.3. Wideband monolithic amplifiers

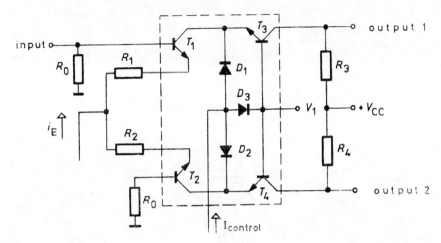

Fig. 11.11. Push-pull gain controlled wideband amplifier with monolithic transistors and diodes

A similar circuit is shown in Fig. 11.11, with three additional regulating diodes within the monolithic circuit. Resistors here are discrete elements. The input is asymmetrical, the output symmetrical. Gain is controlled using current generator $I_{control}$ and diodes D_1–D_3. At full-gain the diodes are off, and increasing the current $I_{control}$ the diodes will be on. Current through diode D_1 is subtracted from the emitter current of transistor T_3 which is thus decreased. The current of transistor T_4 is similarly decreased.

The decrease in emitter currents results in a higher input resistance of the transistors, and a lower resistance of the shunting diodes, therefore, the gain reduces. As soon as the generator current reaches the current value of transistors T_3 and T_4, these are switched off, and thus the complete amplifier is out of order. This is significant if output terminals are connected to several amplifiers according to Fig. 11.11 which are switched on/off alternately.

The amplifier, mentioned above, is used in two-channel oscilloscopes. The elements of both amplifiers may be contained in a single monolithic circuit, which thus contains 8 transistors and 6 diodes. Bandwidth, as before, is 250 MHz.

In Ref. [11.4], a monolithic IF amplifier with limiter, having a cut-off frequency of 170 MHz, is presented. The circuit is made up of a long-tailed pair, emitter followers and common-base stage on the output. In Ref. [6.5], a wideband preamplifier with high input impedance, made up of a monolithic operational amplifier and a FET pair, is investigated.

Figure 11.12 shows a monolithic wideband amplifier. Flat gain within a wide frequency range is assured by monolithic resistances R_C and R_E which are shunted for AC. The low-frequency gain is

$$A_{v0} \cong (R_C + r_C)/(R_E + r_E), \qquad (11.3.2)$$

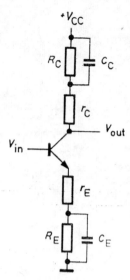

Fig. 11.12. Monolithic wideband amplifier

and the gain at the upper cut-off frequency is

$$A_v(\omega_H) = r_C/r_E. \qquad (11.3.3)$$

The optimum resistance ratio is given by equating these two quantities [11.9]. With this method, a gain of 8 ± 1.5 dB and a rise time of 0.2 ns may be obtained in the frequency range of 0–2.3 GHz.

Reference [11.11] presents a monolithic 1 GHz feedback amplifier compensated by transmission lines; this is treated in Section 18.4.

In Refs. [11.12] and [11.8], symmetrical amplifiers with 500 MHz and 700 MHz bandwidth are presented, consisting of two-stage amplifiers with voltage feedback as shown in Fig. 10.5. The computer-aided design of the three-stage feedback monolithic amplifier shown in Fig. 10.7 is dealt with in Ref. [11.7]. A gain-controlled amplifier with 70 MHz bandwidth is presented in Ref. [11.13]. A high-voltage, wideband monolithic oscilloscope amplifier is investigated in [11.10]. Operational amplifiers having low rise time are treated in [3.20] and [3.21].

11.4. Hybrid integrated wideband amplifiers

A special type of wideband amplifier is represented by DC-coupled two-branch amplifiers having high input impedance and low zero drift. The main application field for these amplifiers is as input amplifier or probe of wideband oscilloscopes where low input impedance would modify the conditions of the test item. However, the low zero drift of the DC transmission is in-

11.4. Hybrid wideband amplifiers

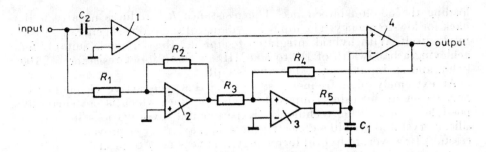

Fig. 11.13. Two-branch wideband hybrid amplifier with high input impedance

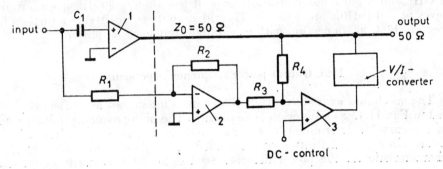

Fig. 11.14. Two-branch wideband hybrid amplifier with high input impedance and output impedance Z_0

compatible with the requirements of the wideband transmission. This is why the low-frequency and high-frequency transmission paths are separated (see Fig. 11.13, Ref. [11.2]). These amplifiers frequently have unity gain and are only used for impedance transforming.

Wideband amplifier *1* is AC coupled, with a lower cut-off frequency of a few kHz. This is also the cross-over frequency of the two branches.

The frequency range below the cut-off frequency, also obviously including DC, is transmitted by operational amplifiers *2* and *3* which have extremely low drifts. Integrator $R_5 C_1$ is designed to set the upper cut-off frequency of this latter branch. The signals of the two branches are added by amplifier *4* by controlling the DC amplifier *3* via R_4. The high input impedance of wideband amplifier 1 is provided by a FET. Amplifier 1 and resistance R_1 are both located in the probe and usually form a hybrid integrated circuit.

Another similar variant of this circuit is shown in Fig. 11.14 [11.5]. The high input impedance, wideband hybrid amplifier has an output impedance of 50 ohms, and is directly coupled to the output cable. DC and low-frequency transmissions are produced via resistance R_1, further operational amplifiers *2* and *3* which are also used for DC adjusting. The DC and low-frequency paths are coupled through a voltage-current converter without loading the 50-ohm impedance. The resistance R_4 produces negativ feedback for low-frequency. The input impedance of the probe is generally given by R_1 itself. With hybrid integrated technique, this circuit is suitable for achieving a bandwidth of DC to 500 MHz, with an input resistance of 100 kohm and an input capacitance of a few pF.

At extremely high frequencies, coupling between active elements can only be set up by using coupled transmission lines which is not currently possible in monolithic technique. (Experimental investigations into using silicon crystal as stripline dielectric material have not yet proved to be successful.) However, in hybrid technology it is very easy to produce striplines on the ceramic substrate, and it results in better cooling and thus higher output power. Hybrid amplifiers with a gain of 40 dB, a cut-off frequency of 2 GHz and a harmonic distortion level of less than -30 dB at an output level of $+10$ dBm are feasible. Hybrid amplifiers are treated further in Section 17.4, in connection with tuned high-frequency amplifiers.

11.5. Output power frequency dependence

The maximum available output power from integrated amplifiers is plotted in Fig. 11.15 as a function of frequency. The high-frequency falling part of the curve is determined by the gain and the compensation circuit. This figure allows the definition of a so-called *power cut-off frequency* which is characteristic for the power decrease, and does not necessarily equal the

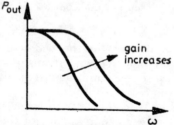

Fig. 11.15. Output power of monolithic wideband amplifiers as a function of frequency

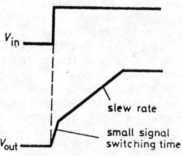

Fig. 11.16. Transient response of monolithic wideband amplifier

11.5. Output power frequency dependence

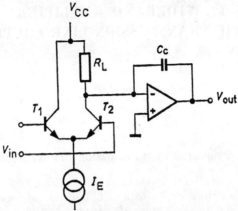

Fig. 11.17. Operational amplifier circuit for slew-rate calculation

cut-off frequency ω_H of the small-signal frequency response. Above the power cut-off frequency increasing the input signal level, the output waveform will approach a triangular waveform. Increasing the compensating capacitances has the effect of decreasing the power cut-off frequency, since these capacitances have to be charged within a finite time interval.

The large-signal transient response is shown in Fig. 11.16. An input step-function signal produces an output signal which has two sections, the first with a short rise time, and the second with a slow rise time. The quick rise is determined by the switching time which, according to Section 5.7, is given by

$$t_r = 2.2/\omega_H . \qquad (11.5.1)$$

Above a certain voltage level, the rate of change is determined by the so-called *slew rate* which is given, as shown in Fig. 11.17, by compensating capacitance C_C and the current I_E charging it:

$$\left.\frac{dV_{out}}{dt}\right|_{max} \cong \frac{I_E}{C_C} = \frac{\omega_u I_E}{g_{m1}} = \omega_u \frac{2kT}{q}, \qquad (11.5.2)$$

where ω_u is the unit-gain frequency, and I_E and g_{m1} are the current and transconductance of the input differential amplifier, respectively. The slew rate is dependent on the *technology* used for the integrated amplifier, which also determines ω_u.

The slew rate may be increased by decreasing the transconductance of the differential amplifier, or by utilizing emitter resistances [3.2], [3.3] or JFET's.

The power bandwidth ω_{max} may be calculated from the slew rate. Assuming a maximum undistorted output voltage $V_{out}(t) = V_0 \sin \omega t$, substitution into (11.5.2) yields the required power bandwidth

$$\omega_{max} = \frac{1}{V_0} \left.\frac{dV_{out}}{dt}\right|_{max} . \qquad (11.5.3)$$

12. WIDEBAND AMPLIFIERS WITH TRANSMISSION-LINE COUPLING

12.1. Properties of transmission-line transformers

In transformers using conventional windings, high-frequency transmission is limited by the winding capacitance. This demands special constructions from the short-wave ranges upwards. In transmission-line transformers, the layout of the coils is chosen so that the winding capacitances become integral parts of a transmission line. With no parasitic resonances, the upper frequency limit is determined by the length of the transmission line. The simplest case is that where the transmission line consists of twisted wire, but flat or coaxial cable may be used equally well. To achieve a low characteristic impedance, two lines may be connected in parallel – this halves the overall characteristic impedance.

The low-frequency bandwidth of the transmission-line transformer is limited by the primary inductance, as is similar to the conventional transformers. The primary inductance may be increased by using high-permeability ferrite toroidal cores which are used exclusively in practical applications.

The permeability of the toroidal core decreases with increasing frequency, and the transformer gradually takes on the characteristics of a transmission line. The 4 : 1 ratio impedance transformer shown in Fig. 12.2 and widely used, has a transmission loss which is given as a function of wave length (assuming matched conditions) by the following expression:

$$a_t = \frac{(1 + \cos \beta l)^2 + 4 \sin^2 \beta l}{4(1 + \cos \beta l)^2}, \qquad (12.1.1)$$

where l is the line length and $\beta = 2\pi/\lambda$ [12.1], [12.7]. According to this expression, attenuation of the line is 1 dB at $l = \lambda/4$, and infinity at $l = \lambda/2$.

The simplest form of line transformer is given by the 1 : 1 ratio polarity changing transformer (Fig. 12.1). Figure (a) shows the conventional form, and Fig. (b) shows the equivalent circuit of the transmission line arrangement. The transmission-line characteristic impedance must be chosen as $Z_0 = R$ where R is the terminating resistance. Practical realization of the toroidal winding is shown in Fig. 12.1 (c), which gives connection points and winding directions. By earthing the middle point of the resistance, a balance output is obtained. Should resistance R differ from the characteristic impedance, the circuit will act as a $\lambda/4$ transformer at the frequency corresponding to line length $l = \lambda/4$, thus causing mismatch. The transmission loss intro-

12.1. Transmission-line transformers

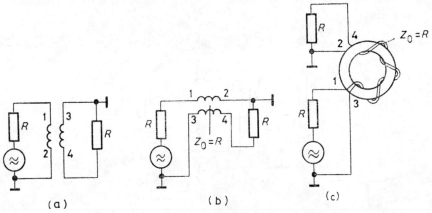

Fig. 12.1. Inverting transmission-line transformer with 1 : 1 ratio. (a) Conventional form, (b) transmission-line form, (c) practical arrangement

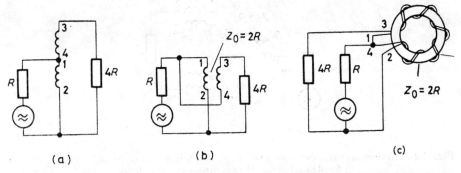

Fig. 12.2. Circuit of 4 : 1 ratio impedance transformer. (a) Conventional form, (b) transmission-line form, (c) practical arrangement

duced by this mismatch is a function of the deviation from the ideal value of $Z_0 = R$. According to measurements, a flat response within 1 dB in the frequency range 0.1–500 MHz can be achieved with a well designed transformer having a high quality core.

Figure 12.2 shows the conventional form and the transmission-line equivalent circuit of the widely used 4 : 1 ratio impedance transformer. Transmission-line characteristic impedance is $Z_0 = 2R$. The practical arrangement is shown in Fig. 12.2 (c). In the ideal case, 1 dB loss occurs at a frequency corresponding to a line length of $l = \lambda/4$. According to measurements, high quality toroidal cores allow a flat response in the frequency range 0.5–300 MHz.

Figure 12.3 shows the conventional form and the transmission-line equivalent circuit of a 4 : 1 ratio impedance transformer having asymmetrical–symmetrical terminals. All three lines are placed on a single toroidal core,

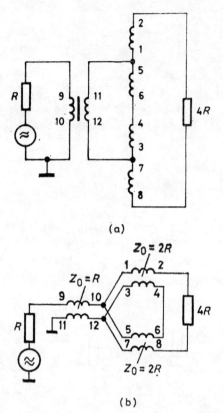

Fig. 12.3. Asymmetrical–symmetrical 4 : 1 ratio impedance transformer. (a) Conventional form, (b) transmission-line form

with winding directions corresponding to the numbering. The resistance $4R$ on the secondary winding may be earthed at any end point or at the middle point, and thus the circuit may also be used for phase splitting. Transmission-line characteristic impedance values are shown in the figure.

The coil of characteristic impedance $Z_0 = R$ in Fig. 12.3 (b) provides an asymmetrical impedance R, as this impedance has a symmetrical appearance at points *10* and *12*. Such a winding acts as a polarity-changing transformer according to Fig. 12.1. By connecting this which has characteristic impedance $Z_0 = 4R$ on the secondary side, to points *2* and *8*, we obtain the phase-splitter of Fig. 12.4. This has a ratio of 8 : 1, but in the downward direction. This configuration is frequently applied in push-pull amplifiers.

In all transformers treated above, the generator shown at the primary side may be placed to the secondary side, i.e. the transformers act in the same way in both directions.

Hybrids are used for combining and branching AC power. Let us consider as an example the hybrid shown in Fig. 12.5. The conventional form is shown

12.1. Transmission-line transformers

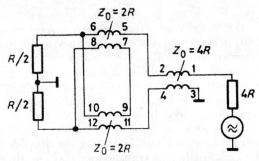

Fig. 12.4. Circuit of 8 : 1 ratio phase splitter

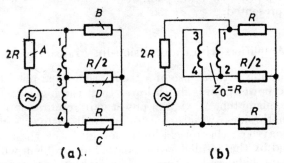

Fig. 12.5. 3-dB hybrid. (a) Conventional form, (b) transmission-line form

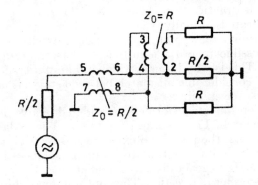

Fig. 12.6. 2 : 1 ratio hybrid using two iron cores

in Fig. (a), and the transmission-line variant in Fig. (b). The power of the generator which has a resistance $2R$ (branch A) is equally divided on the two resistances R. From symmetry, the balance resistance $R/2$ receives no power (branch D). Placing the generator into branch D, the power is also divided between branches B and C, and this time branch A will be without power. Finally, placing the generator into branch B, the power will be divided between branches A and D, and no power will reach branch C. The same is true for the case when the generator is in branch C. Power addition is accomplished in the same way. If generators placed in branches B and C have equal amplitudes and phases, their powers will add in branch D, and branch A will receive no signal.

Placing the generator between points *3* and *4* in the 3-dB hybrid of Fig. 12.5, the generator resistance is decreased to a value of $R/2$. We thus come to the hybrid shown in Fig. 12.6; here resistance $R/2$ is between points *6* and *8*, and is not grounded. It is a shortcoming of this circuit, that the generator and the load resistances have no common ground. However, using a polarity-changing transformer of characteristic impedance $Z_0 = R/2$, the generator may be grounded. This transformer is preferably placed on a separate core in order to give better magnetic conditions.

In Ref. [12.5], practical realization of line transformers with ratios higher than 4 : 1 is presented.

12.2. Amplifiers with transmission-line transformer coupling

If direct coupling without use of a transformer is applied between amplifier stages, the power gain is much lower than the highest attainable value because of the mismatch. Maximum power gain requires complex-conjugate matching at both sides, which can not be realized over a wide frequency range. There may be a theoretical limitation given by the real resistance to be matched and by the parallel capacitance. This problem will be dealt with in the section on wideband matching. If matching is theoretically possible in the frequency range of interest, this does not necessarily mean that the matching circuit is practically realizable (e.g. the reactive network may be extremely complicated). Finally, the wideband impedance matching is critical from the stability viewpoint, and, particularly with multi-stage amplifiers, it is not possible to provide sufficient phase margin over the whole frequency range. This is why in most wideband amplifiers (especially multi-stage amplifiers), considerable mismatch between stages is used giving lower power gain but better stability.

The transmission-line transformers considered above may be used to decrease the mismatch. For instance, for the 4 : 1 impedance transformer according to Fig. 12.2, the power gain is increased by 6 dB in the ideal case. The ratio of transistor-stage output to input impedances in the frequency range below 100 MHz is generally much higher than the ratio of the transmission-line transformer in question. Thus optimum matching is still remote,

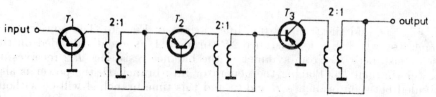

Fig. 12.7. Circuit diagram of three-stage wideband amplifier using transmission-line transformers for inter-stage matching

12.3. Push-pull amplifiers

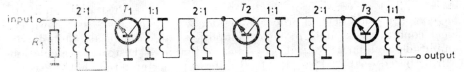

Fig. 12.8. Circuit diagram of three-stage wideband common-base amplifier using transmission-line transformers for inter-stage matching

and no stability problems arise. On the other hand, the gain increase allows much better use of the transistors. Above 100 MHz, these considerations are not always valid. With increasing frequency, output impedance decreases at a higher rate than the input impedance, and thus at higher frequencies instability may occur even in the case for impedance transformation 4 : 1.

A wideband amplifier using transmission-line transformers is shown in Fig. 12.7. Both the common-base and common-emitter stages are coupled by line transformers of 4 : 1 impedance ratio (2 : 1 voltage ratio) to the next stage and the load, respectively. If the applied transistors have transition frequencies of $f_T = 750$ MHz, and collector capacitances $C_c = 1.2$ pF, a flat gain of $A_v = 25 \pm 1$ dB in the frequency range 20–200 MHz may be achieved between 50 ohm terminations [12.2], [12.10], up to an output level of about 10 dBm.

Figure 12.8 shows a three-stage common-base amplifier using wideband line transformers for matching. The 1 : 1 ratio polarity-changing transformers are only used for DC blocking and impedance matching is provided by the 2 : 1 ratio transmission-line transformers. Using transistors of 400-MHz transition frequency and 20-pF collector capacity, a flat gain of 17 dB is possible, up to the 400-MHz cut-off frequency [12.3], [12.4].

12.3. Push-pull amplifiers with transmission-line transformers

The use of transmission-line transformers allows the design of wideband push-pull amplifiers. The opposite phase driving signals are provided by either a phase splitting transformer or a hybrid. Output signals are combined in a similar way. The output power of the push-pull amplifier can be added to the output of further amplifiers by using further hybrids, thus increasing the power level. The push-pull amplifier may be operated in any

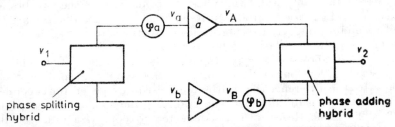

Fig. 12.9. Block diagram of push-pull amplifier using 3-dB hybrids

class, similarly to conventional audio output amplifiers. Push-pull amplifiers have typically low second-order distortion, zero in the ideal case; however, ideal conditions in high-frequency amplifiers are influenced not only by amplitude asymmetry but also by phase differences. Distortion calculation will be based on Fig. 12.9 [12.11]. Taking into account the characteristics of nonlinear amplifiers a and b, the output voltages produced by the input voltages are given as:

$$v_A = v_a a_1 + v_a^2 a_2 + v_a^3 a_3 + \ldots ;$$
$$v_B = v_b b_1 + v_b^2 b_2 + v_b^3 b_3 + \ldots . \qquad (12.3.1)$$

The input voltages in the above expressions can be written as:

$$v_a = v_1 \exp(j\varphi_a)/\sqrt{2} ;$$
$$v_b = v_1/\sqrt{2} , \qquad (12.3.2)$$

where φ_a is the phase shift of the phase splitting hybrid in the upper branch. From summation, the output voltage is given as

$$v_2 = v_A/\sqrt{2} + v_B \exp(j\varphi_b)/\sqrt{2} , \qquad (12.3.3)$$

where φ_b is the phase shift of the combining hybrid in the lower branch. In the spectrum of the output voltage, second-order, third-order etc. components will also be present, in addition to the fundamental frequency:

$$v_2 = v_2^{(1)} + v_2^{(2)} + v_2^{(3)} + \ldots , \qquad (12.3.4)$$

where $v_2^{(n)}$ is the amplitude of the nth order distortion.

The fundamental frequency component, from eqns (12.3.1) to (12.3.4), is given by

$$v_2^{(1)} = v_1 [a_1 \exp(j\varphi_a) + b_1 \exp(j\varphi_b)]/2 . \qquad (12.3.5)$$

The second-order distortion component is

$$v_2^{(2)} = v_1^2 [a_2 \exp(j 2\varphi_a) + b_2 \exp(j\varphi_b)]/2\sqrt{2} , \qquad (12.3.6)$$

and the third-order distortion component is

$$v_2^{(3)} = v_1^3 [a_3 \exp(j 3\varphi_a) + b_3 \exp(j\varphi_b)]/4 . \qquad (12.3.7)$$

Equation (12.3.6) shows that to avoid second-order distortion, the conditions are identical amplitude and phase transmissions in the two branches, i.e. $a_2 = b_2$, and $\varphi_a = \varphi_b = 180°$. In Ref. [12.11], detailed diagrams are presented for determining the second-order distortion introduced by differing phase ($\varphi_a \neq \varphi_b$) and differing gains ($a_2 \neq b_2$).

Figure 12.10 shows a wideband push-pull output amplifier. The input hybrid serving for power branching is the circuit according to Fig. 12.6(a), with the difference that the 50-ohm resistance generator is connected to point 1. Power is divided between two resistances each having the value of $R/2 = 25$ ohms. Resistance R_1 connected to point 4 acts as a balance resistance. The 1 : 1 ratio transformer on a separate core has a characteristic

12.3. Push-pull amplifiers

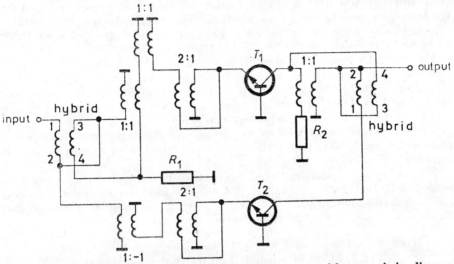

Fig. 12.10. Circuit of wideband push-pull output stage with transmission-line transformer

impedance of 25 ohms, and provides the asymmetrical (earthed) output according to Fig. 12.6. The phase splitting required for push-pull operation is provided by a $1 : -1$ ratio polarity-changing transformer (see Fig. 12.1) situated in the lower branch. As different phase shifts in the two branches must be avoided in order to suppress second-order distortion, a $1 : 1$ ratio transformer is also sited in the upper branch for compensation.

Impedance match is provided in both branches by a $2 : 1$ voltage-ratio transformer (Fig. 12.2).

Thus both base electrodes are driven from an impedance of $R = 25/4 = 6.2$ ohms. Collector signals are added by the output hybrid. The two hybrid generators are connected to points *1* and *4*. Output load is $R/2 = 15$ ohms, and thus collector loads of $R = 30$ ohms are seen by the generators. Balance resistance is also $R = 15$ ohms which is transformed asymmetrically by the $1 : 1$ ratio transformer. The transition frequencies of the applied

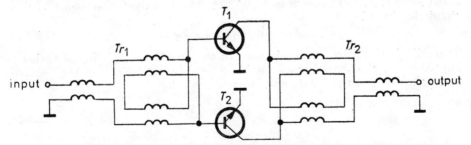

Fig. 12.11. Common-emitter push-pull output stage using transmission-line transformers

transistors are 400 MHz, and high quality ferrite toroidal cores are used for the transformers. The circuit can also be applied as a pulse amplifier; in this case, a 15-ohm load can be fed by 1-A current pulses, at a pulse width of 11 ns and a repetition rate of 80 MHz [12.3], [12.4].

Figure 12.11 shows the simplified circuit diagram of common-emitter push-pull amplifiers. Here the push-pull drive and the collector-voltage addition are provided by a line transformer utilizing a single toroidal core according to Fig. 12.4. The terminating resistances are $4R = 50$ ohms, and thus the input resistance and the load resistance are both $R/2 = 6.2$ ohms. Using power transistors with transition frequencies of 300 MHz, an output power of 50 watts can be obtained in the short-wave range, with better than 30 dB intermodulation distortion.

Further power amplifiers using transmission-line transformers are presented in Refs. [12.9] and [12.12]. A 100-watt amplifier comprising several modules is considered in Ref. [20.15]. A 60-watt amplifier with line transformers is shown in Fig. 20.25 in Section 20.5 [20.10].

12.4. Effect of mismatch on transmission flatness

The overall response of multi-stage amplifiers usually differs from the response calculated from the stage responses. This is due to unwanted inter-stage couplings, typically caused by collector–base feedback. Inter-stage coupling can be reduced by using a high degree of mismatch. This will obviously decrease the gain, and thus mismatch should be chosen so that the transmission requirements are satisfied [10.5], [12.6]. This will be investigated in the following.

The frequency response is primarily determined by input impedance, which vary widely within the transmitted frequency range. The effect of input impedance changes can be reduced if the stage has a flat frequency response stabilized by feedback. If the transfer impedance of the stage, $Z_T = V_{out}/i_{in}$ is frequency-independent, then a current-generator drive has to be used, as encountered in current-feedback amplifiers. In the case of voltage-feedback circuits, the transfer admittance $Y_T = i_{out}/V_1$ is approximately frequency-independent, and so a voltage-generator drive is preferred. Since neither of these ideal cases can be realized, conditions will be investigated using the circuit shown in Fig. 12.12. The voltage transmission of this circuit is given by

$$A_v = V_2/V_1 = Y_T Z_T \eta, \tag{12.4.1}$$

where Y_T is the transfer admittance of the first stage, Z_T is the transfer impedance of the second stage, and the coupling efficiency of the passive four-pole between the ideal generators is given by

$$\eta = \frac{i_{in}}{i_{out}} = \frac{Z_{out}/Z_{in}}{Z_{out}/Z_{in} + 1}. \tag{12.4.2}$$

In the ideal case, the coupling efficiency is $\eta = 1$, and $Z_{out}/Z_{in} = \infty$. The voltage transmission according to (12.4.1) is primarily altered by the change

12.4. Effect of mismatch

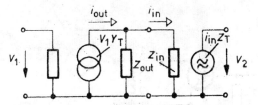

Fig. 12.12. Equivalent circuit for calculating transmission of mismatched amplifier stages

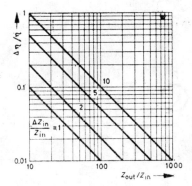

Fig. 12.13. Relative change of coupling efficiency as a function of mismatch, with different values of relative input impedances

in coupling efficiency η and thus, the change in input impedance Z_{in}:

$$\frac{\Delta\eta}{\eta} = -\frac{Z_{in}}{Z_{out}}\frac{\Delta Z_{in}}{Z_{in}}. \qquad (12.4.3)$$

According to this expression, a low ratio Z_{in}/Z_{out} will result in a low relative change of the voltage transmission. This is made possible by a high degree of mismatch between the two stages. Equation (12.4.3) is plotted in Fig. 12.13, showing the change in relative coupling efficiency (and also in transmission) for values of $\Delta Z_{in}/Z_{in}= 1, 2, 5$ and 10. If the input impedance change in the transmitted frequency range is known, the amplifier is designed by determining the ratio Z_{out}/Z_{in} pertaining to the required transmission flatness.

In practical cases, the ratio Z_{out}/Z_{in} for two-stage amplifiers is much higher than required by the transmission flatness. Mismatch can then be decreased and thus the gain increased by applying a wideband impedance transformer. Using a line transformer of ratio $N = 2$, the mismatch Z_{out}/Z_{in} thus produced is sufficient for transmission in most cases, and there is 6-dB increase in gain.

The circuit connection is shown in Fig. 12.14 [12.6]. The transition frequency of the applied transistors is 750 MHz, and with $R_1 = 50$ ohms, $R_E = 24$ ohms, $R_2 = 120$ ohms and $L_2 = 0.05$ μH, the gain is $A_v = 12$ dB,

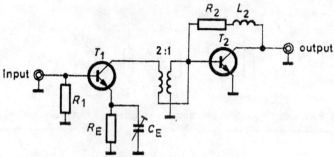

Fig. 12.14. Circuit of two-stage feedback amplifier using a transmission-line transformer for matching

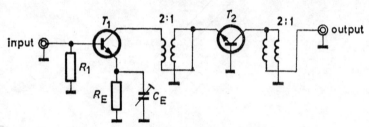

Fig. 12.15. Circuit of two-stage amplifier using a transmission-line transformer for matching

and the transmitted frequency range is 50–500 MHz. Response is flat within ± 0.1 dB. Cascading two amplifiers will produce a gain of 24 dB, leaving the other parameters unaltered without any adjusting or tuning procedures. The mismatch for this circuit is $Z_{out}/Z_{in} = 180$.

Figure 12.15 shows a two-stage amplifier, with the application of a common-base amplifier instead of the compensated current-feedback common-emitter stage. The mismatch in this case is $Z_{out}/4Z_{in} = 670$, the gain $A_v = 14.8$ dB, and the upper cut-off frequency $f_H = 470$ MHz. Inter-stage coupling may be neglected, as in the previous circuit.

Recently the VMOS transistor is also widely used in wideband amplifiers coupled with transmission-line transformers. Reference [12.15] describes an amplifier in the frequency range 40–275 MHz with LR paralell feedback. Reference [12.16] presents push-pull amplifiers for the frequency range 30–88 MHz and 170–230 MHz.

13. DISTRIBUTED AMPLIFIERS

13.1. Principle of operation

All types of amplifier stages investigated in the previous chapters can be characterized by the gain-bandwidth product, which gives an upper limit to high-frequency operation. Above the band limit, no improvement can be attained by cascading amplifier stages, because the less-than-unity stage gains will result in less-than-unity overall gains. Thus the gain-bandwidth product, calculated from the power gain, cannot be surpassed using multiplicative amplification. The cut-off frequency can be increased, however, by additive amplification where stage gains are added. Thus wideband amplifiers utilizing stages with less-than-unity gains are feasible; these are called distributed amplifiers.

The principle of a distributed-amplifier circuit is shown in Fig. 13.1. The generator signal is given on the input delay line which drives the amplifier stages according to the delay times. The output signal of any amplifier stage is propagated in two directions along the delay line. The stage output signals propagated in the direction of load R_L are added in phase with correct adjustment of the delay lines, thus providing an output signal corresponding to the sum of stage amplifications. Signals propagating in the opposite direction are dissipated in resistance R_L at the other end of the line. In the case of correct matching, the terminating resistances are equal to the characteristic impedance for both delay lines, thus eliminating reflections.

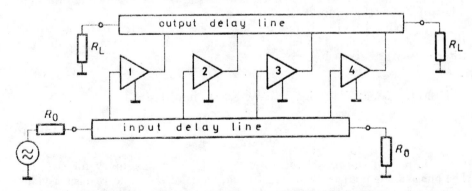

Fig. 13.1. Circuit principle of distributed amplifier

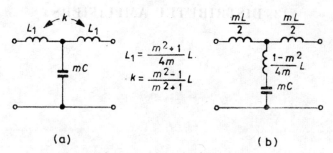

Fig. 13.2. Delay-section element with inductive coupling for use in distributed amplifiers. (a) Circuit diagram, (b) equivalent circuit

The voltage gain of a distributed amplifier is given by

$$A_v = n g_m R_L/2, \qquad (13.1.1)$$

where n is the number of stages, and g_m is the transconductance of the amplifier stages. Thus many stages are needed to achieve high gain. It can be shown that cascading stages above a certain limit is not as advantageous as the usual cascade connection of several distributed amplifiers, each having a number of stages below that limit. The optimum voltage gain for minimum number of stages is $A_v = 2.7$.

The strong frequency dependence of bipolar-transistor input impedance demands the application of compensating elements, in order to decrease the frequency-dependent load of the delay line. For this reason, distributed amplifiers using bipolar transistors have not gained wide acceptance. However, field-effect transistors are frequently used because of the constant input and output capacity which can be included in the delay line.

Theoretical questions of distributed amplifiers and various delay-line designs are considered in Ref. [13.3], in which a detailed list of references dealing with distributed amplifiers is given.

For distributed-amplifier purposes, the type of delay section shown in Fig. 13.2 is most often applied. Figure (a) shows the circuit diagram, and Fig. (b) shows the equivalent circuit.

The coupling between inductances L_1 is easily achieved by correct positioning of the coils. The cut-off frequency of the T network (m-derived filter section) is given by

$$\omega_0 = 2/\sqrt{LC}, \qquad (13.1.2)$$

and the frequency-dependent characteristic impedance is given by

$$Z_0 = \sqrt{\frac{L}{C}\left(1 - \frac{\omega^2}{\omega_0^2}\right)}, \qquad (13.1.3)$$

which is real in the pass band ($0 < \omega < \omega_0$), and imaginary for $\omega > \omega_0$. The phase angle increases from the initial value of zero, and reaches a value at $\pi = \phi$ of $\omega = \omega_0$. Since neither the characteristic impedance nor the group delay time is constant in the complete pass band, only part of the pass band may be utilized. Maximum bandwidth is attained at the optimum value of $m_{opt} = 1.23\text{--}1.27$.

13.2. Distributed amplifiers using transistors

A distributed-amplifier stage using bipolar transistors is shown in Fig. 13.3 (see Ref. [13.1]). To increase the input impedance and decrease the frequency dependence, negative feedback is used in the emitter circuit and series compensation in the base circuit. If the two time constants have a value of $R_E C_E = R_B C_B = 1/\omega_\beta$, then the stage can be regarded as a compensated voltage divider comprising a current generator as active element.

A distributed-amplifier stage containing field-effect transistors is shown in Fig. 13.4 (see Ref. [13.2]). Each amplifier stage is made up of a source-coupled transistor pair (differential amplifier) for decreasing the capacitive feedback. The input capacity of this amplifier stage is increased as a result

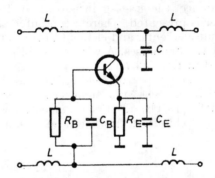

Fig. 13.3. Distributed amplifier stage using bipolar transistors

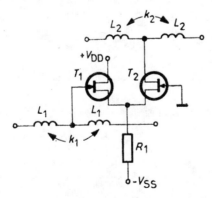

Fig. 13.4. Distributed amplifier stage using field-effect transistors

of the Miller capacity, which has a value of

$$C_{\text{in}} \simeq C_{\text{dg}}(1 + A). \tag{13.2.1}$$

This would require a low-impedance input delay line. The second grounded-gate stage has the effect of decreasing this feedback. In practical cases either the cascaded common-source and common-gate configuration or the long-tailed pair is used, (though the latter has half the transconductance, the input capacity is also halved. This connection was applied in the distributed amplifier outlined above). The input capacity loading the input delay line is given approximately by

$$C_{\text{in}} \simeq C_{\text{gs}}/2 + C_{\text{dg}}. \tag{13.2.2}$$

By using the circuit shown in Fig. 13.4 in a four-stage distributed amplifier, and choosing $R_1 = 1.2$ kohm, $L_1 = L_2 = 0.28$ μH and $k_1 = k_2 = 0.23$, and a gain of 2.8 times, a delay time of 6.5 ns and a 3-dB frequency of approximately 70 MHz have been achieved. For a source-coupled transistor pair, the input and output capacities are 3.4 pF and 1.7 pF respectively, and the overall transconductance is 1.5 mA/V. With an output delay line of 960-ohm characteristic impedance, the stage gain is 0.72. In Ref. [13.6], a 45-MHz distributed amplifier is presented, where a computer-aided design was used to calculate all circuit elements, and element optimization is carried out by taking into account the line losses. A wideband pulse amplifier from DC to 3.6 GHz is described in Ref. [13.7].

With the advent of new types of field-effect transistors which have extremely high cut-off frequencies, the significance of distributed amplifiers has increased. In Refs. [13.4] and [13.5], extremely wideband distributed amplifiers containing field-effect transistors are dealt with.

14. GENERAL DESIGN CONSIDERATIONS OF TUNED AMPLIFIERS

14.1. Building blocks of tuned amplifiers

In tuned amplifiers, the frequency response is implemented by the use of reactive filters placed between active elements as shown in Fig. 14.1. In transistor tuned amplifiers, the active element is a bipolar transistor or a FET, and the reactive filter, also used for impedance transformation, is either a parallel tuned circuit or a double-tuned band-pass filter. Owing to the capacitive feedback of transistors, these amplifiers have to be neutralized in most cases in order to eliminate interaction between stages, which would introduce asymmetry in the frequency response.

Capacitive feedback is reduced if cascaded transistors are used without frequency-selective inter-stage networks. A simple example of this method is a transistor pair made up of common-emitter and common-base stages which have practically no capacitive feedback. Filters at the input and output of such a two-stage amplifier may be calculated separately according to filter-theory methods.

Using high-gain integrated amplifiers, Fig. 14.1 will be modified: active elements are concentrated in a few high-gain integrated amplifiers and, similarly reactive filters are lumped in a single block instead of being distributed. The best solution is the application of high-quality crystal filters which provide high selectivity and frequency stability. The capacitive feedback of high-frequency integrated circuits, caused by the capacity of the package, is extremely low. The frequency response of these amplifiers may be calculated from relationships given by filter theory.

The pole-zero method and the various tables of filter networks are very relevant. However, at the input and output terminals of the filter, the output or input admittance of the active element (both real and imaginary parts) should be taken into account. Since y_{12} is zero, the active element is only characterized by the forward transfer parameter. Transfer of the amplifier is given by the product of active element and filter transfer parameters. In *narrow-band* amplifiers, the parameters of the active elements are fre-

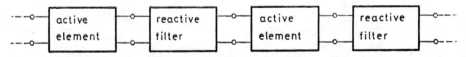

Fig. 14.1. Circuit principle of tuned amplifiers

quency-independent within the pass band, and the frequency response is thus determined exclusively by the reactive filter. In *wideband* amplifiers, the parameters of active elements may vary within the pass band: this problem is dealt with in a later chapter. The frequency response of the amplifier, designed by this method, may differ from the calculated response. In favourable cases, this may be taken into account during the tuning procedure.

For active elements with considerable capacitive feedback, the design of multi-stage amplifiers becomes complicated. From the set of equations applicable to the circuit, the circuit element values providing the transfer function to be achieved may not be determined simply. The exact solution of the problem is only possible using a computer. Multi-stage amplifiers realized by using various filter types have been analyzed by computer, and this enables amplifiers having similar frequency responses to be designed. The design in this case is based on a similar transfer function: a frequency response, characterized by normalized parameters, has to be selected giving the best approximation to the required transfer function [14.8].

14.2. Design by the pole-zero method

If the feedback of active elements is negligible, i.e. $y_{12} = 0$, then the tuned amplifier may be designed using the pole-zero method following from linear four-pole theory. The active element in this case is a single controlled generator, and only the filter need to be designed.

In the linear region of the tuned amplifier the transfer function is characterized by the amplitude response $a(\omega)$ and the phase response $\varphi(\omega)$. When transmitting modulated signals, the group delay, which is defined by the relation

$$\tau(\omega) = \frac{d\varphi}{d\omega}, \tag{14.2.1}$$

and gives the slope of the phase response, is used instead of the phase response. Both the amplitude response $a(\omega)$ and the group-delay response $\tau(\omega)$ may be prescribed in the pass band. In both cases, two types of response may be distinguished, the maximally flat (Butterworth) response shown in Fig. 14.2 (a), and the equal ripple (Chebyshev) response shown in Fig. 14.2 (b).

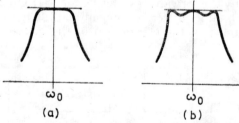

Fig. 14.2. (a) Maximally flat amplitude response, (b) equal-ripple amplitude response

14.2. Pole-zero method

When the pole-zero method is used, the frequency band of the tuned amplifier is transformed into a low-pass transfer function: a low-pass four-pole is first designed, and this can be transformed suitably into the actual frequency range. The method has the advantage that low-pass transfer-function characteristics are available (see Section 5.2). The transformation should enable the pole-zero pattern of the tuned amplifier to be transferred into a low-pass pole-zero pattern. The relation

$$w = \frac{1}{2}\left(p + \frac{\omega_0^2}{p}\right), \tag{14.2.2}$$

known from network theory, transfers the $w = j\Omega$ axis of the low-pass range onto the $p = j\omega$ axis, and point $w = 0$ to the point $p = j\omega_0$ corresponding to the band centre. Because of the conformal mapping, the amplitude and phase responses of the low-pass and band-pass ranges are identical. The relation between the band-pass and low-pass cut-off frequencies is given by the following:

$$\omega_{1,2} = \pm \Omega_b + \sqrt{\omega_0^2 + \Omega_b^2}, \tag{14.2.3}$$

where Ω_b is the low-pass cut-off frequency, and ω_1 and ω_2 are symmetrically placed 3-dB cut-off frequencies of the band-pass filter; thus the bandwidth is given by

$$\omega_b = \omega_2 - \omega_1 = 2\Omega_b. \tag{14.2.4}$$

From eqn. (14.2.2), the group-delay response of the band-pass filter is given by

$$\tau(\omega) = \frac{d\varphi}{d\Omega}\frac{d\Omega}{d\omega} = \frac{\tau(\Omega)}{2}\left[1 + \frac{\omega_0^2}{\omega^2}\right], \tag{14.2.5}$$

where $\tau(\Omega)$ is the group delay in the low-pass range.

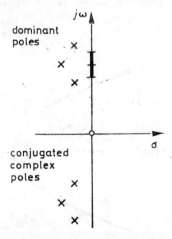

Fig. 14.3. Pole-zero pattern of narrow-band amplifiers

The frequency transformation given by eqn. (14.2.2) may also be used for wideband cases. The method of dominant poles, as explained below, may be used to facilitate the design of narrow-band amplifiers. As shown in Fig. 14.3, which illustrates the pole-zero pattern of narrow-band amplifiers, the dominant poles are near the band-centre frequency $p = j\omega_0$, whereas the complex-conjugate poles are further away. In narrow-band amplifiers, these latter may be neglected to good approximation, and thus the transformation relation reduces to a simple frequency shift, i.e. $w = p - j\omega$ and $\Omega = \omega - \omega_0$. The 3-dB bandwidth is given by $\omega_b = 2\Omega_b$.

14.3. Stability

From the stability standpoint, single-stage and multi-stage amplifiers must be investigated separately. The stability of single-stage amplifiers is given by the *Stern factor* (see eqn. (4.3.4)):

$$k = \frac{2 g_1 g_2}{|y_{12} y_{21}| (1 + \cos \varphi)} = \frac{2}{A^*(1 + \cos \varphi)} \geq 1, \qquad (14.3.1)$$

where the notations $g_{11} + g_g = g_1$ and $g_{22} + g_L = g_2$ are used, $\varphi = \varphi_{12} + \varphi_{21}$ and A^* is the open-loop gain of the terminated four-pole. In Fig. 14.4, the loop gain A^* is plotted on the complex plane. According to relation (14.3.1), the criterion for stability is a loop-gain curve within the parabola.

The stability of the single-stage amplifier depends on the types of reactive networks at the input and output side. By generalization relation (14.3.1) stability is given by an expression of the form

$$k = a_F/A^*, \qquad (14.3.2)$$

where a_F is a factor depending on the type of filter applied.

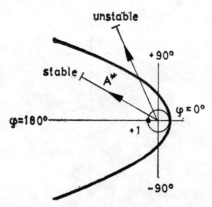

Fig. 14.4. Open-loop gain plotted on the complex plane

14.3. Stability

If single-tuned circuits are applied at both the input and the output, then $a_{F1} = 2/(1 + \cos \varphi)$, yielding the value given by (14.3.1).

For *multi-stage* amplifiers applying single-tuned circuits between stages, stability is given by

$$k = a_{F1} a_n / A^*, \qquad (14.3.3)$$

where a_n is a factor depending on the number of stages, given in Table 14.1. It is seen that stability reduces rapidly when the number of stages is increased. The factor a_F for double-tuned filters is given by Table 14.2 for various phase angles φ, and for various kQ coupling factors of the band-pass filters. It can be seen from the table that the factor a_F is higher for double-tuned filters than it is for single-tuned circuits, and thus the former solution is better as regards stability.

Reference [14.8] contains a detailed investigation on the stability calculation of multi-stage double-tuned amplifiers.

Table 14.1. Factor a_n characterizing the decrease in stability factor, for n-stage single-tuned amplifier

n	a_n
1	1.0
2	0.5
3	0.38
4	0.33
5	0.31
6	0.29
7	0.28
8	0.28
9	0.27
10	0.27
∞	0.25

Table 14.2. The factor a_F of double-tuned amplifiers

φ	$\dfrac{2}{1+\cos \varphi}$	one-stage amplifier with double-tuned filter			two-stage amplifier with double-tuned filter		
		$kQ = 0.7$	1.0	1.41	0.7	1.0	1.41
0	1	2.3	4.0	4.0	1.5	1.7	1.9
15	1.1	2.0	2.9	3.0	1.4	1.6	1.9
30	1.2	2.0	2.6	2.9	1.4	1.6	1.9
45	1.3	2.0	2.6	2.9	1.5	1.7	2.0
60	1.4	2.1	2.6	3.0	1.6	1.8	2.0
75	1.5	2.4	2.8	3.3	1.8	2.0	2.3
90	2.0	2.7	3.2	3.8	2.3	2.5	2.7
105	2.7	3.5	4.0	4.6	3.0	3.2	3.5
120	4.0	5.5	6.0	7.0	4.3	4.7	5.0
135	7.0	8	8.5	9.5	9.2	9.5	10
150	15	13	15	18	13	14	16

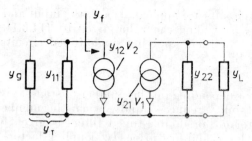

Fig. 14.5. Interpretation of the stability factor

The stability factors above apply for *operational* conditions of the amplifier. In fact, the amplifier may be operated under a variety of conditions, e.g. during tuning. A more characteristic parameter is the minimum stability factor given by

$$k_{min} = \frac{\mathrm{Re}(y_I)_{min}}{\mathrm{Re}(y_f)_{max}}. \tag{14.3.4}$$

In this relation, $y_I = y_{11} + y_g$ is the total admittance at the input, and y_f is the admittance appearing at the input resulting from the feedback (see Fig. 14.5):

$$y_f = -\frac{y_{12}\, y_{21}}{y_{22} + y_L}. \tag{14.3.5}$$

Admittances y_I and y_f are connected in parallel at the input, and the real part of their resultant value may be zero or negative in an unfavourable case. Stability is characterized by the ratio of these two admittances, according to (14.3.4). When calculating the minimum stability factor, the minima and maxima of the possible real values of admittances y_I and y_f have to be determined; the admittances depend on the terminations and filter types.

14.4. Neutralization

The instability of high-frequency tuned amplifiers is caused by the feedback given by the four-pole parameter which has subscript 12. Feedback may be reduced or eliminated by neutralization, which is a cancelling of the output-to-input transfer path in some way.

For the case of perfect neutralization, no signal reaches the transistor input from the transistor output, thus reducing the subscript-12 parameter to zero. This kind of four-pole having zero-valued h_{12}, y_{12}, z_{12} or s_{12} parameters is called a unilateral four-pole. The unilateral four-pole is thus non-transparent, its loop gain is zero, and the output signal does not reach the input terminal [14.2].

The neutralizing circuit is essentially a bridge, and the feedback signal depends on its balance. Since a perfectly balanced high-frequency bridge

14.4. Neutralization

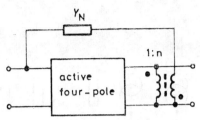

Fig. 14.6. *y*-type neutralization of an active four-pole

cannot be realized permanently, the ideal unilateral four-pole cannot be either.

It is important to determine the worst-case residual feedback value. If this does not interfere esessentially with the operation of the circuit, the neutralizing circuit can be considered good.

The most generally used neutralizing method is the so-called *y* neutralization in which a phase-reversing transformer of ratio $1:n$ and an admittance Y_N are applied (Fig. 14.6). According to relation (4.6.1), the four-pole becomes unilateral if the condition

$$y_{12} + Y_N/n = 0, \qquad (14.4.1)$$

is satisfied. Expressing Y_N from the above equation and assuming a given turns ratio n, the other three overall y parameters may be calculated.

For the case of a narrow bandwidth, the admittance y_{12} at the band centre is relevant. Y_N is proportional to this middle value and may be achieved by either parallel or series RC configuration. Within a narrow bandwidth, the variation of y_{12} is small, so the neutralization is sufficient for both designs. At higher frequencies, the series resistance is usually omitted because of the imaginary nature of y_{12}. Within a wide bandwidth, the variation of y_{12} may be substantial, so the admittance Y_N has to be set-up by several circuit elements, with combination of resistors and capacitors, in order to simulate the frequency dependence of parameter y_{12} (see Ref. [14.5]). The phase-reversal transformer is used simultaneously for matching the next stage or the termination.

Resistive elements of the neutralizing circuit dissipate power and thus decrease the power gain available with the active four-pole. A higher power gain can be achieved with a neutralizing circuit using reactive elements exclusively.

Figure 14.7 shows an active four-pole neutralized by reactive elements. Susceptance B_N is connected in parallel with the output terminal, and a parallel-series connected ideal transformer is also used. The following equations apply to this circuit:

$$i_1 = y_{12} V_2 = \frac{y_{12}}{y_{22} + jB_N} i_2', \qquad (14.4.2)$$

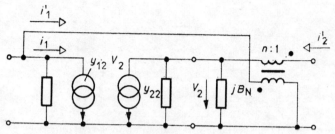

Fig. 14.7. Neutralization of an active four-pole through the use of reactive elements

and $i_1' n = i_2'$. The circuit is unilateral if the algebraic sum of the two input currents is zero, i.e. $i_1 + i_1' = 0$. Substituting the two current values, the susceptance and the turns ratio are calculated as:

$$B_N = b_{12} g_{22}/g_{12} - b_{22}; \qquad (14.4.3)$$

$$n = -g_{22}/g_{11}.$$

Let us now return to the neutralized amplifier shown in Fig. 14.6 and calculate the power transfer. The four-pole is unilateral if condition (14.4.1) is met. Using the expression for neutralizing admittance Y_N taken from the equation, and substituting this into expression (4.6.1) to yield the overall y parameters (also taking into account the effect of the phase reversal transformer), we get

$$g_{11R} = g_{11} - n g_{12};$$
$$y_{21R} = y_{21} - y_{12}; \qquad (14.4.4)$$
$$g_{22R} = g_{22} - g_{12}/n.$$

Substituting these values into expression (4.4.1) yielding the matched transfer, the following relation is calculated:

$$MAG = \frac{|y_{21R}|^2}{4 g_{11R} g_{22R}}. \qquad (14.4.5)$$

According to eqns. (14.4.4) and (14.4.5), the power gain is somewhat decreased by the neutralization. Since the power gain is dependent on the turns ratio n, the extremum value giving the maximum power gain is determined by differentiating expression (14.4.5) with respect to n:

$$\frac{d(MAG)}{dn} = 0. \qquad (14.4.6)$$

From this, the optimum turns ratio is given by $n_{opt} = \sqrt{g_{11}/g_{22}}$, and the power gain is

$$MAG = \frac{|y_{21} - y_{12}|^2}{4(g_{11} g_{22} - 2 g_{12} \sqrt{g_{11} g_{22}} + g_{12}^2)} \simeq \frac{|y_{21}|^2}{4 g_{11} g_{22}}, \qquad (14.4.7)$$

14.5. Loop-gain limit

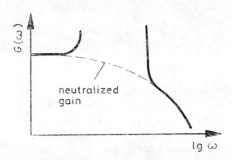

Fig. 14.8. Frequency dependence of power gain

where the approximate expression on the right-hand side may be used for small g_{12} values.

Figure 14.8 shows the power-gain frequency dependence of transistor-tuned amplifiers. Above a given frequency limit, the power gain increases sharply as a result of positive feedback and the circuit becomes unstable. In this frequency range, the amplifier may only be used with neutralization. Above a further frequency, the amplifier is again stable, which can be explained by the decreased loop gain and by the more suitable phase angle of the loop gain. In this frequency range, the amplifier may be operated without neutralization; in any case, the design of the neutralizing circuit at higher frequencies would present difficulties.

14.5. Design based on the loop-gain limit. Linvill design

The instability due to capacitive feedback within the active device may be decreased by forfeiting power gain. As the amplifier stability is characterized by the loop gain, this design method is called design for loop-gain limit [14.8]. The starting-point data are: the so-called intrinsic loop gain of the four-pole terminated by open circuits:

$$A_i^* = \frac{|y_{12} y_{21}|}{g_{11} g_{22}}, \qquad (14.5.1)$$

and also the phase angle of the product of transfer parameters, and the stability value as given by (14.3.2).

For an n-stage amplifier with single tuned circuits without neutralization, the permitted loop gain per stage, based on eqn. (14.3.3), is

$$A^* = \frac{2a_n}{k(1 + \cos \varphi)}, \qquad (14.5.2)$$

where a_n is the reduction factor given in Table 14.1, and k is the stability factor which, for practical amplifiers, is within the range of 2–4.

For band-pass amplifiers, the permitted loop gain per stage, based on eqn. (14.3.2), is similarly

$$A^* = a_F/k, \qquad (14.5.3)$$

where a_F is a factor given in Table 14.2, depending on the number of stages and type of band-pass.

With knowledge of the loop gain, the product of the input and output mismatches, based on eqn. (14.5.1), is given by

$$\frac{g_{11}}{g_1} \frac{g_{22}}{g_2} = \frac{A^*}{A_i^*}. \qquad (14.5.4)$$

In the following chapters, the choice of mismatch and its effects are treated in detail for a variety of basic circuits.

Loop-gain limit design may also be used for neutralized amplifiers. This is justified by the fact that because of imperfect neutralization, a residual uncompensated internal feedback y_{12} will be present. The loop gain for design purposes will then be

$$A_N^* = \frac{|\Delta y_{12} y_{21}|}{g_1 g_2}, \qquad (14.5.5)$$

where Δy_{12} is the maximum un-neutralized feedback capacity during operation (e.g. during gain control).

A special design for tuned amplifiers is called the *Linvill design* which utilizes the modified Smith diagram, also known as the Linvill diagram. The following modification is carried out. The Smith diagram is rotated by 180°, so that $\varphi = 0$ corresponds to infinite impedance. The diagram shows the admittance $y_2 = y_{22} + y_L$ at the active four-pole output. The unity real circle corresponds to the value $g_2 = 2g_{22}$, and the unity imaginary circles correspond to the value $b_2 = \pm g_{22}$.

Let us assume a potentially unstable four-pole, i.e. the stability factor as given by eqn. (4.3.7) will be $K < 1$. In the diagram shown in Fig. 14.9, the y_2 admittances pertaining to constant G_T power transfer are located on a circle. The centre of this circle is on a straight line passing through the origin at an angle $-(180 + \varphi)$ at a distance from the origin given by

$$r = G_T \left| \frac{y_{21}}{y_{12}} \right|. \qquad (14.5.6)$$

The radius of the circle is given by

$$R = \sqrt{1 + r^2 - 2rK}. \qquad (14.5.7)$$

With increasing power gain, the circles approach the unstable region shown by the hatched area, within which no admittance y_2 is permitted. During output tuning, the imaginary part b_2 is varied, and the admittance

14.6. Calculation by scattering parameters

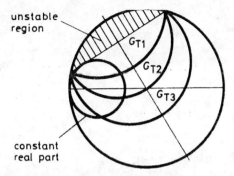

Fig. 14.9. Circles of terminating admittances on the Linvill diagram, resulting in constant power gains for potentially unstable four-poles ($G_{T_1} > G_{T_2} > G_{T_3}$)

y_2 is thus changed along a real circle. Stability requires that the admittance y_2 should not be transferred into the hatched area.

The task is thus to choose y_2 so as to provide maximum power gain with suitable stability. If the power gain is specified, then the real circle which is tangent to the specified power gain circle should be chosen as a value for g_2. The tangent point will give the value of the imaginary part b_2. From the expression $g_2 = g_{22} + g_L$, the terminating conductance g_L may be found.

For unconditionally stable four-poles ($K > 1$), the constant power-gain circles lie entirely within the diagram, and there is no unstable range. The power-gain circle corresponding to the conjugated matching case as given by (4.4.5) degenerates into a single point which gives the admittance $y_2 = = y_{22} + y^*_{out}$ according to (4.4.2).

14.6. Stability calculation by scattering parameters

During circuit design by scattering (reflection) parameters (s parameters), stability check is also necessary. This is simple with the aid of the scattering parameters and the Smith diagram [14.9]. The active four-pole is unconditionally stable if both the s_{11} and s_{22} parameters are less than unity, and the factor K according to (4.3.7) exceeds unity. Expressing this in terms of scattering parameters, we obtain

$$K = \frac{1 - |s_{11}|^2 - |s_{22}|^2 + |\Delta|^2}{2|s_{12}s_{21}|} > 1 , \qquad (14.6.1)$$

where we have used the notation:

$$\Delta = s_{11}s_{22} - s_{12}s_{21}. \qquad (14.6.2)$$

For maximum power gain, the unconditionally stable four-poles may be matched in a conjugated manner at both ends. As in expressions (4.4.3), the matched terminations, referred to the characteristic impedance Z_0, are expressed by the scattering parameters as follows:

$$r_{g,\text{opt}} = C_1^* \frac{B_1 \pm \sqrt{B_1^2 - 4|C_1|^2}}{2|C_1|^2} ; \qquad (14.6.3)$$

$$r_{L,\text{opt}} = C_2^* \frac{B_2 \pm \sqrt{B_2^2 - 4|C_2|^2}}{2|C_2|^2} ,$$

where

$$B_1 = 1 + |s_{11}|^2 - |s_{22}|^2 - |\Delta|^2 ;$$
$$B_2 = 1 - |s_{11}|^2 + |s_{22}|^2 - |\Delta|^2 ; \qquad (14.6.4)$$
$$C_1 = s_{11} - s_{22}^* \Delta ;$$
$$C_2 = s_{22} - s_{11}^* \Delta ,$$

and the asterisks denote complex-conjugate value. In both expressions, the sign of the square-root should be negative for positive B_1 and B_2 values, and vice versa. Using the above expressions, the power gain of the stage with complex-conjugate matching is, as for (4.4.5), given by

$$G_{T,\max} = \left| \frac{s_{21}}{s_{12}} \right| \left(K \pm \sqrt{K^2 - 1} \right) , \qquad (14.6.5)$$

where the sign of the square-root should be negative for positive B_1, and vice versa.

For potentially unstable four-poles ($K < 1$), the terminations providing stable operation may be determined using the impedance diagram. For both the input and output sides, terminations, resulting in unstable operation are situated within a circle (see Fig. 14.10). The centre points r_{s1} and r_{s2} and radii R_{s1} and R_{s2} of these circles may be calculated from the following expressions:

$$r_{s1} = \frac{C_1^*}{|s_{11}|^2 - |\Delta|^2} ;$$

$$r_{s2} = \frac{C_2^*}{|s_{22}|^2 - |\Delta|^2} ;$$

$$R_{s1} = \frac{|s_{12} s_{21}|}{|s_{11}|^2 - |\Delta|^2} ; \qquad (14.6.6)$$

$$R_{s2} = \frac{|s_{12} s_{21}|}{|s_{22}|^2 - |\Delta|^2} .$$

The stability of the circuit may be controlled directly by drawing the instability region on the impedance diagram, on the basis of the above expressions.

14.6. Calculation by scattering parameters

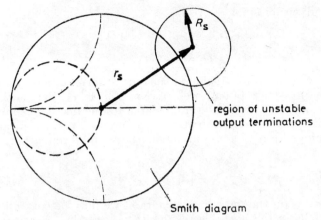

Fig. 14.10. Region of unstable output terminations on the impedance diagram

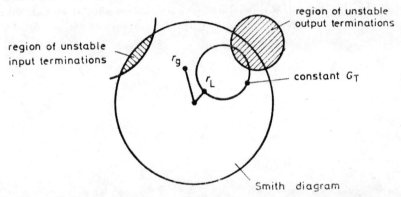

Fig. 14.11. Region of unstable terminations on the impedance diagram

The circuit-design procedure taking into account the stability requirements has several steps. The first step is to draw the circles of terminations corresponding to constant power gains G_T on the impedance diagram. For the output-impedance diagram, the distance of the circle centre from the origin is given by

$$r_0 = \frac{G_s}{1 + D_2 G_s} C_2^*, \qquad (14.6.7)$$

and the radius of the circle is given by

$$R_0 = \frac{\sqrt{1 + 2K|s_{12}s_{21}|G_s + |s_{12}s_{21}|^2 G_s^2}}{1 + D_2 G_s}, \qquad (14.6.8)$$

where C_2 is given by (14.6.4). Also,

$$D_2 = |s_{22}|^2 - |\Delta|^2 ; \qquad (14.6.9)$$

$$G_s = G_T/|s_{21}|^2 .$$

By drawing the unstable region of the output terminations, on the impedance using expressions (14.6.6), we arrive at Fig. 14.11. Part of the terminations pertaining to constant power gains fall in the unstable region and cannot be used. The output termination r_L is preferably chosen far from the unstable region. From this requirement, the input termination is given by

$$r_g = \left(\frac{s_{11} - r_L \Delta}{1 - r_L s_{22}}\right)^* . \qquad (14.6.10)$$

By drawing this on the impedance diagram, we may be controlled whether or not this falls in the unstable region. If it does, the procedure has to be repeated with a further output termination.

For wide frequency ranges, changes in scattering parameters should be expected. In this case, the procedure outlined above should be carried out with scattering parameters relevant to the two band limits.

15. SINGLE-TUNED AMPLIFIERS

15.1. Power gain

Narrow-band frequency response characteristics are most frequently produced by parallel resonant circuits. This solution is simple and inexpensive, though the response may have the disadvantage of too sharp resonance and low bandwidth.

Figure 15.1 shows the principle of the single-tuned amplifier. The loss of the parallel resonant circuit is represented by the conductance g'_k. The parallel resonant circuit is loaded by further real and reactive elements from both sides, as a result of the connected active four-poles.

In Fig. 15.2 the driving active four-pole is replaced by a transistor characterized by y parameters, and the loading active four-pole is represented by the load conductance g'_L. The output capacity $\mathrm{Im}(y_{22}) = C_{22}$ of the driving transistor and the reactive part of the load is included in the resonant circuit capacity C_0; the resonant frequency is given by $\omega_0 = 1/\sqrt{L_0 C_0}$.

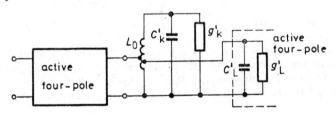

Fig. 15.1. Principle of single-tuned amplifier

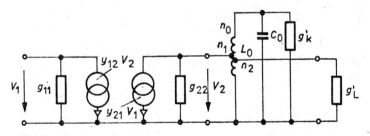

Fig. 15.2. Equivalent circuit of single-tuned amplifier stage characterized by y parameters, with all capacities included in the resonant circuit

15. Single-tuned amplifiers

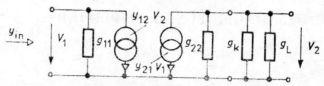

Fig. 15.3. Equivalent circuit for calculating the power gain at midband of a single-tuned amplifier stage

Let us now calculate the power gain. At the resonant frequency, the reactive elements are just omitted, and in order to simplify the calculation, the conductance g'_k of the resonant circuit loss and the conductance g'_L of the load are transformed to the output. The transformed values are denoted by g_k and g_L (see Fig. 15.3). The power gain of the single-tuned amplifier, given by the ratio of power P_2 delivered to the load g_L, to the input power P_1, may now be determined. Note that the power P_1 is not the power dissipated in conductance g_{11} but in the real part of the input admittance, i.e. $g_{in} = \mathrm{Re}(y_{in})$. Owing to the internal feedback, g_{in} is not identical to g_{11}, and may be calculated from the expression

$$y_{in} = y_{11} - \frac{y_{12} y_{21}}{y_{22} + y_L}, \tag{15.1.1}$$

where y_L is the resultant of all admittances loading the output. Rewriting this equation for the real parts, the following relation holds at the resonant frequency:

$$g_{in} = g_{11} - \frac{|y_{12} y_{21}| \cos (\varphi_{21} + \varphi_{12})}{g_2}, \tag{15.1.2}$$

where g_2 is the overall conductance at the output, its value being $g_2 = g_{22} + g_k + g_L$.

Let us now introduce a quantity derived from the unloaded and loaded bandwidths of the resonant circuit, B_0 and B_L. The unloaded bandwidth is proportional to the circuit loss, i.e. $B_0 = g_k/2\pi C_0$, and the loaded bandwidth is proportional to the overall conductance loading the resonant circuit:

$$B_L = \frac{g_{22} + g_k + g_L}{2\pi C_0}. \tag{15.1.3}$$

From the above relations,

$$1 - \frac{B_0}{B_L} = \frac{g_{22} + g_L}{g_{22} + g_L + g_k}. \tag{15.1.4}$$

Using the above notations, the power gain at resonance can thus be expressed as follows:

$$G_0 = \frac{P_2}{P_1} = \frac{|y_{21}|^2}{4 g_{11} g_{22}} \left(1 - \frac{B_0}{B_L}\right)^2 \frac{4\nu}{(1+\nu)^2} \frac{1}{1 - \frac{|y_{12} y_{21}| \cos (\varphi_{21} + \varphi_{12})}{g_{11}(g_{22} + g_L + g_k)}}, \tag{15.1.5}$$

15.1. Power gain

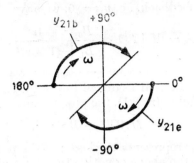

Fig. 15.4. Frequency dependence of the transconductance for common-emitter and common-base configuration

where $\nu = g_L/g_{22}$ is the mismatch factor expressing the deviation of the load conductance g_L from the matched value resulting in maximum power gain. The power gain has a maximum value at $\nu = 1$.

Let us consider eqn. (15.1.5). The first factor is the maximum available gain MAG, giving the power gain of a four-pole with conjugated matching and ideal neutralization. The power gain as given by (15.1.5) differs from this ideal value for three reasons. The resonant circuit introduces a loss as expressed in the second factor by which a decrease of gain, proportional to the square of expression (15.1.4), is introduced. The second cause of gain decrease is the mismatch ($\nu \neq 1$). The third cause is the internal feedback of the four-pole which has the effect of changing the input conductance g_{in} and thus the input power. In contrast with the two factors above which cause a decrease, this may lead to either an increase or to a decrease in power gain, depending on the phase angle $\varphi_{21} + \varphi_{12}$. At high frequencies, the imaginary value of the admittance y_{12} representing the feedback is generally much higher than the real part, i.e. $\omega C_{12} \gg g_{12}$, and thus the phase angle of the admittance y_{12} is $\varphi_{12} \sim -90°$.

The situation is more complicated for the admittance y_{21} responsible for the transconductance. The phase angle of the transconductance at low frequencies is given by $\varphi_{21b} = +180°$ for the common-base or common-gate configuration. With increasing frequency, the phase angle is reduced and may even fall below $+90°$ at very high frequencies (Fig. 15.4). The situation is just the contrary for the common-emitter or common-source configuration. At low frequencies, the phase angle is given by $\varphi_{21e} = 0$, and increases in the negative direction with increasing frequency.

For common-base or common-gate configurations, the quantity $\cos(\varphi_{12} + \varphi_{21})$ is positive, and so in expression (15.1.5) the denominator is smaller because of the subtraction, thus increasing the power gain.

For the common-emitter configuration, the opposite situation exists. The quantity $\cos(\varphi_{12} + \varphi_{21})$ is always negative, the two terms in the denominator add thus decreasing the power gain.

For the case of zero feedback, i.e. $y_{12} = 0$, we have

$$G_0 = MAG\left(1 - \frac{B_0}{B_L}\right)^2 \frac{4v}{(1+v)^2}, \qquad (15.1.6)$$

where the value of MAG is given by expression (14.4.7).

For the case of zero internal feedback ($y_{12} = 0$), assuming complex-conjugate impedance matching ($v = 1$), the power gain at resonance is given by

$$G_0 = MAG(1 - B_0/B_L)^2, \qquad (15.1.7)$$

where the second factor represents the losses of the parallel tuned circuit.

It may be seen from the above equation that for a given (and practically achievable) unloaded bandwidth B_0, the power gain is increased by an increase in loaded bandwidth B_L, and vice versa.

15.2. Stability

It has been shown in the previous chapter that the power gain at resonance is changed by the feedback. In extreme cases, the positive feedback may lead to instability, so the internal feedback should be taken into account precisely at circuit design. However, the conditions at resonance do not provide a true picture of the instability of the stage. In other words, the evaluation of eqn. (15.1.5) in itself is not sufficient to determine the circuit instability. Taking into account the phase response which has a high slope in the vicinity of resonance, it is clear that oscillations may occur at frequencies other than the resonant frequency. It thus follows that stability investigations should be extended to the whole frequency range in question.

Consideration of the stability of a single-tuned transistor amplifier will be carried out using Fig. 15.5, and the relations of the stability factor as given in Section 4.3.

In Fig. 15.5, both the input and output ports of the four-pole are terminated by single-tuned circuits. In fact, the transistor is coupled to a tapping of both resonant circuits, although this has been ignored, and all external elements have been transformed down to the appropriate transistor connections. Furthermore, admittances y_{11} and y_{22} have been excluded from the transistor, and included in the parallel resonant circuit. The imaginary

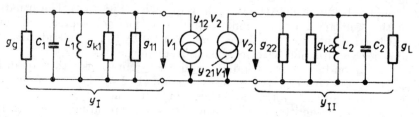

Fig. 15.5. Equivalent circuit for calculating the stability of a four-pole terminated by a single-tuned circuit at each ends

15.2. Stability

parts of the admittances are included directly in the resonant circuit of frequency ω_0, and all conductances are also included in the external conductances. The overall input conductance is expressed by $g_1 = g_g + g_{k1} + g_{11}$, where g_g is the conductance of the generator or of the previous stage acting as a generator, and g_{k1} is the loss of the input resonant circuit. Similarly, at the output side, $g_2 = g_{22} + g_{k2} + g_L$, where g_{k2} is the loss of the output resonant circuit, and g_L is the load conductance.

Let the resonant frequency of the input and output resonant circuits be ω_0, and the relative detuning (assumed small) for the input circuit be

$$x_1 = 2\Delta\omega_1 Q_1/\omega_0, \qquad (15.2.1)$$

and for the output circuit

$$x_2 = 2\Delta\omega_2 Q_2/\omega_0, \qquad (15.2.2)$$

where Q_1 and Q_2 are the loaded Q values of the two resonant circuits, and $\Delta\omega_1$ and $\Delta\omega_2$ are the detunings from the resonant frequency. The admittance of the input resonant circuit can then be expressed in the form:

$$y_I = g_1(1 + jx_1), \qquad (15.2.3)$$

and similarly the admittance of the output resonant circuit is given by

$$y_{II} = g_2(1 + jx_2). \qquad (15.2.4)$$

The feedback admittance as given in (14.3.5), may be expressed using eqn. (15.2.4) as:

$$y_f = -\frac{y_{12} y_{21}}{g_2(1 + jx_2)}. \qquad (15.2.5)$$

Using the absolute value and the phase of the transfer parameters, the above expression is modified as follows:

$$y_f = -\frac{|y_{12} y_{21}| e^{j\varphi}}{g_2(1 + jx_2)}, \qquad (15.2.6)$$

where $\varphi = \varphi_{12} + \varphi_{21}$. According to (14.3.4), the minimum stability factor is determined from the minimum real value of admittance y_I as given by (15.2.3) and from the maximum real value of admittance y_f as given by (15.2.6).

Figure 15.6 shows a plot of above two admittances on the complex plane. Admittance y_I is represented by a straight line parallel to the imaginary axis, at a distance g_1 from that axis in the right half-plane, the intersection with the real axis corresponding to $x_1 = 0$.

Admittance y_f is represented by a circle crossing the origin, with an angle φ between the positive real axis and the circle diameter pertaining to the origin. The figure shows the maximum real value of admittance y_f. Instability occurs if this maximum exceeds the minimum real value of admittance y_I:

$$\text{Re}(y_f)_{\max} > \text{Re}(y_I)_{\min}. \qquad (15.2.7)$$

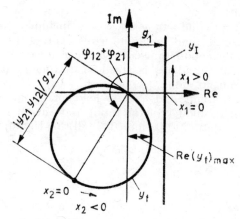

Fig. 15.6. Plot of the admittances y_I and y_f determining the stability on the complex plane

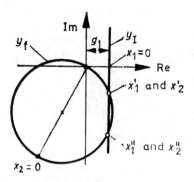

Fig. 15.7. The intersection of admittance curves y_I and y_f causes instability

According to Fig. 15.6, instability will be characterized by an intersection between the admittance circle and the admittance line; the tangential position marks the threshold of instability. The case of intersection is shown in Fig. 15.7, together with the tuning frequencies corresponding to the two intersection points: x_1' and x_1'' for amittance y_I, and x_2' and x_2'' for admittance y_f. Between the intersection points, condition (15.2.7) is obviously not fulfilled. If the frequency ranges determined by the above sections ($x_1' - x_1''$, and $x_2' - x_2''$, respectively) have at least one common frequency oscillation will be present.

Stability without oscillation prevails if the straight line and the circle have no intersection. The minimum stability factor tells whether or not instability will occur in the worst case, when detuning is least favourable.

15.2. Stability

The real parts of the admittances are

$$\mathrm{Re}(y_1)_{min} = g_1; \qquad (15.2.8)$$

$$\mathrm{Re}(y_f)_{max} = |y_{12}y_{21}|(1 + \cos \varphi)/2g_2. \qquad (15.2.9)$$

On the basis of the above two values, the minimum stability factor for single-tuned amplifiers is given by

$$k_{min} = \frac{2g_1 g_2}{|y_{12}y_{21}|(1 + \cos \varphi)}. \qquad (15.2.10)$$

Let us rearrange this expression similarly to (15.1.4) by introducing the factors which apply to the bandwidths of the input and output resonant circuits:

$$\frac{g_{11} + g_g}{g_{11} + g_g + g_{k1}} = 1 - \frac{B_0}{B_1}; \qquad (15.2.11)$$

$$\frac{g_{22} + g_L}{g_{22} + g_L + g_{k2}} = 1 - \frac{B_0}{B_2}. \qquad (15.2.12)$$

Utilizing these expressions, we have

$$k_{min} = \frac{2(g_{11} + g_g)(g_{22} + g_L)}{|y_{12}y_{21}|(1 + \cos \varphi)} \cdot \frac{1}{(1 - B_0/B_1)(1 - B_0/B_2)}. \qquad (15.2.13)$$

The instability depends to a large extent on the quantity $|y_{12}y_{21}|/g_2$ giving the circle diameter. The danger of instability increases with increasing circle diameter. Phase angle φ determines the position of the circle. If this angle is around $180°$ then the circle is on the left-hand side of the origin, and thus the possibility of intersection is smaller. If the angle is around zero then the circle is to the right of the origin, thus making intersection possible with even a relatively small circle diameter.

The two limiting cases above are represented well by the common-base and common-emitter configurations. For the common-base configuration, the phase angle $\varphi = \varphi_{12} + \varphi_{21}$ approximates to zero with increasing frequency. This means that the circle representing admittance y_f shifts right and downwards, in the direction of the positive real axis, thus increasing the danger of instability.

For the common-emitter configuration, the phase angle $\varphi = \varphi_{12} + \varphi_{21}$ approximates to $-180°$, the initial value being $-90°$. At higher frequencies at which the phase angle approximates to $180°$, the common-emitter tuned amplifier may definitely be regarded as stable since the circle on the left-hand side cannot intersect the straight line on the right.

If there is no intersection, i.e.

$$\mathrm{Re}(y_1)_{min} > \mathrm{Re}(y_f)_{max}, \qquad (15.2.14)$$

then $k_{min} > 1$ and instability cannot arise.

In most neutralized amplifiers, the residual non-neutralized feedback is small and may thus be neglected when considering the minimum stability factor. However, it may be significant when dealing with the frequency response, as will be shown in later chapters.

15.3. Design for maximum power gain in the case of ideal neutralization

For the case where the internal feedback is perfectly eliminated, i.e. $y'_{12} = 0$ for the resultant four-pole, the power gain is given by relation (15.1.6). Let us now determine the load conductance and turns ratio yielding maximum power gain for this case.

First of all let us quote those parameters which are predetermined. Normally, the loaded bandwidth B_L of the amplifier is given. For an n-stage amplifier, where each stage has identical resonant circuits with the same tuning frequency and bandwidth, and the overall bandwidth of the complete amplifier is B_r, then the bandwidth of a single stage is calculated from the usual formula:

$$B_L = \frac{B_r}{\sqrt{2^{1/n} - 1}}. \qquad (15.3.1)$$

The unloaded bandwidth B_0 is determined by the unloaded Q factor Q_0 of the resonant circuit: $B_0 = f_0/Q_0$, where Q_0 depends on the capacity C_0 and inductance L_0 of the resonant circuit. The inductance value is limited by the capacity which cannot be chosen very low since the detuning effect of the changing transistor capacities which are transformed into the resonant circuit must be avoided.

The power gain given in eqn. (15.1.6) is highest at $\nu = 1$, when there is exact impedance match between load conductance g_L and output conductance g_{22}. This maximum power gain is given by

$$G_{\max} = \frac{|y_{21}|^2}{4 g_{11} g_{22}} \left(1 - \frac{B_0}{B_L}\right)^2. \qquad (15.3.2)$$

The four-pole parameters in the above expression apply to the neutralized transistor.

The transformer ratio will be calculated using the notations given in Fig. 15.2. In the figure, the number of turns of inductance L_0 is n_0, with n_1 and n_2 turns referring to tappings connected to the four-pole output and to the load respectively. Turns ratio $n_1/n_2 = \sqrt{g'_L/g_{22}}$ corresponds to matching as expressed by $\nu = 1$. The ratio n_1/n_0 is determined by the loaded bandwidth B_L given by

$$B_L = \frac{1}{2\pi C_0} \left[\left(\frac{n_1}{n_0}\right)^2 g_{22} + g'_k + \left(\frac{n_2}{n_0}\right)^2 g'_L \right]. \qquad (15.3.3)$$

15.3. Design for given stability factor

From this expression, the turns ratio to be determined is given by

$$\frac{n_1}{n_0} = \sqrt{\frac{\pi C_0(B_L - B_0)}{g_{22}}}. \tag{15.3.4}$$

The above design procedure usually results in a rather high power gain which may cause instability in the case of imperfect neutralization. The design procedure given should therefore only be applied with perfect neutralization.

15.4. Design for given stability factor

In order to achieve stable operation, the minimum stability factor should always exceed unity. However, even more stringent requirements on the stability factor may be necessary for maintaining the frequency response. A mismatch results in decreased power gain but increased stability. In the following design procedure, the turns ratios will be calculated from an assumed stability factor k_{min}.

Let us rearrange expression (15.2.13) in the following way:

$$k_{min} = k_0(1 + \nu_1)(1 + \nu_2)/4, \tag{15.4.1}$$

where k_0 is the stability factor determined by the transistor parameters and the bandwidth values:

$$k_0 = \frac{8g_{11}g_{22}}{|y_{12}y_{21}|(1 + \cos\varphi)} \frac{1}{(1 - B_0/B_1)(1 - B_0/B_2)}. \tag{15.4.2}$$

The input and output mismatch factors are $\nu_1 = g_g/g_{11}$ and $\nu_2 = g_L/g_{22}$, respectively.

Stability may be improved by applying mismatch if the value of k_0 is too low for the matched case. Let us assume identical mismatch factors at input and output. For a required stability factor k_{min}, the mismatch needed is given by

$$\nu_1 = \nu_2 = 2\sqrt{k_{min}/k_0} - 1. \tag{15.4.3}$$

Thus the generator conductance transformed to the input and the load conductance transformed to the output are given by the following expressions:

$$g_g = g_{11}(2\sqrt{k_{min}/k_0} - 1); \tag{15.4.4}$$

$$g_L = g_{22}(2\sqrt{k_{min}/k_0} - 1). \tag{15.4.5}$$

The loaded bandwidth of the output resonant circuit is calculated to be

$$B_L = \frac{1}{2\pi C_0}\left[\left(\frac{n_1}{n_0}\right)^2 g_{22} + g'_k + \left(\frac{n_1}{n_0}\right)^2 \nu_2 g_{22}\right], \tag{15.4.6}$$

and thus the turns ratio is given by

$$\frac{n_1}{n_0} = \sqrt{\frac{2\pi C_0(B_L - B_0)}{(1+\nu_2)g_{22}}}. \qquad (15.4.7)$$

Finally, the turns ratio pertaining to the tapping connected to the load conductance is

$$n_1/n_2 = \sqrt{g'_L/\nu_2 g_{22}}. \qquad (15.4.8)$$

The value of k_{min} is normally chosen by meeting the requirement of $k_{min} > 2$, provided that no restriction from the frequency response is given.

The application of mismatch in multi-stage amplifiers requires special attention inasmuch as an input mismatch applied to any stage results in a mismatch of opposite sense in the previous stage. This will decrease the load conductance of the previous stage and thus also the stability factor. Therefore, the stage stability factors in multi-stage single-tuned amplifiers should be distributed so that each stage may have suitable stability factors.

15.5. Frequency-response distortion due to internal feedback

An annoying consequence of the internal feedback (y_{12}) is the distortion of the frequency response. This is due to the sharp change of feedback resulting from the abruptly changing phase response. This results in a substantial asymmetry even in a relatively narrow pass band.

The distortion of the frequency response also depends on the method of tuning procedure. Two methods of tuning may be applied for single-tuned single-stage amplifiers which will be considered using the equivalent circuit shown in Fig. 15.8. The input and output admittances of the transistor are included in the input and output resonant circuit respectively, and thus the active four-pole only consists of two current generators.

When applying the tuning procedure which uses loading, one of the resonant circuits is heavily loaded while the other resonant circuit is tuned to resonance. Feedback is eliminated by the loading, and thus both circuits are tuned to the correct resonant frequency. After tuning in the input circuit the output load is removed, thus producing the known admittance y_f at the input as a result of the feedback. Thus asymmetry is due to the imaginary part of this admittance by detuning the input circuit.

The admittance appearing at the input is calculated from eqn. (15.1.1):

$$y_{in} = g_1 \left[1 + jx_1 - \frac{y_{12} y_{21}}{g_1 g_2 (1 + jx_2)} \right], \qquad (15.5.1)$$

where x_1 and x_2 are the relative detunings of the input and output resonant circuits respectively. Let us introduce the following notation:

$$H = \frac{y_{12} y_{21}}{g_1 g_2} = \text{Re}(H) + j\text{Im}(H). \qquad (15.5.2)$$

15.4. Frequency-response distortion

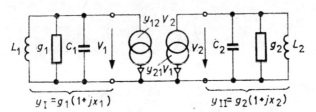

Fig. 15.8. Equivalent circuit of a single-stage single-tuned amplifier for calculating the frequency-response distortion

Thus the admittance appearing at the input is

$$y_{in} = g_1\left(1 + jx_1 - \frac{H}{1 + jx_2}\right). \tag{15.5.3}$$

During the tuning procedure, the output circuit is tuned to resonance with a loaded input circuit; in practice, this loading is given by a low-impedance driving generator. In this way, the condition $x_2 = 0$ is fulfilled, and there is no feedback.

The second step of the tuning procedure is to tune the input circuit to resonance with a loaded output circuit. Again there is no feedback, thus satisfying condition $x_1 = 0$ for the input circuit. After removing the output circuit load, an admittance given by

$$y_{in1} = g_1[1 - \text{Re}(H) - j\,\text{Im}(H)], \tag{15.5.4}$$

will appear at the input, and the imaginary part of this admittance detunes the input circuit.

The *dynamic* method of tuning procedure is carried out as follows. First, the output circuit is tuned to resonance with loaded input, i.e. condition $x_2 = 0$ is satisfied. Next, the input circuit is tuned to resonance with unloaded output, i.e. with the feedback being present. During the input circuit tuning, the imaginary part of y_{in} is set to zero, and thus the adjusted input admittance is given by

$$y_{in2} = g_1[1 - \text{Re}(H) + jx_1 - j\,\text{Im}(H)] = g_1[1 - \text{Re}(H)], \tag{15.5.5}$$

since for the imaginary part, $x_1 - \text{Im}(H) = 0$.

Figure 15.9 shows frequency responses of a single-stage single-tuned amplifier with a stability factor of $k_{min} = 4$. The ful lline response corresponds to the case without internal feedback ($y_{12} = 0$), while the dashed line responses are results of various tuning methods with feedback. A substantial asymmetry may be noted, a peak at negative detunings is present, and the level at resonance is lowered.

For the dynamic-tuning method, the peak is somewhat lower, and the level at resonance is somewhat higher. The asymmetry is thus reduced, and the pass band is flatter.

The single-tuned amplifier may also be designed by the shape of the frequency response. This requires that responses relating to different stability

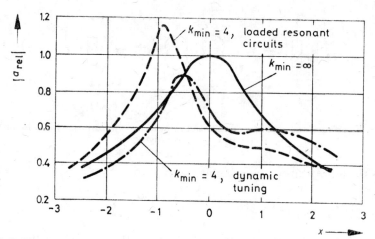

Fig. 15.9. Frequency responses of single-stage single-tuned amplifier for different tuning procedures, and a stability factor of $k_{\min} = 4$

factors and transistor parameters are available. By selecting the most suitable frequency response, the required stability factor is determined, and the design according to the previous chapter is thus possible.

Several frequency responses are possible, depending on the stability factor and the phase angle $\varphi = \varphi_{12} + \varphi_{21}$. In the following, the most characteristic responses will be investigated.

Figure 15.10 shows the distortion of responses at a phase angle of $\varphi = -120°$ for different stability factors. The responses of Fig. 15.10 (a) apply to the tuning method which uses loading. The asymmetry increases with decreasing stability factor, and a sharp peak is shown at negative detunings. For comparison, the response without feedback and with infinite stability factor is also shown in the figure.

Figure 15.10 (b) shows the responses achieved by dynamic tuning, with the same parameters. When applying this tuning procedure, the input circuit is tuned to resonance with the internal feedback effective, and thus the reactive term due to feedback is tuned out. This has the effect of increasing the resonance level and decreasing the peak, resulting in a more even pass band. For comparison, the response corresponding to infinite stability factor is again shown. Figure 15.11 shows responses corresponding to various phase angles for a stability factor of $k_{\min} = 4$, assuming the dynamic tuning method. The dependence of the response shape on the sum of transfer-parameter phases is illustrated for a given stability factor.

For an arbitrary stability factor and phase angle, the response of a single-tuned amplifier can be calculated as follows. From eqs. (15.2.10) and (15.5.2), the real and imaginary parts of the factor H are given by

$$\mathrm{Re}(H) = 2 \cos \varphi / k_{\min}(1 + \cos \varphi); \tag{15.5.6}$$

$$\mathrm{Im}(H) = 2 \sin \varphi / k_{\min}(1 + \cos \varphi). \tag{15.5.7}$$

15.4. Frequency-response distortion

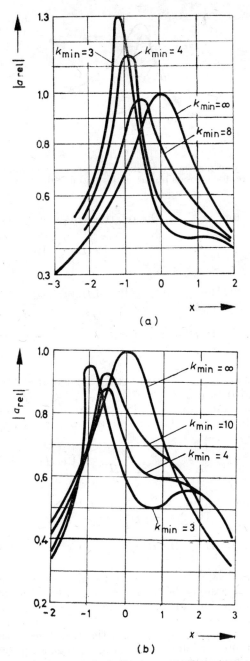

Fig. 15.10. Frequency responses of single-stage single-tuned amplifier for various stability factors and a phase angle of $\varphi_{12} + \varphi_{21} = -120°$. (a) Loaded-tuning procedure, (b) dynamic-tuning procedure

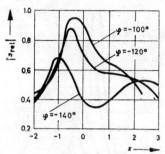

Fig. 15.11. Frequency responses of single-stage single-tuned amplifier for dynamic-tuning procedure, for various phase angles and a stability factor of $k_{min} = 4$

For the loaded-tuning method, the relative frequency function is expressed by (15.5.3):

$$\frac{1}{a_{rel}} = \frac{y_{in}}{g_1} = 1 + jx - \frac{\mathrm{Re}(H) + j\mathrm{Im}(H)}{1 + jx}. \qquad (15.5.8)$$

After rearrangement, the absolute value of the relative response is given by

$$|a_{rel}| = \frac{1 + x^2}{\sqrt{[1 + x^2 - \mathrm{Re}(H) - \mathrm{Im}(H)x]^2 + [x + x^3 - \mathrm{Im}(H) + \mathrm{Re}(H)x]^2}}. \qquad (15.5.9)$$

This relation enables the response to be calculated for arbitrary cases, assuming the loaded-tuning method.

For the dynamic-tuning method, i.e. the tuning process being carried out with internal feedback, the following relation is obtained using eqn. (15.5.5) for the relative response:

$$\frac{1}{a_{rel}} = \frac{y_{in}}{g_1} = 1 + j[x + \mathrm{Im}(H)] - \frac{\mathrm{Re}(H) + j\mathrm{Im}(H)}{1 + jx}. \qquad (15.5.10)$$

By rearranging, the absolute value of the relative response is given by

$$a_{rel} = \frac{1 + x^2}{\sqrt{[1 + x^2 - \mathrm{Re}(H) - \mathrm{Im}(H)x]^2 + [x + x^3 + \mathrm{Im}(H)x^2 + \mathrm{Re}(H)x]^2}}. \qquad (15.5.11)$$

From the above expression, the frequency response of stages using dynamic tuning can be obtained for arbitrary cases.

The frequency response as given by the above relation applies to the input circuit alone. In single-stage amplifiers, the response of the output circuit must also be taken into account, and in multi-stage amplifiers, the responses of the additional filter elements should be considered.

15.6. Stagger tuning

A relatively higher bandwidth may be achieved by placing stagger-tuned resonant circuits between active elements (see Fig. 15.12).

By correct choice of resonant frequencies and damping factors, a wide frequency range can be attained. In the following, stagger-tuned amplifiers with two and three resonant circuits will be investigated.

Let the overall bandwidth required be B_r. Relating this to the band-centre frequency f_0, the relative bandwidth is given by $\delta = B_r/f_0$.

Let us determine the resonant frequencies and loaded Q factors needed for a required relative bandwidth. Two-stage stagger-tuned amplifiers will be investigated first. The two resonant circuits are stagger-tuned symmetrically, and thus the two resonant frequencies are

$$f_1 = f_0/\alpha; \qquad (15.6.1)$$
$$f_2 = \alpha f_0,$$

where α is a factor depending on the relative bandwidth, plotted in Fig. 15.13. The reciprocal values of the loaded Q factors are plotted, also as a function of relative bandwidth.

For small bandwidths, i.e. $\delta < 0.3$, $\alpha = 1 + 0.35\delta$ and the two resonant frequencies are given by

$$f_1 = f_0 - 0.35 B_r; \qquad (15.6.2)$$
$$f_2 = f_0 + 0.35 B_r.$$

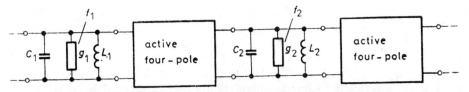

Fig. 15.12. Principle of multi-stage stagger-tuned amplifier

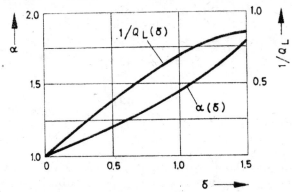

Fig. 15.13. Resonant circuit parameters of two-stage stagger-tuned amplifier as a function of relative bandwidth

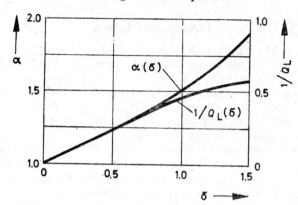

Fig. 15.14. Resonant circuit parameters of three-stage stagger-tuned amplifier as a function of the relative bandwdith

The loaded Q factor is given by

$$1/Q_L = 0.71\delta. \tag{15.6.3}$$

For both circuits, the loaded bandwidth is $B_L = 0.71 B_r$, i.e. 0.71 times the overall bandwidth.

Let us now investigate a three-stage stagger-tuned amplifier. In this case, one circuit is tuned to the band centre, and the other two are symmetrically stagger-tuned:

$$\begin{aligned} f_1 &= f_0/\alpha; \\ f_2 &= f_0; \\ f_3 &= \alpha f_0. \end{aligned} \tag{15.6.4}$$

The factor α is plotted in Fig. 15.14 as a function of relative bandwidth.

The circuit tuned to band centre should have a loaded bandwidth just equal to B_r. The loaded Q of the other two circuits is plotted as a function of relative bandwidth in Fig. 15.14. For convenience, the reciprocal values of the loaded Q factors are again given.

For $\delta < 0.3$, i.e. for low relative bandwidth values, the situation is simplified, and the resonant frequencies are given as follows:

$$\begin{aligned} f_1 &= f_0 - 0.43 B_r; \\ f_2 &= f_0; \\ f_3 &= f_0 + 0.43 B_r. \end{aligned} \tag{15.6.5}$$

The bandwidth of the circuit tuned to f_0 is B_r as before, and for the other two circuits, $1/Q_L = 0.5\delta$, i.e. the loaded bandwidth is half the overall bandwidths for both circuits, i.e. $B_L = 0.5 B_r$.

It was assumed in the above computation that all stages were perfectly neutralized, i.e. $y_{12} = 0$. Substantial internal feedback would result in a distorted response, and the above design equations would lead to incorrect results.

16. DOUBLE-TUNED AMPLIFIERS

16.1. Voltage transfer

In order to satisfy stringent requirements on bandwidth, flatness or selectivity double-tuned amplifiers are applied. In the following, several methods for designing transistor double-tuned amplifiers are presented.

Let us first investigate the voltage transfer of a single-stage amplifier shown in Fig. 16.1. Band-pass filters comprising two inductively coupled resonant circuits are applied at the input and output port of the active four-pole. Inductive coupling will be assumed throughout, but the results may be directly applied to other types of couplings, e.g. capacitive coupling.

In Fig. 16.1, g_{k1} and $g_{\varkappa 2}$ are the primary and secondary conductances due to losses in the output band-pass filter. L_1 and L_2 are the two inductances, between which the inductive coupling is

$$M = k \sqrt{L_1 L_2}. \tag{16.1.1}$$

Tappings and transformers which may in fact be applied are not shown, as the resonant circuits and external conductances are transformed to the appropriate points of the active four-pole.

In Fig. 16.2, the active four-pole is characterized by y parameters. This allows direct inclusion of the real and imaginary parts of the output admit-

Fig. 16.1. Single-stage amplifier comprising band-pass filters with two inductively coupled resonant circuits at the input and the output

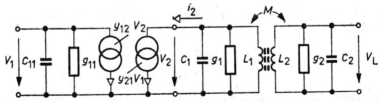

Fig. 16.2. Admittance parameter equivalent circuit of an active four-pole and of an output band-pass filter, after combining the parallel admittances

tance in the resonant circuit elements. Capacity C_1 of the primary circuit includes both the capacity of the resonant circuit and the output capacity C_{22} of the four-pole. It is assumed that both circuits are tuned to the same frequency, so that, for both the primary and secondary circuits,

$$\omega_0 = 1/\sqrt{L_1 C_1} = 1/\sqrt{L_2 C_2}. \qquad (16.1.2)$$

The conductances at resonance are $g_1 = g_{22} + g_{k1}$, and $g_2 = g_L + g_{k2}$ for the primary and secondary circuits respectively.

Circuit losses are expressed by the unloaded Q factors. Let us assume that these are identical for the two circuits:

$$Q_0 = \omega_0 C_1/g_{k1} = \omega_0 C_2/g_{k2}. \qquad (16.1.3)$$

The loaded Q factors are given by

$$Q_1 = \omega_0 C_1/g_1; \qquad (16.1.4)$$

$$Q_2 = \omega_0 C_2/g_2. \qquad (16.1.5)$$

In the following, the voltage transfer expressed by the ratio V_L/V_1 is to be determined, where V_L is the voltage appearing at the output load. The voltage transfer is given by

$$V_L/V_1 = -y_{21} Z_T, \qquad (16.1.6)$$

where y_{21} is the active four-pole transfer parameter and Z_T is the band-pass filter transfer parameter. Using the notations of Fig. 16.2, we have

$$Z_T = -\frac{V_L}{i_2} = \frac{1}{\sqrt{g_1 g_2}} \frac{p}{2\sqrt{f(\eta)}}, \qquad (16.1.7)$$

where $f(\eta)$ is the frequency dependence of the band-pass filter transfer impedance:

$$f(\eta) = \frac{(1 - \eta^2 Q_1 Q_2 + k^2 Q_1 Q_2)^2 + 4\eta^2 Q_1 Q_2}{(1 + k^2 Q_1 Q_2)^2}, \qquad (16.1.8)$$

the relative detuning being expressed by $\eta = 2\Delta\omega/\omega_0$. The coupling factor is given by

$$p = \frac{2k\sqrt{Q_1 Q_2}}{1 + k^2 Q_1 Q_2}. \qquad (16.1.9)$$

Writing the amplifier voltage transfer as a product of the transfer parameters and taking into account eqn. (16.1.7), we obtain

$$\frac{V_L}{V_1} = -\frac{y_{21}}{\sqrt{g_1 g_2}} \frac{p}{2\sqrt{f(\eta)}}. \qquad (16.1.10)$$

At resonance ($\eta = 0$), the factor $f(\eta)$ is unity. Amplification at any frequency is given by $f(\eta)$ which is a function of the quantity $k\sqrt{Q_1 Q_2}$.

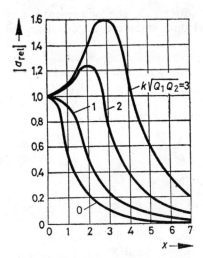

Fig. 16.3. Relative frequency response of double-tuned band-pass filter, by plotting the function $1/\sqrt{f(\eta)}$ for various coupling factors $k\sqrt{Q_1Q_2}$

Figure 16.3 shows frequency responses for various coupling factors as a function of relative detuning η. Only response parts pertaining to positive detuning are shown as the responses are symmetrical. For convenience, the horizontal axis is calibrated in units of $x = \eta\sqrt{Q_1Q_2}$.

16.2. Power gain at resonance

For the determination of power gain, the real part of the input admittance is needed. As a result of internal feedback the input admittance is expressed by

$$y_{\text{in}} = y_{11} - y_{12}y_{21}/y_{\text{II}}, \qquad (16.2.1)$$

where y_{II} is the admittance at the active four-pole output terminal, which, according to Fig. 16.2 is the input admittance of the equivalent band-pass filter. In order to determine the power gain at resonance, the real part of the admittance y_{in} at frequency $\omega = \omega_0$ will be calculated. Relationships away from resonance will be dealt with in the next chapter, in connection with the stability factor.

The input admittance of the band-pass filter at resonance, using the notations of Fig. 16.2, is $y_{\text{II}} = g_1(1 + k^2Q_1Q_2)$, assuming that both the primary and the secondary circuits have been tuned to resonance. Thus the input conductance at resonance is given by

$$g_{\text{in}} = g_{11} - \frac{\text{Re}(y_{12}y_{21})}{(g_{22} + g_{k2})(1 + k^2Q_1Q_2)}. \qquad (16.2.2)$$

From the input conductance, the power gain may be calculated as the ratio of the delivered output power $P_{out} = V_L^2 g_L$ to the dissipated input power $P_{in} = V_1^2 g_{in}$:

$$G_0 = \frac{P_{out}}{P_{in}} = \left(\frac{V_L}{V_1}\right)^2 \frac{g_L}{g_{in}}. \tag{16.2.3}$$

Using the relation derived for the ratio V_L/V_1, we have

$$G_0 = \frac{|y_{21}|^2}{4g_1 g_2} p^2 g_L \bigg/ \left[g_{11} - \frac{\text{Re}(y_{12} y_{21})}{g_1(1 + k^2 Q_1 Q_2)}\right]. \tag{16.2.4}$$

Let us introduce the following:

$$1 - B_0/B_1 = g_{22}/g_1; \tag{16.2.5}$$

$$1 - B_0/B_2 = g_L/g_2, \tag{16.2.6}$$

where B_0 is the unloaded bandwidth of the primary and secondary circuits, B_1 is the loaded bandwidth of the primary circuit, and B_2 is the loaded bandwidth of the secondary circuit. The power gain thus becomes

$$G_0 = \frac{|y_{21}|^2}{4g_{11} g_{22}} p^2 \frac{(1 - B_0/B_1)(1 - B_0/B_2)}{1 - \dfrac{|y_{12} y_{21}| \cos \varphi}{g_{11} g_1 (1 + k^2 Q_1 Q_2)}}. \tag{16.2.7}$$

The first factor of this expression is the *MAG* value defined in (4.4.1), which is the maximum available gain of the active four-pole. This value is reduced by the coupling loss expressed by the factor p and by the factors (16.2.5) and (16.2.6) accounting for the primary and secondary resonant circuit losses. The denominator may be less or more than unity, depending on the phase angle φ, increasing or decreasing the power gain at resonance. The value of $\cos \varphi$ is always positive for a common-base configuration, thus raising the gain. The contrary is true for the common-emitter configuration: $\cos \varphi$ is negative, reducing the power gain as a result of internal feedback.

For zero feedback ($y_{12} = 0$), the denominator is unity, and thus the power gain at resonance is given by

$$G_0 = MAG \; p^2 (1 - B_0/B_1)(1 - B_0/B_2). \tag{16.2.8}$$

The factor p is dependent on the coupling factor $k\sqrt{Q_1 Q_2}$, which is equal to unity at critical coupling. Thus the power gain at critical coupling is given by

$$G_0 = MAG \; (1 - B_0/B_1)(1 - B_0/B_2). \tag{16.2.9}$$

For above-critical and below-critical coupling values, the power gain is reduced as shown by Fig. 16.4.

The relation for power gain is similar in many respects to the power gain of single-tuned amplifiers as given by (15.1.7). The difference is in the factor

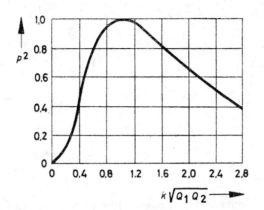

Fig. 16.4. The dependence of factor p^2, determining the power gain, on the coupling factor $k\sqrt{Q_1Q_2}$

p expressing the band-pass type of coupling; however, this disappears at critical coupling.

Let us rearrange eqn. (16.2.8) in order to express the Q factors rather than the bandwidth values, using the relations $B_0/B_1 = Q_1/Q_0$ and $B_0/B_2 = Q_2/Q_0$. The power gain at resonance, without feedback ($y_{12} = 0$) thus becomes

$$G_0 = MAG\; p^2(1 - Q_1/Q_0)(1 - Q_2/Q_0). \qquad (16.2.10)$$

16.3. Stability

The minimum stability factor of band-pass amplifiers is given, based on relation (14.3.4). Firstly let us determine the admittance y_I at the input terminal of the transistor. For this calcuation, only admittance y_{11} will be taken into account and internal feedback will be neglected.

In Fig. 16.1, the elements of the input band-pass filter are primed, to distinguish them from those of the output band-pass filter. Admittance y_I is the parallel connection of the input band-pass filter admittance and the admittance y_{11}:

$$y_I = (g'_{k2} + g_{11})\left(1 + jx'_2 + \frac{k^2 Q'_1 Q'_2}{1 + jx'_1}\right), \qquad (16.3.1)$$

where g'_{k2} is the conductance of the input band-pass filter secondary circuit at resonance, and Q'_1 and Q'_2 are the loaded Q factors. The relative detuning of the two circuits are $x'_1 = 2\Delta\omega'_1 Q'_1/\omega_0$ and $x'_2 = 2\Delta\omega'_2 Q'_2/\omega_0$. Further, it has been assumed that the coupling factors of the input and output filters are identical. The minimum value of the real part of admittance y_I

is given by

$$\text{Re}(y_I)_{\min} = g'_{k2} + g_{11}, \qquad (16.3.2)$$

which is valid at a detuning of $|x'_1| = \infty$, i.e. for a very large detuning of the primary circuit.

Secondly, we shall consider the admittance y_f due to the internal feedback given by $y_f = y_{12}y_{21}/y_{II}$. Admittance y_{II} is the parallel resultant of the output band-pass filter input admittance and admittance y_{22} of the active four-pole:

$$y_{II} = (g_{22} + g_{k1})\left[1 + jx_1 + \frac{k^2 Q_1 Q_2}{1 + jx_2}\right], \qquad (16.3.3)$$

where g_{k1} is the output band-pass filter primary-circuit conductance at resonance, and x_1 and x_2 are the relative detunings of the two circuits. Substituting the value of y_{II} we have

$$y_f = \frac{y_{12}y_{21}}{(g_{22} + g_{k1})[1 + jx_1 + k^2 Q_1 Q_2/(1 + jx_2)]}. \qquad (16.3.4)$$

The real part of this admittance is now analyzed to find the maximum value as a function of the detunings. Considering detuning x_2, the real part will be maximum for a completely detuned secondary, i.e. $|x_2| = \infty$. The maximum value may be determined by extremum calculation:

$$\text{Re}(y_f)_{\max} = |y_{12}y_{21}|(1 + \cos\varphi)/2(g_{22} + g_{k1}), \qquad (16.3.5)$$

and thus the minimum stability factor is given by

$$k_{\min} = \frac{\text{Re}(y_I)_{\min}}{\text{Re}(y_f)_{\max}} = \frac{2(g_{11} + g'_{k2})(g_{22} + g_{k1})}{|y_{12}y_{21}|(1 + \cos\varphi)}. \qquad (16.3.6)$$

Let us rearrange eqn. (16.3.6) by introducing the Q factors:

$$g_{11}/(g_{11} + g'_{k2}) = 1 - Q'_2/Q_0; \qquad (16.3.7)$$

$$g_{22}/(g_{22} + g_{k1}) = 1 - Q_1/Q_0, \qquad (16.3.8)$$

where Q'_2 and Q_1 are the Q factors of the input band-pass filter secondary circuit and of the output band-pass filter primary circuit respectively. With this substitution, the minimum stability factor is given by

$$k_{\min} = \frac{2g_{11}g_{22}}{|y_{12}y_{21}|(1 + \cos\varphi)(1 - Q'_2/Q_0)(1 - Q_1/Q_0)}. \qquad (16.3.9)$$

The above stability factor must exceed unity in order to avoid instability even with detuned resonant circuits. Stability properties may be investigated by using Fig. 16.5 which shows admittances y_I and y_f on the complex plane [15.3]. Admittance y_I is a vertical line at a distance $g_{11} + g'_{k2}$ from the

16.3. Stability

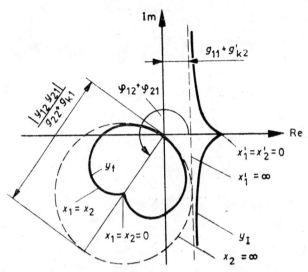

Fig. 16.5. Plot of admittances y_I and y_f, determining the stability of double-tuned band-pass amplifiers, on the complex plane

imaginary axis for a completely detuned primary, i.e. $|x_1'| = \infty$. With no detuning ($x_1' = x_2' = 0$), the real part is given by $y_I = (g_{22} + g_{k2}')(1 + k^2 Q_1' Q_2')$.

Admittance y_f is respresented by a circle for a completely detuned output secondary circuit ($|x_2| = \infty$), shown by a dashed line, and with its axis at angle φ to the positive real axis. For a tuned-in secondary circuit ($x_1 = x_2$) the circle changes into a cardioid shown by a full line. For the case shown, the coupling factor is $k\sqrt{Q_1 Q_2} = 1$, i.e. the output band-pass filter has critical coupling. The involution of the curve on the axis depends on the coupling factor but always remains within the dashed-line circle. For the minimum stability factor, the worst case is considered so that stability is governed by the two dashed lines of Fig. 16.5. If the two dashed lines have an intersection or, in the limiting case, are tangential (this latter case corresponds to $k_{\min} = 1$), then oscillation might occur during tuning.

Stability is primarily influenced by the phase angle φ. For the common-emitter configuration, the circle is on the left of the origin, thus the probability of intersection between the two curves is reduced. For the common-base connection, however, the circle is on the right, near the vertical line, so the stability is worse.

Stability may only be improved by decreasing the loaded Q factors Q_2' and Q_1. For a given coupling factor this will intolerably increase the transmitted frequency band. In Ref. [15.3], the dependence of the response curves on the stability factor and the phase angle φ is investigated in detail for band-pass amplifiers. It may be stated that for a given stability factor and phase angle φ, the band-pass filter responses show much less asymmetry than single-tuned amplifiers.

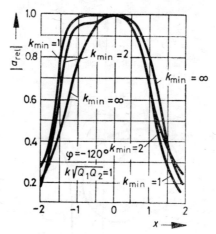

Fig. 16.6. Frequency responses of double-tuned band-pass amplifier with critical coupling, for various stability factors, for a phase angle of $\varphi_{12} + \varphi_{21} = -120°$

In Fig. 16.6, the frequency responses of a single-stage band-pass amplifier are shown for various stability factors, for a phase angle of $\varphi = -120°$. It is seen that the asymmetry is much smaller compared with single-tuned amplifiers of the same stability factors. The response curves correspond to the dynamic tuning process considered earlier, during which the output band-pass filter is tuned in with the input short-circuited, and the input band-pass filter is tuned to maximum with effective feedback present.

16.4. Design procedure with ideal neutralization

Since designing can be substantially simplified by using neutralization, amplifiers with ideal neutralization will be considered in the following, with $y_{12} = 0$.

There are several design methods, the use of which depends on the specified parameter, which may be power gain, bandwidth or selectivity and response shape. The design is complicated by the strong interdependence of the above parameters. For instance, bandwidth is determined by both the coupling factor and the loaded Q value. Two main design procedures may be applied:

(a) *Design for a given coupling factor kQ_L and bandwidth $\Delta\omega_B$*

According to Fig. 16.3, the response shape is determined by the coupling factor $kQ_L = k\sqrt{Q_1 Q_2}$. However, the bandwidth $\Delta\omega_B$ is also specified and this determines the loaded Q factors. The relative bandwidth is given by $\eta_B = 2\Delta\omega_B/\omega_0$ and this must be taken into account for the determination of

16.4. Design procedure

the unknown loaded Q factors Q_1 and Q_2. Power gain is optimum if the equality $Q_1 = Q_2 = Q_L$ is valid. Q_L may be computed from the relation (16.1.8), by substituting the value of $\eta_B Q_L$:

$$f(\eta_B Q_L) = 2. \tag{16.4.1}$$

The loaded Q factor may be calculated from the above expression:

$$Q_L = \sqrt{(kQ_L)^2 - 1 + \sqrt{2} \cdot \sqrt{(kQ_L)^2 + 1}} \Big/ \eta_B. \tag{16.4.2}$$

The design process is simplified by using the plot of the above relation as shown in Fig. 16.7. From this figure, the quantity $\eta_B Q_L$ for different Q_L values may be determined.

The primary and secondary resonant circuit conductances, corresponding to the losses, as transformed to the active four-pole output terminal, may be expressed by the loaded Q factors:

$$g_{k1} = \frac{g_{22}}{Q_0/Q_L - 1}; \tag{16.4.3}$$

$$g_{k2} = \frac{g_L}{Q_0/Q_L - 1}, \tag{16.4.4}$$

assuming identical unloaded Q_0 factors for both circuits.

For known values g'_k of the two circuit losses, the matching is determined by using the ratios g_{k1}/g'_k and g_{22}/g'_k. Thus for the primary circuit we have

$$\frac{n_0}{n_1} = \sqrt{\frac{g_{k1}}{g'_k}} = \sqrt{\frac{g_{22}}{\omega_0 C_1} \frac{Q_0 Q_L}{Q_0 - Q_L}}, \tag{16.4.5}$$

where n_0 is the total number of turns of the resonant circuit inductance, n_1 is the turns number of the tapping connected to the active four-pole output terminal, and C_1 is the total capacity of the primary circuit.

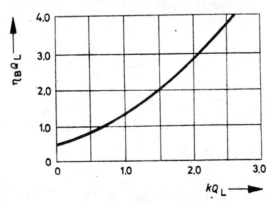

Fig. 16.7. Calculation of the loaded Q factor from the coupling factor kQ and the relative bandwidth η_B

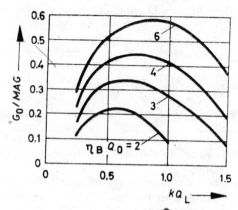

Fig. 16.8. The ratio determining the power transfer of the double-tuned band-pass filter as a function of coupling factor kQ_L, for various values of $\eta_B Q_0$

A similar condition holds for the secondary circuit. If the secondary circuit capacity is also C_1, then the tapping n_2 of a coil having n_0 turns may be calculated from the following equation:

$$\frac{n_0}{n_2} = \sqrt{\frac{g_{k2}}{g'_k}} = \sqrt{\frac{g_L}{\omega_0 C_1} \frac{Q_0 Q_L}{Q_0 - Q_L}}. \tag{16.4.6}$$

(b) *Design for a given bandwidth and maximum power gain*

The previous design procedure will not give maximum power gain because of a specified value of kQ_L. This is easily verified by the dependence of factor p^2 on the coupling factor. The curve has a maximum at the critical coupling $kQ_L = 1$. If substantially different kQ_L values are chosen, the power gain will be greatly reduced because of the factor p^2. If the response shape is of no interest, it is expedient to choose the coupling kQ_L which results in maximum power gain. However, the optimum will not necessarily be at the critical coupling value.

Let us investigate the relation (16.2.10) which gives the power gain. The power transfer of the band-pass filter is given by

$$\frac{G_0}{MAG} = \left(\frac{2kQ_L}{1 + k^2 Q_L^2}\right)^2 \left(1 - \frac{Q_L}{Q_0}\right)^2. \tag{16.4.7}$$

This equation is plotted in Fig. 16.8, which shows that for small relative bandwidth (or more precisely, for small values of $\eta_B Q_0$), maximum power gain is attained below critical coupling, though the value of p^2 is not maximum for this case. This is explained by the factor containing Q_L. By increasing the relative bandwidth or the value of $\eta_B Q_0$, the maximum point is shifted to the value of $kQ_L = 1$. This means that for high relative bandwidth values, maximum power gain is actually given by the critical value of coupling.

Figure 16.8 is used to assist in designing. Using the specified relative bandwidth η_B and the realizable unloaded Q factor Q_0, the kQ_L value yielding the maximum G_0/MAG value is determined for the product $\eta_B Q_0$. Knowing the coupling factor, the design is performed according to point (a).

Both design procedures may only be applied theoretically to amplifiers with ideal neutralization. For non-ideal neutralization, the frequency response will be distorted, and the actual power gain will differ from the calculated value.

The design of multi-stage band-pass amplifiers is dealt with in Ref. [14.8], which also gives the exact frequency responses, taking into account the feedback. By selecting the most suitable normalized frequency response, a multi-stage amplifier may be designed.

17. INTEGRATED-CIRCUIT TUNED AMPLIFIERS

17.1. General considerations

Considering the application of integrated circuits in tuned amplifiers, the two main problems are the provision of the frequency response and the realization of a small internal feedback to ensure stability. We shall investigate these problems in more detail.

Selective high-frequency amplifiers, especially intermediate frequency amplifiers, have stringent demands on the frequency response which may only be satisfied by using filter networks with steep response slopes. In this respect, special attention should be given to *crystal filters* which can be used to satisfy the highest demands. Though crystal filters cannot be included within a monolithic integrated circuit, they are still more suitable for use with integrated circuits than conventional LC filter networks with discrete elements.

In transistor amplifiers, one or two resonant circuits are applied between amplifier stages, and the overall frequency response is thus determined by separated filter networks. However, the power gain of integrated amplifiers is usually much higher than the gain of transistor stages, requiring a smaller number of amplifying stages. This means that the required filter networks can be minimized and applied within a single block — this is advantageous from the aspects of both design and tuning. In this case, the integrated amplifier has to have a large bandwidth in order to avoid any effect on the frequency response of the filter network.

These considerations have been taken into account in the selective amplifier shown in Fig. 17.1 [3.5]. The lumped filter network which may be either a crystal filter or a conventional LC filter network is at the input. This is followed by an amplifier with gain control which can be adjusted over a wide range by the control signal feedback via resistor R_2. The gain-controlled amplifier is followed by a fixed-gain wideband amplifier. The bandwidth of both amplifiers is much larger than the pass band of the filter network. The bias stability of the fixed-gain amplifier is provided by a DC feedback network made up of elements R_1 and C_3. The amplifiers are followed by a detector providing both the demodulated signal and the gain-control signal. The separation of blocks with different functions can be seen clearly.

Special consideration should be given to the internal feedback in integrated amplifiers. As the IC amplifier has a high gain (y_{21}), the internal feedback y_{12} should be kept small because of the loop-gain limitation. In multi-stage

17.1. General considerations

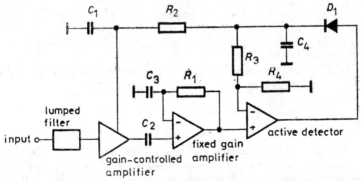

Fig. 17.1. Monolithic gain-controlled amplifier with external filter network and detector

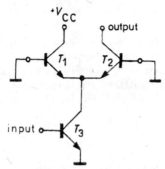

Fig. 17.2. Monolithic integrated differencial amplifier used as common-emitter–common-base two-stage amplifier

amplifiers, the feedback due to internal capacitances can be decreased by correct choice of configurations for the stages. On the other hand, the case capacitance may have a significant effect. To illustrate these parameters, let us investigate a typical monolithic amplifier consisting of a common-emitter and a common-base stage (Fig. 17.2). Denoting the parameters of the common-emitter transistor T_3 by y_e and those of the common-base transistor T_2 by y_b, the overall admittances will be given by [3.4], [11.4]

$$y_{11} = \frac{y_{11e}(y_{12e} + y_{11b}) - y_{12e} y_{21e}}{y_{11b} + y_{22e}} \approx y_{11e}; \qquad (17.1.1)$$

$$y_{12} = - \frac{y_{12e} y_{12b}}{y_{11b} + y_{22e}} \approx 0; \qquad (17.1.2)$$

$$y_{21} = - \frac{y_{21e} y_{21b}}{y_{11b} + y_{22e}} \approx - y_{21e}; \qquad (17.1.3)$$

$$y_{22} = \frac{y_{22b}(y_{22e} + y_{11b}) - y_{12b} y_{21b}}{y_{11b} + y_{22e}} \approx y_{22b}. \qquad (17.1.4)$$

According to eqn. (17.1.2), the parameter y_{12} in the two-stage amplifier is nearly zero, and is only caused by the stray capacitance between input and output. The arrangement is thus equivalent to a common-emitter stage without internal feedback.

Monolithic integrated amplifiers are frequently characterized by the ratio of the transfer parameters. According to eqn. (4.4.5) we have

$$G = \left| \frac{y_{21}}{y_{12}} \right| (K - \sqrt{K^2 - 1}), \tag{17.1.5}$$

where K is the stability factor. In the case of $K < 1$, the circuit becomes unstable, and the expression is meaningless. For the case of $K > 1$, the circuit is stable, but the power gain is lower than in the limiting case of $K = 1$. The factor

$$G_{max} = \left| \frac{y_{21}}{y_{12}} \right|, \tag{17.1.6}$$

is therefore called the *maximum stable gain*, and is frequently used to characterize monolithic integrated amplifiers as the theoretical upper limit of realizable gain. The actually realizable power gain is somewhat lower.

Monolithic integrated amplifiers are sometimes characterized by the so-called k parameters, which are defined by the following set of equations:

$$i_1 = k_{11} V_1 + k_{12} i_2; \tag{17.1.7}$$

$$V_2 = k_{21} V_1 + k_{22} i_2. \tag{17.1.8}$$

Internal feedback is expressed by parameter k_{12} which is the ratio of the output to input current and is usually expressed in dB.

17.2. Application of monolithic differential amplifiers

A typical high-frequency monolithic amplifier is shown in Fig. 17.3. This is a tuned amplifier having a common-collector and a common-base stage. The overall four-pole parameters of the two-stage amplifier are given to good approximation (assuming identical transistors) by the following equations:

$$y_{11} \approx y_{11e}/2; \tag{17.2.1}$$

$$y_{12} \approx - y_{11e} y_{22e}/2 y_{12e}; \tag{17.2.2}$$

$$y_{21} \approx - y_{21e}/2; \tag{17.2.3}$$

$$y_{22} \approx y_{22e}/2. \tag{17.2.4}$$

Since the parameter y_{12} is decreased according to eqn. (17.2.2) the circuit is generally unconditionally stable. This makes a complex-conjugated match possible at both sides, and the power gain can be calculated from eqn. (17.1.5). The terminations $y_{g\,opt}$ and $y_{L\,opt}$ necessary for this match can be

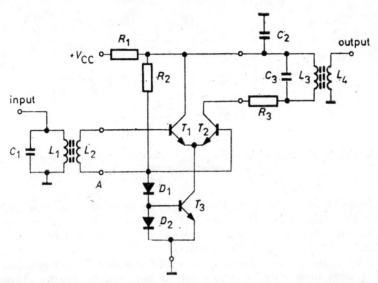

Fig. 17.3. Circuit diagram of a typical high-frequency monolithic integrated circuit

provided by suitable turns ratios of inductive couplings $L_1 - L_2$ and $L_3 - L_4$. A gain of 35 dB at an intermediate frequency of 30 MHz may be implemented with this kind of circuit. The bias stability of the transistors is provided by diodes D_1 and D_2. The circuit is simpler and provides better temperature stability compared with a circuit using discrete components. The amplifier has low distortion in the linear region, especially for even harmonics.

The tuned amplifier shown in Fig. 17.3 has good limiting properties. In the quiescent state, the collector current of transistor T_3 is divided equally between transistors T_1 and T_2. The driving has the effect of upsetting this balance and causing one transistor to draw more current at the expense of the other. Limiting occurs at both sides by the alternating turn-off of transistor currents in T_1 and T_2. By correct choice of output load resistance, saturation of transistor T_2 can be avoided during complete turn-off of transistor T_1. The load resistance required to satisfy this condition requirement is given by

$$R_L \leq 2(V_{CC} - V_{BE})/I_{c3}, \qquad (17.2.5)$$

where V_{CC} is the supply voltage, V_{BE} is the base voltage of the transistors and I_{c3} is the collector current of transistor T_3. The power gain will be somewhat lower as the load resistance thus calculated is less than the matched load resistance needed for maximum power gain.

Figure 17.4 shows a VHF amplifier containing the monolithic circuit shown in Fig. 17.3. The zero-voltage point A of the input terminal has to be earthed with capacitance C_3 as the impedance of the forward-biased

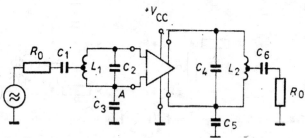

Fig. 17.4. Circuit diagram of high-frequency selective amplifier using a monolithic integrated circuit

diodes $D_1 - D_2$ is significant. Using 50-ohm terminations, this amplifier can provide a gain of 14 dB at 200 MHz and a gain of 21 dB at 100 MHz.

A common-collector–common-base pair with gain control is shown in Fig. 17.5, using the monolithic circuit shown in Fig. 7.6. Gain control is achieved by regulating the base voltage of transistor T_1 via resistance R_4. Assuming zero base current, the relation between transistor currents is given by

$$I_1 + I_2 = I_3 . \tag{17.2.6}$$

Because of identical transistor parameters, the collector current of transistor T_1 is given by

$$I_1 = \frac{I_3}{1 + e^{q\Delta V/kT}}, \tag{17.2.7}$$

where ΔV is the voltage appearing between the base electrodes of transistors T_1 and T_2. Under normal conditions, $\Delta V = 0$, and identical currents $I_1 = I_2$ will flow through the two transistors of the differential amplifier. During gain control, current I_1 is decreased substantially.

As shown in the figure, many internal points are brought out to assist different applications. The circuit shown in Fig. 17.5 provides a gain of 26 dB at an intermediate frequency of 10.7 MHz, and can also be operated as a limiter without saturation.

Figure 17.6 shows the common-emitter–common-base (cascade) variant of the previous monolithic circuit; C_3 serves for earthing the base of transistor T_1 which thus works only in order to decrease the DC current of transistor T_2 during control. The y parameters are given by the relations (17.1.1)–(17.1.4). Internal feedback is practically zero, making the circuit unconditionally stable, and the complex-conjugate match can be implemented. The value of the power gain is given by (17.1.6). From a knowledge of the admittance parameters, the optimum loading resistances and the reactive transforming elements may be calculated. The cascade connection may be applied advantageously for achieving high power gain and low noise figure. However, its limiting properties are worse than those of the previously investigated differential amplifier as saturation may occur. On the other hand, the input impedance is practically constant during control (see

17.2. Monolithic differential amplifiers

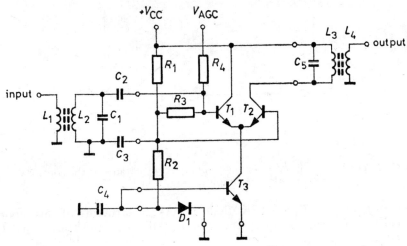

Fig. 17.5. Gain-controlled high-frequency common-collector – common-base amplifier using a monolithic integrated circuit

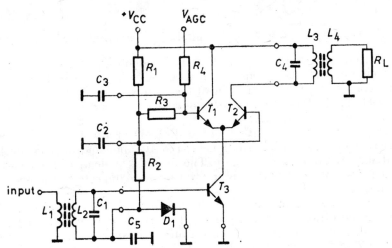

Fig. 17.6. Gain-controlled high-frequency common-emitter – common-base amplifier using a monolithic integrated circuit

Fig. 17.7) which is due to the fact that the current of transistor T_3 depends only slightly on the control voltage.

The circuit shown in Fig. 17.6 provides a gain of 30 dB at zero control voltage, at a frequency of 60 MHz and a bandwidth of 0.5 MHz. In the minimum-gain condition at a control voltage of 6 V, the gain is decreased to -25 dB. According to Fig. 17.7, the dependence of the input resistance and input capacitance on the control voltage is less than 10%. The base

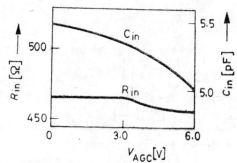

Fig. 17.7. Dependence of input resistance and input capacitance on the control voltage, for the selective amplifier shown in Fig. 17.6

voltage of transistor T_3 is set by diode D_1. As the structure of this diode is identical to the structure of the emitter–base diode of the transistor, they have equal temperature behaviour, resulting in only a slight change of current I_3 as a function of temperature. This kind of temperature compensation has the effect of stabilizing the high-frequency gain to ± 1.5 dB within a temperature range of $-55 -- +125$ °C.

17.3. Application of high-complexity monolithic circuits

In addition to the previously described monolithic circuits, high-complexity circuits are also used in tuned amplifiers. These highly complex circuits, comprising a large number of transistors, were partially treated in Chapter 11. In the following, variants developed primarily for tuned-amplifier applications are investigated.

Figure 17.8 shows a simplified circuit diagram of a wideband monolithic amplifier. Applying suitable filter networks at input and output, we have a tuned amplifier of high power gain. The circuit is made up of three emitter-coupled pairs (differential amplifier stages), interconnected by emitter-follower stages.

It is worthwhile dealing with the DC adjustment of the circuit. The base voltage of the differential amplifiers is half the supply voltage, and thus the collector resistance R_2 is twice the emitter resistance R_1. Since the base voltage of T_2 and the emitter voltage of T_3 are identical, the collector-base voltage of T_2 is equal to the turn-on voltage V_{BE} of T_3 and is thus fairly low. The base voltage is set by resistance R_9 and transistor T_9, whose current is determined by the voltage V_B. This is generated by a diode voltage-divider chain contained in the monolithic circuit but not shown in the figure. Voltage V_A which sets the base voltage of transistor T_{10} is also provided by this diode chain. Thus the first two differential amplifiers are supplied by a regulated supply voltage. The third differential amplifier is also suitable for limiting according to the factors mentioned previously.

Figure 17.9 shows a two-stage application of this circuit, denoted in this figure in the conventional manner. Connection pins can be identified by the

17.3. High-complexity monolithic circuits

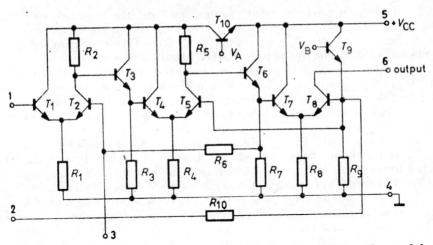

Fig. 17.8. Circuit of high-frequency monolithic integrated amplifier made up of three emitter-coupled pairs and emitter followers

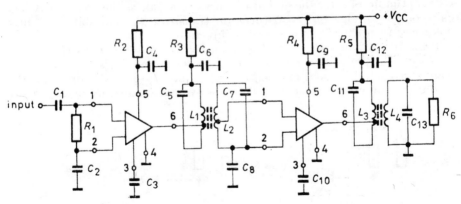

Fig. 17.9. Two-stage selective amplifier with double-tuned band-pass filters, using the monolithic integrated circuits shown in Fig. 17.8

numbering. Double-tuned band-pass filters with inductive coupling are applied. The pins *1* and *2* are connected for DC and pin *3* is AC grounded. The gain of the monolithic circuit is 60 dB at a frequency of 10.7 MHz, and the insertion loss of the filter is about 8 dB. Since the overall gain of the two stages is higher than required by the IF amplification of receivers, the second stage can also be used for limiting. It is essential that the frequency response should not change as a function of the input signal level. According to measurements, the coupling factor kQ_L of the band-pass filter changes within the range 0.5–1.0, which introduces some changes into the frequency response. Further, the selectivity requirements cannot generally be met by the application of two band-pass filters. By applying band-pass filters made up of three inductively coupled resonant circuits, the required selectivity can be assured, and the dependence of the response on the signal level will be negligible. Insertion loss of about 12–17 dB of this band-pass filter presents no problem due to the high reserve in gain.

Figure 17.10 shows an improved version of the monolithic amplifier shown in Fig. 17.8. Subsequent differential amplifiers are inter-connected by two emitter followers, which make the amplifier symmetrical. In the figure, only two identical stages are shown, but three- and four-stage symmetrical amplifiers are also used. Gain control is provided by voltage V_A and thus by the current change of the last differential amplifier. As in the previous circuit, pins *1* and *2* should be connected for DC and pin *3* should be earthed for signal frequencies. Voltage V_B, which sets the current of the uncontrolled differential amplifiers as usual, is generated by a voltage-divider diode chain.

Figure 17.11 shows the tuned-amplifier application of a high-gain monolithic circuit. Transistors T_1, T_2 and T_3 operate in the common-emitter, emitter-follower and common-emitter configurations, respectively. The remaining elements are present for biasing and gain control. For adjusting the current of the common-emitter stages, the base of T_1 and pin *2* should be connected for DC. This introduces a DC negative feedback which provides optimum bias for the amplifier stages. The gain is then 80–90 dB. The gain-control voltage on pin *3* turns on transistor T_6, and shifts the base voltage of transistor T_1 in the reverse direction.

The efficiency of gain control is improved if, in addition to turning off transistor T_1, the collector voltage of T_1 is decreased. This is provided by transistor T_5 and T_4, and resistance R_9; the increase in collector voltage of T_1 during gain control may be avoided or even decreased, by proper choice

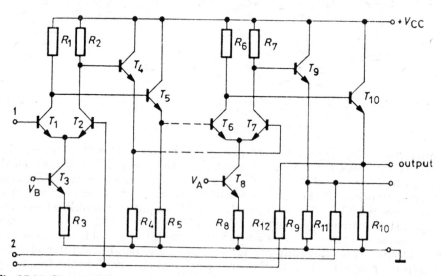

Fig. 17.10. Circuit of high-gain monolithic integrated amplifier made up of symmetrical emitter-coupled pairs

17.3. High-complexity monolithic circuits

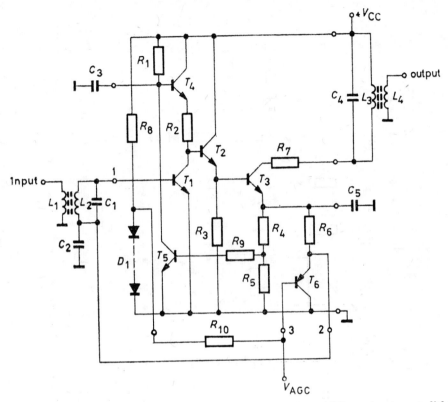

Fig. 17.11. Circuit of high-gain selective high-frequency amplifier using a monolithic integrated circuit

of resistances. With this circuit, a gain control range of about 60 dB may be achieved.

A simplified circuit diagram of a monolithic circuit developed for IF amplifiers is shown in Fig. 17.12. The gain-controlled amplifier is between input *1* and output *2*. The filter network needed to achieve the required frequency response is at the input (it may be either a crystal or an LC-type filter network). The output of the controlled amplifier is connected through external capacitor C_2 to the fixed-gain stage. At the output of this stage (collector of transistor T_7) the IF voltage is provided.

The controlled amplifier basically consists of emitter follower T_1. During gain control, the decrease in gain is provided by a current decrease of T_1 and a simultaneous current increase of T_2 which had been originally turned off. Since the turning on of transistor T_2 has the effect of increasing the output loading, the control is of the so-called *series–parallel* type. The DC control of transistor T_2 is provided by control voltage at pin *3* via emitter follower T_3. In the minimum gain condition, T_1 is turned off, yielding a gain

17. Integrated-circuit tuned amplifiers

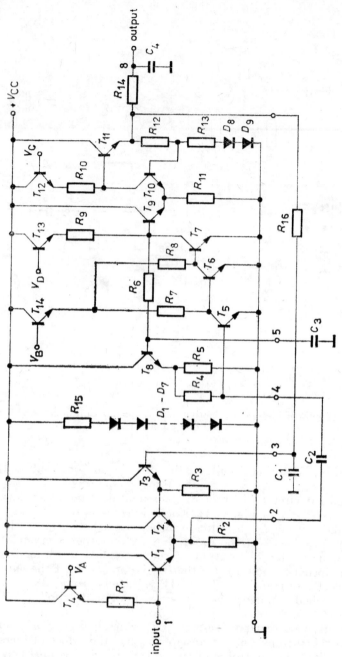

Fig. 17.12. Monolithic integrated circuit intended for IF amplifiers

17.3. High-complexity monolithic circuits

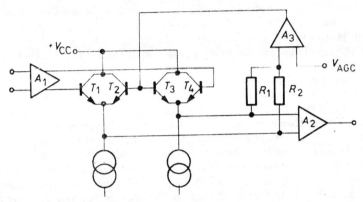

Fig. 17.13. Fast-response gain-controlled amplifier using a monolithic integrated circuit

decrease of about 60 dB. The base voltage of T_1 is set by transistor T_4 which receives voltage V_A from a suitable tapping of the voltage divider comprised of resistance R_{15} and diodes D_1–D_7. Other tappings of this divider provide voltages V_B, V_C and V_D required for further level adjustments.

The fixed-gain stage is made up of the cascaded common-emitter transistors T_5, T_6 and T_7. The DC bias is stabilized by negative feedback which is effected via resistance R_6 and emitter follower T_8. AC feedback is eliminated by external capacitor C_3 connected to pin 5. There are three common-emitter stages: the collector-emitter voltages of the first two stages provide the voltage V_{BE} for transistor T_7, so that transistors with low saturation voltages are needed. DC stability is increased by transistors T_{13} and T_{14} which act as series-type regulators, and thus decrease the dependence on the supply voltage.

The high-gain part is followed, according to Fig. 17.1, by an active detector made up of differential amplifier T_9, T_{10} and T_{11} working as a high-level detector. As a result of the feedback to the base of T_{10}, the circuit is an ideal peak rectifier, with the demodulated signal appearing at pin 8 (after high-frequency filtering). After further filtering via R_{16} and C_1, the demodulated signal is used for gain control at pin 3 of the controlled amplifier.

The monolithic amplifier shown in Fig. 17.12 is a good example for producing the active elements of an IF amplifier in a simple and economic block, to which only the input filter network (preferably a crystal filter) and a few capacitors have to be added externally. This is not only inexpensive, but has the advantage of easy servicing by exchange of factory-made units. This circuit has a sensitivity of 50 μV at 450 kHz, with a control range of about 60 dB.

Figure 17.13 shows a fast-response, symmetrical gain-controlled amplifier without transients. The first stage, differential amplifier A_1, has unity gain. Control is achieved through emitter-coupled pairs T_1–T_4 which act as differential amplifiers; transistors T_2 and T_3 are driven by control amplifier A_3 which itself is driven by the difference between the output signal and the control signal V_{AGC}. The gain control of T_1 and T_4 is implemented by the usual series–parallel method. For more efficient control, the emitter current of the emitter-coupled pairs is provided by current generators. The symmetrical voltage between the emitters is amplified by A_1 which has a gain of 100.

In the foregoing, wideband monolithic integrated amplifiers with externally connected filter networks were studied. In most cases, this method is applied as high-frequency filter networks are difficult to design into monolithic amplifiers. However, a monolithic circuit is presented below which can be used as a wideband tuned amplifier in the microwave region.

It is well known that the common-collector configuration has an inductive output impedance as shown in Fig. 6.12. This type of inductance may be utilized as a resonant circuit tuning element to give selective transmission [17.11], [17.20]. The Q factor of this inductance is very small, less than 3, giving a flat response over a wide frequency range. In the circuit, shown in Fig. 17.14, T_1 acts as amplifier and T_2 acts as inductance. The latter has a common-collector configuration, and the generator resistance is $R_g = R_b$. Figure 17.15 shows the simplified equivalent circuit of the output impedance of T_2. The emitter-circuit RC elements have been replaced by a single resistance r_e, and the additional phase shift has been taken as zero.

Let us now consider the current transfer of this circuit, for the case where the load is given by the input impedance of a common-emitter transistor T_3, taken into account at high frequencies by the series connection of base resistance r_b and emitter capacitance C_e. At sufficiently high frequencies ($\omega > \omega_\beta$), the current gain of the transistor T_1 is given by

$$\beta = \frac{\beta_0}{1 + j\omega/\omega_\beta} \approx -j\frac{\omega_T}{\omega}. \qquad (17.3.1)$$

The common-collector transistor T_2 has an output impedance appearing parallel with the output, and may be written as:

$$Y_c = \frac{1}{r_b + R_b} - \frac{j\omega_T}{\omega(r_b + R_b)}. \qquad (17.3.2)$$

The input capacitance C_e of the next stage T_3 and the coupling capacitance C_0 may be combined. Thus the conductance appearing at the collector of transistor T_1, due to the loading of the next stage, is given by

$$Y_2 = \frac{j\omega C_2}{1 + j\omega r_b C_2}, \qquad (17.3.3)$$

where the contracted capacitance is $C_2 = C_0 C_e/(C_0 + C_e)$. According to these equations, the current gain is given by

$$\frac{i_2}{i_1} = A(\omega) = \frac{\omega_T}{j\omega}\frac{Y_2}{Y_2 + Y_c}. \qquad (17.3.4)$$

17.3. High-complexity monolithic circuits

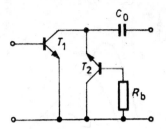

Fig. 17.14. Monolithic selective amplifier stage using the inductive output impedance of the common-collector configuration

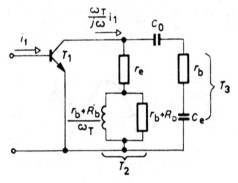

Fig. 17.15. The circuit of Fig. 17.14 using the equivalent circuit of transistor T_2

Substituting the values of Y_2 and Y_c we obtain

$$A(\omega) = \frac{-j\omega C_2(R_b + r_b)}{1 + j\omega(1 + \omega_T r_b C_2)/\omega_T - \omega^2 C_2(R_b + 2r_b)/\omega_T}. \qquad (17.3.5)$$

From the frequency dependence of the current gain given by (17.3.5), we can calculate resistance R_b and capacitance C_0 which are needed for the specified response. The calculation may be performed either by the pole-zero method or by using the curves given in Chapter 5. For maximally flat response the maximum value of the upper cut-off frequency is

$$\omega_H = \frac{1 + \omega_T r_b C_2}{\sqrt{2C_2(R_b + 2r_b)}}. \qquad (17.3.6)$$

This value is higher for the equal-ripple (Chebyshev-type) response.

We are primarily interested in the selective response. Figure 17.16 shows the frequency responses of the circuit presented in Fig. 17.14 for various values of C_0 and for $R_b = 100$ ohms. It is assumed here that the output load is presented by similar circuits. The *iterative gain* thus defined is the gain of a single stage of an amplifier chain comprised of a number of identical

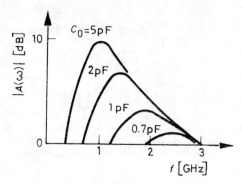

Fig. 17.16. Frequency response of the circuit shown in Fig. 17.14, for various capacity values C_0

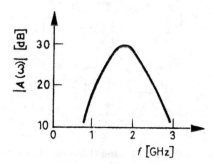

Fig. 17.17. Frequency response of six-stage selective monolithic amplifier

stages. Expressing this by the admittance parameters, we have

$$A_{it} = \frac{y_{21} y_{it}}{y_{11}(y_{22} + y_{it}) - y_{12} y_{21}}, \qquad (17.3.7)$$

where y_{it} is the iterative input admittance, i.e. the input admittance of a single stage taken out of an infinitely long chain:

$$y_{it} = [y_{11} - y_{22} + \sqrt{(y_{11} + y_{22})^2 - 4y_{12}y_{21}}]/2 . \qquad (17.3.8)$$

The iterative gain has a selective character according to Fig. 17.16, though the bandwidth is rather high. The bandwidth may be decreased by increasing the number of stages. Figure 17.17 shows the frequency response of a six-stage amplifier in which two cascaded selective stages according to Fig. 17.14 are followed by a common-emitter stage, and this configuration is repeated once. This kind of amplifier can easily be produced in monolithic form and is still suitable for high frequencies.

17.4. Application of hybrid integrated circuits

Hybrid integrated circuits used at high frequency have mostly a multichip construction, where several semiconductor chips are die-bonded to the ceramic substrate and interconnected to each other and to passive components by thin gold wire. The semiconductor chips may be bipolar transistors, FETs, MOS capacitors or even monolithic integrated circuits. The passive components are capacitors and resistors made by thin-film technique. The great advantage of hybrid circuits in high-frequency applications is the possibility to use simultaneously active device chips made by different technologies (e.g. bipolar and MESFET transistors), the superior isolation between the components reduces hereby the parasitics, and the higher power-handling capability. In addition, hybrid circuits are inexpensive when used in small quantities, in contrast with the monolithic integrated circuits [17.19].

Hybrid integrated circuits can be used to advantage in a frequency range which is too high for circuits applying lumped circuit elements (coils, capacitors), but too low for waveguide or stripline circuits. It is well known that there is a lower limit to inductance and capacitance which can be achieved with conventional elements, because of geometrical dimensions: this limits resonance frequency. However, with hybrid circuits, evaporated, helical metal stripes allow the production of extremely low inductances which may be used in the GHz frequency range. The dimensions of the lumped element reactances in hybrid circuits are about a tenth of the dimension of microstrip circuits operating at about 1 GHz since in the latter type of circuit, the stripline length is a function of the wavelength. Use of lumped inductances and capacitances allows smaller dimensions, particularly important in the VHF range. The Q factor of inductances and capacitances which are built with hybrid technology is generally not too high, at most about 50. This results in a wide pass band of the selective amplifier, which is a certain limitation.

A general cross-sectional view of a hybrid integrated circuit is shown in Fig. 17.18, and the top view of a specific circuit is illustrated by Fig. 17.19 [17.14]. Various metal and dielectric layers are deposited in several steps onto the insulating substrate. Inductances are normally formed by helical metal layers, prepared by a photolithographic process. The inductance value may be calculated from the geometrical dimensions of the helix and of the thickness of the evaporated metal layer, by using the following expression:

$$L \cong \frac{0.04 a^2 n^2}{8a + 11b}, \qquad (17.4.1)$$

where L is the inductance in nH, n is the number of turns, $a = (r_1 + r_2)/2$ and $b = r_1 - r_2$, where r_1 and r_2 are the outer and inner radii of the helix, respectively, in mm. The inductance of the single-turn coil shown in Fig. 17.19 is $L_2 \approx 2$ nH, which is extremely low and cannot be achieved using conventional methods. The inductance L_1 of the radio-frequency choke is

17. Integrated-circuit tuned amplifiers

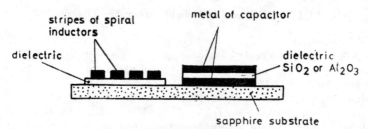

Fig. 17.18. Cross-sectional view of hybrid integrated circuit

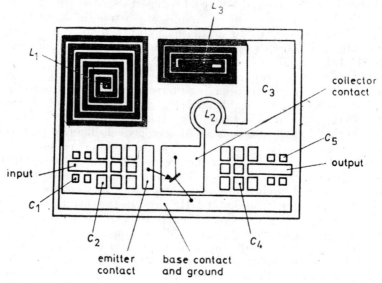

Fig. 17.19. Top view of single-stage hybrid integrated selective amplifier

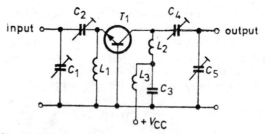

Fig. 17.20. Circuit of hybrid integrated amplifier shown in Fig. 17.19

produced by a multi-turn helix, its value being 25 nH. Adjustment is possible by applying a short-circuit to sections of the helix.

It is well known that to avoid energy radiation, the dimension of the inductance should be less than a thirtieth of the wavelength. This condition is met by helical inductances produced by hybrid technology, inasmuch as

17.4. Hybrid integrated circuits

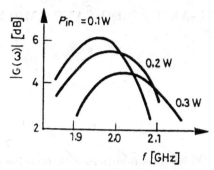

Fig. 17.21. Frequency responses of the hybrid integrated selective amplifier shown in Fig. 17.19, for various input power levels

the outer diameter of the inductances used in the 0.4–2 GHz frequency range is $2r_1 \approx 1$–2 mm. The practical value of the unloaded Q factor of the inductances is 30–50, which allows selective inter-stage networks to be designed with reasonably small losses.

The capacity of the thin-film capacitor depends on its surface area, and the material and thickness of the dielectric layer. Usual methods allow extremely small surfaces in order to produce a circuit with small dimensions. Capacitors are made up of several cells (see C_2 in Fig. 17.19), and the required capacity is achieved by connecting a suitable number of cells. This allows circuit tuning. The unloaded Q factors which can be obtained are around 100.

Figure 17.20 shows the diagram of the circuit, of which the top view is shown in Fig. 17.19 [17.14]. The selective amplifier stage is matched to 50-ohm characteristic impedance at the input by an L network made up of capacitances C_1 and C_2, and at the output by a π network made up of circuit elements L_2, C_4 and C_5. All matching capacitances are made up of several cells and are thus adjustable. Inductances L_1 and L_3 have a high number of turns and act as radio-frequency chokes. Capacitor C_3 has a large surface for earthing. The dimensions of the substrate comprising the complete circuit are 2.6×3 mm. The class C power amplifier provides an output power of 0.5 watts at a frequency of 2 GHz with a gain of about 6 dB. The power loss introduced by the matching elements is about 0.8 dB. The bandwidth may be determined from Fig. 17.21, which also shows the level dependence of the frequency response.

Reference [17.15], contains a description of a three-stage hybrid amplifier for an intermediate frequency of 500 MHz and a bandwidth of 400 MHz. Gain of the wideband RC amplifier is 24 dB. It is coupled to the driving circuit and the load by a matching circuit which has lumped L and C elements produced by hybrid technology.

In Ref. [17.16], a hybrid integrated power amplifier is presented, which can deliver a power of 20 W in the frequency range 225–400 MHz at a gain of 12 dB ± 1 dB. We shall consider this amplifier in a later chapter.

18. UHF AND MICROWAVE AMPLIFIERS

18.1. Wideband amplifiers with reactive filters

Figure 14.1 shows the principle of amplifiers using reactive matched networks to achieve a specified frequency response. Single-tuned and double-tuned resonant circuits were dealt with earlier, and it was shown that the transmission band attainable with these filters is generally small, so their main application field is in narrow-band (intermediate frequency) amplifiers. Several reactive elements are needed to increase the transmission bandwidth, and this is most easily achieved using the ladder network. In filter catalogues tables are presented for various types of ladder networks which are suitable for the design of both maximally flat low-pass filters (Butterworth filters) and equal-ripple low-pass filters (Chebyshev filters).

High-pass and band-pass filters may be derived from low-pass filter, using transformation. However, most filter catalogues present little data for filters with high transforming ratio, despite the fact that rather high ratios are required for transforming the transistor output impedance to the conventional characteristic impedance of coaxial cables. This problem will be detailed below.

In Ref. [18.1], the design of equal-ripple low-pass filters having up to ten sections is studied, with transforming ratios in the range 1.5–50. The circuit of the low-pass filter is shown in Fig. 18.1, and the frequency response is given in Fig. 18.2. Between the lower and upper cut-off frequencies ω_1 and ω_2, the response shows an equal-ripple fluctuation of a_r. Since the filter is antimetric and performs a resistance transformation, there is substantial DC reflection and thus a DC loss a_{DC}. An important characteristic is the relative bandwidth:

$$\Omega = 2 \frac{\omega_2 - \omega_1}{\omega_2 + \omega_1}. \tag{18.1.1}$$

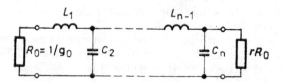

Fig. 18.1. Circuit diagram of low-pass filter

18.1. Amplifiers with reactive filters

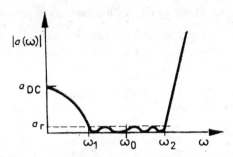

Fig. 18.2. Frequency response of equal-ripple (Chebyshev-type) low-pass filter

Reference [18.1] contains tabulation of the values of ripple a_r pertaining to various relative bandwidths and transforming ratios, and element values. In the context of transistor amplifiers, four-element filters are most significant, so we present the data of these filters in Tables 18.1, 18.2 and 18.3. These tables give normalized element values

$$\omega_0 L_1/R_0 = \gamma_1; \qquad (18.1.2)$$

$$\omega_0 C_2/g_0 = \gamma_2, \qquad (18.1.3)$$

and similarly,

$$\omega_0 L_{n-1}/R_0 = \gamma_{n-1}; \qquad (18.1.4)$$

$$\omega_0 C_n/g_0 = \gamma_n, \qquad (18.1.5)$$

where ω_0 is the centre frequency of the pass band:

$$\omega_0 = (\omega_2 + \omega_1)/2. \qquad (18.1.6)$$

Table 18.1. Equal-ripple (Chebyshev-type) low-pass filter parameters for an element value of $n = 4$. Ripple amplitude a_r for different transforming ratios r and relative bandwidth values Ω

r	$\Omega = 0.1$	0.2	0.3	0.4	0.6	0.8	1.0
1.50	0.00	0.00	0.00	0.00	0.00	0.00	0.03
2.00	0.00	0.00	0.00	0.00	0.01	0.05	0.11
2.50	0.00	0.00	0.00	0.00	0.03	0.09	0.21
3.00	0.00	0.00	0.00	0.00	0.04	0.13	0.30
4.00	0.00	0.00	0.00	0.01	0.07	0.23	0.50
5.00	0.00	0.00	0.00	0.02	0.10	0.32	0.70
6.00	0.00	0.00	0.00	0.02	0.14	0.41	0.90
8.00	0.00	0.00	0.01	0.04	0.20	0.60	1.26
10.00	0.00	0.00	0.01	0.05	0.27	0.78	1.60
15.00	0.00	0.00	0.02	0.08	0.43	1.19	2.36
20.00	0.00	0.00	0.03	0.12	0.58	1.58	3.00
25.00	0.00	0.00	0.05	0.15	0.73	1.93	3.57
30.00	0.00	0.01	0.06	0.18	0.87	2.25	4.06

Table 18.2. Equal-ripple (Chebyshev-type) low-pass filter parameters for an element value of $n = 4$. First element relative value γ_1 for different transforming ratios r and relative bandwidth values Ω

r	$\Omega = 0.1$	0.2	0.3	0.4	0.6	0.8	1.0
1.50	0.65	0.65	0.66	0.66	0.67	0.68	0.69
2.00	0.81	0.82	0.82	0.83	0.84	0.87	0.89
2.50	0.93	0.93	0.94	0.95	0.97	1.00	1.03
3.00	1.01	1.02	1.03	1.14	1.07	1.11	1.16
4.00	1.15	1.16	1.17	1.19	1.23	1.29	1.36
5.00	1.26	1.27	1.28	1.30	1.36	1.36	1.44
6.00	1.34	1.35	1.37	1.40	1.47	1.57	1.67
8.00	1.49	1.50	1.52	1.56	1.65	1.78	1.92
10.00	1.60	1.62	1.65	1.69	1.81	1.97	2.14
15.00	1.82	1.84	1.88	1.94	2.12	2.34	2.60
20.00	1.98	2.01	2.07	2.14	2.36	2.66	2.98
25.00	2.11	2.15	2.22	2.31	2.58	2.93	3.31
30.00	2.23	2.27	2.35	2.46	2.77	3.18	3.61

Table 18.3. Equal-ripple (Chebyshev-type) low-pass filter parameters for an element value of $n = 4$. Second element relative value γ_2 for different transforming ratios r and relative bandwidth values Ω

r	$\Omega = 0.1$	0.2	0.3	0.4	0.6	0.8	1.0
1.50	0.87	0.87	0.87	0.86	0.85	0.83	0.80
2.00	0.86	0.86	0.85	0.84	0.82	0.80	0.76
2.50	0.83	0.82	0.82	0.81	0.79	0.76	0.72
3.00	0.80	0.79	0.79	0.78	0.75	0.72	0.68
4.00	0.74	0.74	0.73	0.72	0.69	0.65	0.61
5.00	0.70	0.70	0.69	0.68	0.65	0.61	0.56
6.00	0.67	0.67	0.66	0.64	0.61	0.57	0.52
8.00	0.62	0.62	0.61	0.59	0.56	0.51	0.47
10.00	0.59	0.58	0.57	0.55	0.52	0.47	0.42
15.00	0.53	0.52	0.51	0.49	0.45	0.40	0.36
20.00	0.49	0.48	0.47	0.45	0.40	0.36	0.31
25.00	0.46	0.45	0.44	0.42	0.37	0.32	0.28
30.00	0.44	0.43	0.41	0.39	0.35	0.30	0.26

In the tables, normalized γ values of elements in the first filter half are presented. The second half is the inverse of the first, so these element values can be calculated from the table as follows:

$$\gamma(n/2 + 1) = r\gamma(n/2); \qquad (18.1.7)$$

$$\gamma(n/2 + 2) = \frac{1}{r}\gamma(n/2 - 1); \qquad (18.1.8)$$

$$\gamma(n/2 + 3) = r\gamma(n/2 - 2), \qquad (18.1.9)$$

18.1. Amplifiers with reactive filters

etc. For the case presented of $n = 4$, based on the γ_1 and γ_2 values from the tables, we have

$$\gamma_3 = r\gamma_2; \qquad (18.1.10)$$

$$\gamma_4 = \gamma_1/r. \qquad (18.1.11)$$

The first step in the filter design is the determination of the relative bandwidth by using (18.1.1). Whether or not the ripple a_r is suitable with the chosen filter degree number n should also be evaluated. Should the ripple prove to be excessive, the filter degree should be increased. Following this, the element values L_1, C_2 etc. may be calculated by determining the relative values γ_1, γ_2 from tables (for $n = 4$, using Tables 18.2 and 18.3). The driving generator may be connected to any of the two filter terminals, so the circuit is also suitable for transforming down resistances. DC attenuation a_{DC} follows from the transformation value:

$$a_{DC} = 10 \lg \frac{(r+1)^2}{4r}. \qquad (18.1.12)$$

Let us now consider the problem of matching in transistor amplifiers. For correct operation of the previously designed filter we need real and constant values of resistances R_0 and rR_0 throughout the frequency range $\omega_2 - \omega_1$. However, this is not at all the case with transistor input and output impedances. If the reactive component of the impedance to be matched is constant within the band then the reactance may be combined with the first or nth reactive section of the filter. Figure 18.3 shows an inductive input impedance and a capacitive output admittance. We may contract L_{in} and L_1 in the first case, and in the second case C_{out} and C_n, or if necessary, the nth circuit element may consist of C_{out} itself. This is often the situation at extremely high frequencies because of the high output capacities.

If the impedance elements shown in Fig. 18.3 are not constant over the frequency range, the filter element values must also be modified. However, a much poorer flatness should be expected. This, of course, is highly dependent on the transmitted bandwidth. In this case, after designing the filter by considering the input and output impedances at band centre, the best response should be adjusted experimentally by tuning the filter elements.

Figure 18.4 (a) shows the circuit diagram of a wideband amplifier with Chebyshev filters of fourth degree [18.9]. In the base-circuit of the com-

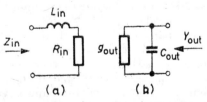

Fig. 18.3. Equivalent circuits for bipolar transistor impedances. (a) Inductive input impedance, (b) capacitive output admittance

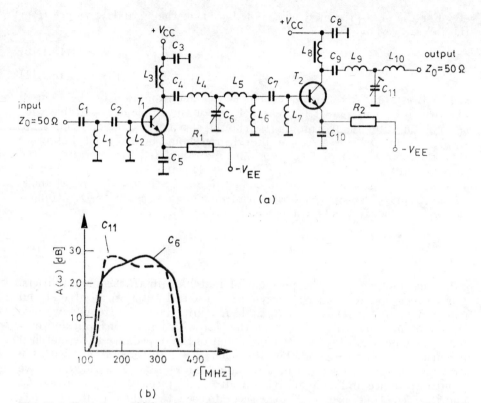

Fig. 18.4. Two-stage wideband amplifier with Chebyshev filters. (a) Circuit diagram, (b) frequency response for two different tuning methods

mon-emitter transistors, there are high-pass filters made up of elements $L_1 - C_2 - L_2$ and $L_6 - C_7 - L_7$. The fourth filter element, the series capacitance, is given by the capacitive part of the transistor input impedance (in this frequency range, the applied transistor type has a capacitive input impedance). Output points are also capacitive and feed low-pass filters made up of elements $L_4 - C_6 - L_5$ and $L_9 - C_{11} - L_{10}$. In both cases, the first filter element is the output capacitance of the transistor. Capacitors $C_1 - C_4 - C_9$ serve for DC blocking, and capacitors $C_3 - C_5 - C_8 - C_{10}$ for high-frequency earthing. Coils L_3 and L_8 are high-frequency chokes. The frequency response measured between 50-ohm terminations is shown in Fig. 18.4 (b) using two separate tuning methods.

The full curve corresponds to the case when the filter at the first-stage output is tuned to maximum flatness with trimmer capacitor C_6. For the dashed curve maximum flatness is adjusted by C_{11} at the output of the second stage. It is seen that the transmitted band is very wide but flatness is difficult to achieve, especially in multi-stage amplifiers and in the wideband case.

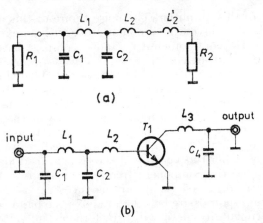

Fig. 18.5. (a) Transistor input impedance matched by reactive filter, (b) circuit diagram of wideband power amplifier

Reference [18.20] contains tables which give data on reactive filters with high-frequency peaking. The low-pass filters shown in Fig. 18.5 (a) are present for matching the transistor input impedance replaced by elements L'_2 and R_2. In the tables, element values $L_1 - L_2 - C_1 - C_2$ are given for various pass-band ripples and relative bandwidths, for a slope of 4, 5 and 6 dB/octave and for transforming ratios of $R_1/R_2 = 20-100$. Figure 18.5 (b) shows the circuit diagram of a class C hybrid amplifier for the frequency range of 225–400 MHz, designed according to the tables.

In Ref. [20.14] a power amplifier delivering 50 W output power in the range 100–160 MHz is presented. In this amplifier, Chebyshev filters utilizing thin-film inductances are applied.

18.2. Wideband amplifiers with transformers

It was shown in Chapter 12 that by using transmission-line transformers as coupling elements, high amplifier bandwidths may be produced, by eliminating the stray capacitance which would otherwise be present. This has not to be considered in narrow-band amplifiers, so in these cases, conventional tuned high-frequency transformers are often used. However, the high operating frequency requires careful layout as stray capacitances would limit the transmitted frequency band. Good results may be achieved by application of miniature ferrite toroidal cores. Stray inductance may be reduced by the close spacing of primary and secondary windings.

In Ref. [18.19], an eight-stage amplifier operating at 600 MHz is presented. The high internal feedback would render the common-emitter stage impractical at this frequency, so emitter-coupled transistor pairs are applied in each stage (Fig. 18.6). This also has the advantage of limiting — an essential aspect of this amplifier. The noise figure of this amplifier is poor, so it is preceded by a conventional preamplifier with an extremely low noise figure made up of a common-emitter and a common-base stage and also having transformer output coupling.

The transformer at the output of the emitter-coupled stage utilizes a ferrit toroidal core which has a shape similar to that used in balun transformer cores. Small dimensions and small number of turns are used to attain the high resonant frequency. The primary inductance of the transformer and the resonant frequency determined by the output capacitance of the emitter-coupled stage can usually only be determined by measurement. The resonance peak is rather flat because of the transformed load value.

Figure 18.7 shows the component frequency responses. Responses a and b result from transformers having various turns ratios. The resonant frequencies of these responses are below and above the band centre. By applying these experimentally designed transformers in suitable numbers a relatively wide bandwidth around 600 MHz may be attained, as shown by curve c. Should the two resonance frequencies be distributed symmetrically around the band centre, then both transformers should be applied in equal amounts. The impedance match required for the high power amplification generally requires transformer ratios between 2 : 1 and 4 : 1.

Figure 18.8 shows the frequency response of the eight-stage amplifier. The high gain demands careful shielding and suitable filtering of the power-supply voltages. In order to avoid parasitic oscillations, 100-ohm resistors are connected in series with all transistor collectors. Midband gain is about 85 dB, and bandwidth is 50 MHz.

Figure 18.9 shows one stage of a transformer-coupled microwave amplifier [18.6]. A stripline is used to connect the input signal to the base of T_1, operating in the common-emitter configuration. The stripline impedance is 50 ohms. An air-core inductance having a few turns is in the collector circuit, its tap feeding another 50-ohm stripline. The low-inductance ceramic capacitor C_1, built into the stripline, produces the DC decoupling. The base of the next stage is biased via inductance L_2. The stage is essentially a single-tuned amplifier, whose resonant frequency is determined by inductance L_1 and the transistor output capacitance. The extremely high loading of the resonant circuit results in a frequency response which is substantially different from the response of a conventional single-tuned amplifier, and the transmitted band is extremely large.

The practical design of this circuit is shown schematically in Fig. 18.10. As shown in Fig. 18.9, this is the first stage of a multi-stage amplifier with the place of the coaxial connector, the input stripline, and the decoupling capacitor C_3. The output is connected to the base of the next stage T_2. The transistors have special packages suitable for connection to the striplines. Use of striplines produces lower stray capacitances and negligible radiation of interconnections also results. The optimum number of turns of coil L_1 for 800 MHz band-centre frequency is about 3–4 turns, and tapping near the centre results in a fairly good impedance match to the next stage input. The inductance value and thus the resonant frequency are adjustable by bending the coil windings.

18.2. Amplifiers with transformers

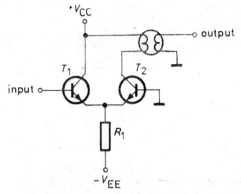

Fig. 18.6. Circuit diagram of emitter-coupled amplifier using wideband transformer matching

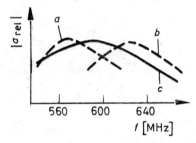

Fig. 18.7. Frequency responses of the circuit shown in Fig. 18.6, for various transformer turns ratios

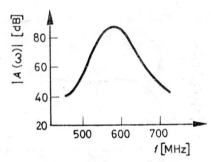

Fig. 18.8. Frequency response of an eight-stage amplifier made up of stages shown in Fig. 18.6

Figure 18.11 shows the frequency response of an eleven-stage amplifier, using stages according to Fig. 18.9. The power gain as measured between 50-ohm terminations in the frequency range 0.5–1.1 GHz is 60 dB ± 2 dB, and the noise figure is 7 dB at 1 GHz. The transition frequency of the applied germanium mesa-transistor is $f_T = 1.4$ GHz, the internal feedback

18. UHF and microwave amplifiers

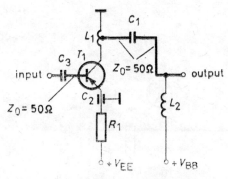

Fig. 18.9. Circuit diagram of microwave amplifier stage with transformer coupling

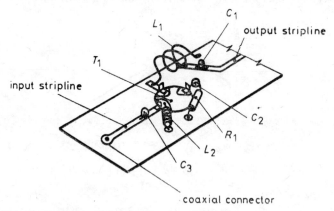

Fig. 18.10. Practical design of microwave amplifier stage with transformer coupling

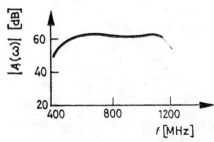

Fig. 18.11. Frequency response of an eleven-stage wideband microwave amplifier made up of stages as shown in Fig. 18.9

time-constant is $r_b C_c = 5$ ps. A similar three-stage wideband amplifier applying hybrid technology provides a gain of 20 dB $\pm$ 0.5 dB in the frequency range 1.1–2.1 GHz.

18.3. Narrow-band amplifiers with transmission lines

The highest resonant frequency which can be achieved with conventional inductances is about 200–300 MHz. Above 300 MHz, resonant circuits cannot be designed with this method, so resonant circuits made up of transmission-line sections are used. Inductive elements using transmission-line sections have the advantage over conventional inductances of having higher Q factors and good reproducibility, especially regarding tappings. In the following, various transmission-line layouts and properties will be investigated from the aspect of narrow-band amplifier application.

Figure 18.12 shows the cross-sectional views of eight transmission-line configurations, with approximate expressions given for the characteristic impedances. It has been assumed that the dielectric is air ($\varepsilon_r = 1$), and that the diameter or width of the inner conductor is small compared with the dimensions of the outer conductor ($d/D < 0.5$ and $b/D < 0.5$).

The first two transmission lines in the figure have cylindrical inner conductors. It is seen that by removing to infinity the horizontal outer conductor planes the characteristic impedance is increased. This arrangement is advantageous because practically, the cover of the metallic box which forms the outer conductor has less significance in generating the field of the transmission line. In the next three arrangements, it is assumed that the inner conductor is formed by striplines of zero thickness. In most cases, a thin stripline parallel to two infinite conductor planes is applied. The sixth drawing shows a rectangular transmission line with an inner conductor of finite dimensions. In the last two drawings, the outer conductor is formed

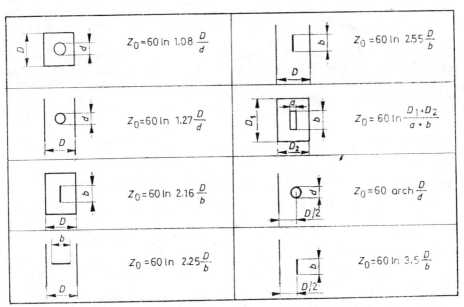

Fig. 18.12. Transmission-line configurations and characteristic impedances

by a single conducting plane of infinite dimension. The characteristic impedance of an inner conductor situated asymmetrically in a cavity can best be approximated by these arrangements. The characteristic impedance relations given in Fig. 18.12 may be generalized for any dielectric material by division by $\sqrt{\varepsilon_r}$.

One of the most important transmission line equation is that providing the input resistance. Terminating the output of a transmission line of length l and characteristic impedance $Z_0 = 1/y_0$ by admittance y_2, the input admittance y_1 is given by

$$y_1 = y_0 \frac{y_2 + jy_0 \tan \beta l}{y_0 + jy_2 \tan \beta l}, \qquad (18.3.1)$$

where the wave length is expressed through $\beta = 2\pi/\lambda$.

In narrow-band amplifiers, the inductances are realized by transmission-line sections short-circuited at one end. In this case $y_2 = \infty$, and for lengths shorter than quarter wavelength, the input admittance is inductive and is given by

$$y_1 = -jy_0 \cot \beta l. \qquad (18.3.2)$$

Connecting a capacitance C in parallel to admittance y_1 at the transmission-line section input, a typical UHF resonant circuit is formed whose resonance may be obtained from the equation:

$$\omega C = y_0 \cot \beta l. \qquad (18.3.3)$$

The unloaded Q factor of the resonant circuit is in most cases much higher than that of a conventional resonant circuit. Coupling may be arranged by tapping, by an inductive loop or by a capacitive probe. When tapping is used, any point of the inner conductor may be connected — it should be remembered that the impedance increases according to the function $\tan \beta l$ when moving away from the short circuit. The inductive-loop coupling is effective at a high-current point, i.e. near to the short-circuit. The coupling value may be calculated approximately from the geometrical dimensions of the loop. The capacitive coupling is effective at a high impedance point, i.e. near the transmission-line input; it may be accomplished either by a capacitor of low value or a capacitive probe.

Figure 18.13 shows the circuit of a common-base narrow-band amplifier with a wideband input coupling. The collector is connected to a tapping of a transmission-line resonant circuit. The circuit is tunable by capacitor C_4 in the UHF band, and another tapping at a substantially lower impedance point is used for output coupling. Elements L_1 and C_1 at the input form a high-pass filter thus preventing signals below 400 MHz from reaching the emitter circuit and producing intermodulation. Capacitor C_1 also provides impedance matching, primarily from the noise aspect, by adjusting the minimum noise figure.

Figure 18.14 shows the circuit of a tunable UHF amplifier with three resonant circuits [18.11]. The emitter circuit of the common-base transistor is tuned out by a transmission-line resonant circuit tunable by adjustable capacitor C_1. Both the input and emitter points are coupled to the trans-

18.3. Narrow-band amplifiers with transmission lines

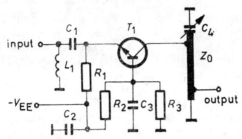

Fig. 18.13. Circuit diagram of narrow-band amplifier tuned by transmission-line section

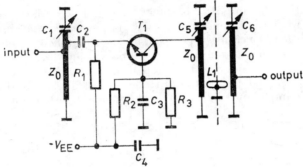

Fig. 18.14. Circuit diagram of tunable UHF amplifier using three transmission-line sections

mission line by tappings. A double-tuned band-pass filter is applied at the output, with the coupling adjusted by loop L_1. The loop is placed at a high-current point near the short-circuit. The coupling coefficient is changed by adjusting the loop surface and the distance from the inner conductor. Rotary condensers $C_1 - C_5 - C_6$ have a common shaft and are thus aligned when tuning is accomplished. Construction is simplified by applying varicap diodes instead of rotary condensers. Assuming identical diodes, the adjustment of the reverse bias would provide simultaneous tuning of the three resonant circuits. With this connection, a power gain 10–40 dB at a bandwidth 15–20 MHz is possible in the frequency range 450–800 MHz. This is proof of the high Q factors attainable using transmission lines.

Several different solutions and applications are known in the field of narrow-band amplifiers utilizing transmission-line resonant circuits. They all originate from lumped element circuits dealt with earlier, by replacing the inductance by a transmission-line section. Stability may be similarly calculated, though its significance is not as high as at lower frequencies, because of the smaller loop gain in the UHF range. Even so, solutions for transistor neutralizations are known. The neutralizing signal is coupled by a loop placed into the output transmission-line section and is fed to the input via a small capacitor. Neutralization may be adjusted by changing the loop positions.

The transmission line is used to advantage in antenna output amplifiers of medium power if the collector and case of the transistor are connected. In a conventional resonant circuit, transistor heat-sinks introduce substantial parallel capacity. However, the transistor may be connected directly to the inner conductor if this is sufficiently thick for efficient heat-sink. The direct metallic contact provides good heat transfer, and no parasitic reactances are formed.

In another group of transmission line amplifiers, the task is not to achieve high-Q resonant circuits but to match the complex transistor input/output impedances to the real generator/load impedances. In these amplifiers which usually provide substantial output power, the bandwidth is of only minor importance.

The circuit of the amplifier is shown in Fig. 18.15. Chokes L_1 and L_2 are present for DC circuit completion. Low-frequency oscillation is prevented by resistance R_1. The transistor impedances are matched to the terminations by correct choice of characteristic impedances Z_{01} and Z_{02} and transmission-line lengths. Let us assume that one end of the transmission line is terminated by admittance y_2 and an input admittance y_1 is to be achieved. The required characteristic impedance $Z_0 = 1/y_0$ and transmission-line length l are given through

$$y_0 = \sqrt{g_1 g_2 - b_1 b_2 + \frac{b_1 - b_2}{g_1 - g_2}(g_1 b_2 - g_2 b_1)} \; ; \qquad (18.3.4)$$

$$\tan \beta l = \frac{y_0(g_1 - g_2)}{g_1 b_2 + g_2 b_1} , \qquad (18.3.5)$$

where the two admittances to be matched are given by $y_1 = g_1 + jb_1$ and $y_2 = g_2 + jb_2$. In the case of real admittances ($b_1 = b_2 = 0$), eqn. (18.3.4) gives the admittance of the well known quarter-wave transformer. The general condition for setting up the match is a positive radical; if otherwise it is impossible to achieve the impedance transformation using a single transmission-line section.

To achieve maximum transistor power gain, there must be a complex impedance match at both input and output, and this requires the unconditional stability of the transistor ($K > 1$). This requirement is normally met in the UHF and microwave frequency range by most transistors, so transmission-line sections can be designed by using eqn. (18.3.4) if the admittances to be matched and the generally real terminations y_g and y_L are known. The transmission-line sections of given characteristic impedance can most conveniently be realized by microstrip techniques as the characteristic impedance of microstrips can be chosen over a wide range. A disadvantage of a transmission-line match, however, is the fact that the adjustments required by the spread of transistor parameters are not possible. For this reason, trimmer capacitors are usually applied at the input or output of the transmission-line sections, providing an accurate impedance match within a narrow frequency range.

At the output of the amplifier stage shown in Fig. 18.16, two trimmer capacitors are used to give an accurate match [18.14]: capacitor C_4 in series

18.3. Narrow-band amplifiers with transmission lines

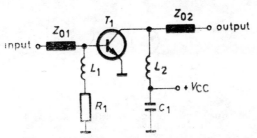

Fig. 18.15. Circuit diagram of common-emitter amplifier matched by transmission-line sections of different characteristic impedances

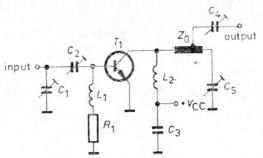

Fig. 18.16. Circuit diagram of common-emitter amplifier with output transmission-line matching and small tuning range

with the transmission line tapping, and capacitor C_5 terminating the transmission line. The inductive input admittance of the transistor is matched to the generator by an L section comprising lumped elements C_1 and C_2. Inductances L_1 and L_2 are high-frequency chokes, and resistor R_1 is present to eliminate the low-frequency oscillation. With this connection, a gain of 6 dB and an output power of 1 W can be obtained at a frequency of 1 GHz.

For the case where the transmission-line characteristic impedance is given either from calculation or from requirements on mechanical layout, it is most convenient to use the Smith diagram to visualize impedance relations. Special attention should be given to the transmission-line section length of $\lambda/8$: by choosing a proper characteristic impedance, a real input impedance can be obtained. If the termination is $Z_2 = R_2 + jX_2$ and a characteristic impedance of

$$Z_0 = \sqrt{R_2^2 + X_2^2}, \qquad (18.3.6)$$

is chosen, then the input impedance is real and has a value of

$$Z = R = \frac{R_2}{1 - X_2/Z_0}. \qquad (18.3.7)$$

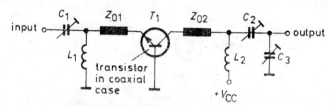

Fig. 18.17. Circuit diagram of microwave amplifier matched by coaxial line sections of different characteristic impedances

Figure 18.17 shows the circuit diagram of an amplifier with coaxial mount [18.14]. Transistor T_1 has a coaxial case, the cylindrical emitter and base contacts are connected directly to the cylindrical inner conductor of the transmission line. Ceramic trimmers C_1, C_2, C_3 are needed for accurate tuning. The high-power transistor is cooled by a ring made of beryllium oxide which has good heat-transfer properties; this is placed at the output end of the coaxial line of characteristic impedance Z_{02}. In class B biasing, the circuit will deliver an output power of 1.2 W with a gain of 6 dB at a frequency of 2 GHz.

In Ref. [18.27], the design procedure of a microwave amplifier operating in the frequency range 2.3–3.5 GHz and using microstrip matching is given. The initial design requirement is the unconditional stability of the transistor in the operating frequency range. Instability at lower frequencies is eliminated by matching pads which present an open circuit to the input and a short-circuit to the output. The input is matched for minimum noise.

In Ref. [18.24], a microwave amplifier operating in the frequency range 1.75–1.95 GHz, also using microstrip matching, and having a gain of 13 dB is presented. A specially arranged power amplifier consisting of 12 GaAs FETs is described in Ref. [18.32]. Another microwave amplifier is considered in Ref. [18.17].

In Refs. [1.40] and [1.41], microwave amplifiers using TRAPATT and IMPATT diodes are presented.

18.4. Wideband amplifiers with transmission lines

Transmission lines may be used conveniently as impedance transformers in wideband amplifiers, especially if their frequency range extends into the UHF region. Wideband amplification is provided by the following procedure: maximum power transfer is adjusted by complex-conjugated impedance matching at the upper limit of the frequency range, and the increase in power gain at lower frequencies is compensated by mismatch. This compensation is often imperfect, resulting in rather high fluctuations in the pass band. The circuit design aims to minimize this power-gain fluctuation.

A single-stage wideband amplifier circuit using a transmission-line section is shown in Fig. 18.18. The design procedure for this circuit manifests the design aspects of multi-stage amplifiers [18.2]. The voltage gain is expressed

18.4. Wideband amplifiers with transmission lines

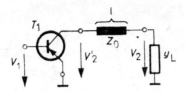

Fig. 18.18. Circuit diagram of single-stage wideband amplifier with transmission-line section

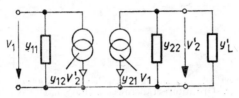

Fig. 18.19. Equivalent circuit of single-stage wideband amplifier with transmission-line section

through the equivalent circuit of Fig. 18.19; the transformed value at the transmission-line input (i.e. at the collector) of the load admittance y_L is denoted by y'_L. Calling the collector voltage V'_2, the voltage gain is

$$A_v = \frac{V_2}{V_1} = \frac{|y_{21}|}{y_{22} + y'_L} \frac{V_2}{V'_2}. \qquad (18.4.1)$$

Voltage V'_2 is calculated using the relation of the potential wave along the transmission line:

$$V(l) = V_2(\cos \beta l + jy_L \sin \beta l/y_0), \qquad (18.4.2)$$

where V_2 is the voltage appearing across load y_L at the distance $l = 0$, and l is the distance measured from the terminated transmission-line end. Calculating the value of voltage V'_2 and substituting it into eqn. (18.4.1), we have

$$|A_v| = \frac{|y_{21}|}{|(y_{22}+y_L)\cos \beta l + j(y_0 + y_{22}y_L/y_0) \sin \beta l|} = \frac{|y_{21}|}{|H|}, \qquad (18.4.3)$$

where the denominator has been denoted by H. In the following, the condition resulting in maximum voltage gain or minimum denominator value H will be determined. The denominator depends on two quantities, the characteristic impedance $Z_0 = 1/y_0$ and length l. The minimum value may be obtained by partial differentiation of function H with respect to both variables. The characteristic impedance and line length resulting in

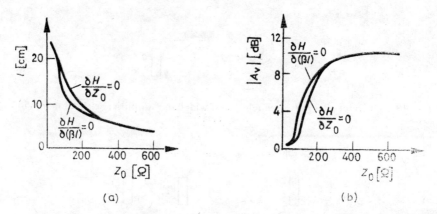

Fig. 18.20. Dependence of single-stage amplifier parameters on the characteristic impedance. (a) Optimum transmission-line length, (b) voltage gain

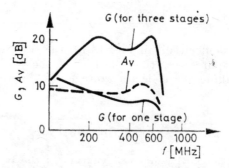

Fig. 18.21. Frequency response of the power gain and voltage gain for a wideband amplifier with transmission-line section, for one stage and three stages

maximum voltage gain may be calculated from the equation

$$\partial H/\partial Z_0 = 0, \qquad (18.4.4)$$

for constant line length l, and from the equation

$$\partial H/\partial(\beta l) = 0, \qquad (18.4.5)$$

for constant characteristic impedance Z_0. The result of this calculation for a specific case is shown in Fig. 18.20 (a). For large characteristic impedances, identical results are obtained by the two methods, and the two curves coincide. However, for small characteristic impedances, the results are different, and the choice between the two methods should be based on Fig. 18.20 (b). This figure shows the dependence of the maximum voltage gain on the characteristic impedance, calculated for the two cases. From eqns. (18.4.4) and (18.4.5), the one yielding higher voltage gain should be chosen.

18.4. Wideband amplifiers with transmission lines

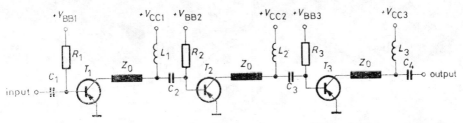

Fig. 18.22. Circuit diagram of three-stage wideband amplifier with transmission-line sections

The calculation should be carried out at the upper cut-off frequency ω_H; at lower frequencies, the voltage gain will be lower because of the higher mismatch.

The resulting voltage gain A_v is shown in Fig. 18.21. This also shows the power gain G which decreases with frequency for a single-stage amplifier. The cut-off of the range below the lower frequency limit can be accomplished by a high-pass filter.

A similar process may be used to design multi-stage amplifiers; however, in this case the measured data will show deviations from the calculated values because of interaction between the stages. Figure 18.22 shows the circuit diagram of a three-stage wideband amplifier. In all stages, the transistor output impedances are matched by transmission lines to the next stage input impedances or the terminating resistance. The high-pass filter is made up from the series capacitances. The supply voltage is fed via resistances and chokes to the low-impedance points of the transmission lines. The amplifier power gain G is shown in Fig. 18.21, according to which the gain is flat within about 3 dB in the frequency range of more than one octave. The transmission-line characteristic impedance is 230 ohm, with a helical inner conductor.

A wideband impedance match can be implemented using the tapered transmission lines shown in Fig. 18.23 (a). The stripline is made up of two conducting planes which have a width b changing exponentially along the line, thus also changing the characteristic impedance, given by:

$$Z_0 = 377 a/b \sqrt{\varepsilon_r} \,. \tag{18.4.6}$$

Here a is the width of the dielectric layer and ε_r is the dielectric constant. Increasing width b will decrease the characteristic impedance, and as shown in Fig. 18.23 (b), the impedance transformation is achieved over a wide frequency range. The length of the exponential transmission line is a quarter of the wavelength corresponding to two-thirds of the upper cut-off frequency. However, a length which is 40% below this value should actually be used, according to experimental results.

Figure 18.24 shows the circuit of a wideband amplifier which comprises an input tapered line [18.29] for matching the low input impedance of the transistor to the 50-ohm generator resistance [18.15]. In order to obtain wideband flat transmission, the transistor gain decreasing with frequency

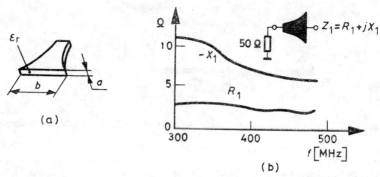

Fig. 18.23. (a) Sketch of an exponential stripline, (b) real and imaginary parts of the input impedance versus frequency for 50-ohm termination

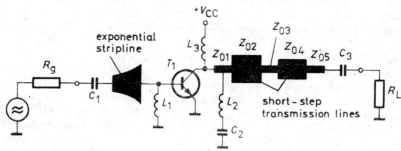

Fig. 18.24. Circuit diagram of a common-emitter wideband amplifier, with exponential stripline matching at the input, and stepped transmission-line matching at the output

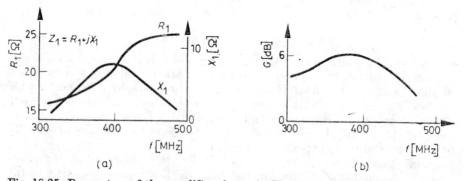

Fig. 18.25. Parameters of the amplifier shown in Fig. 18.24. (a) Input impedance of the five-section transmission line versus frequency for 50-ohm termination, (b) the frequency response of gain

18.4. Wideband amplifiers with transmission lines

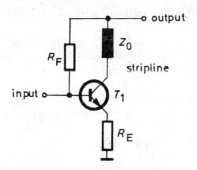

Fig. 18.26. Feedback amplifier with transmission-line compensation

should be compensated by the output matching circuit made up of a five-section short-step transmission line. By correct choice of characteristic impedances of the various sections, a Chebyshev-type response is attained over a wide frequency range [18.8]. Terminating the five-section stepped transmission-line output by a 50-ohm resistor, the input impedance Z_1 as seen from the collector side is shown in Fig. 18.25 (a). The higher resistance R_1 at higher frequencies allows a higher power gain.

The frequency response of the amplifier is shown in Fig. 18.25 (b). The collector capacitance is tuned out by inductance L_3 at about 450 MHz, and capacitances C_1 and C_3 are present for DC decoupling. High-frequency chokes L_1 and L_3 are applied for supply voltage connection. The inductive component of the transistor input impedance is tuned out by the capacitive component of the input tapered transmission-line impedance.

In Ref. [11.11], a feedback amplifier with transmission-line compensation is investigated. The circuit diagram of this amplifier is shown in Fig. 18.26. Because of the delay of the transmission line of characteristic impedance Z_0, feedback is decreased at high frequencies, and thus a peaking at the upper cut-off frequency may be produced, depending on the values of Z_0 and the line length. With this method, a bandwidth of 0–1.0 GHz has been achieved by a two-stage symmetrical amplifier using $f_T = 4$ GHz transistors.

The broadband microwave amplifier, described in Ref. [18.34], uses transmission-line impedances in the source lead of GaAs FET as negativ feedback.

In Ref. [18.30], an amplifier for the frequency range of 2–6.5 GHz is considered; this amplifier makes use of a thin-film circuit optimized by computer. The optimization was carried out by simultaneously accounting for the gain and the voltage standing-wave ratio (VSWR) by use of a weighting function. Small-signal s parameters were used for the design; these assure sufficiently good approximation up to the signal level corresponding to a 1-dB gain compression. In Refs. [18.28], [18.31], [18.33], [18.35] and [23.22], further wideband microwave amplifiers employing transmission-line matching are dealt with.

18.5. Wideband amplifiers with directional couplers

In the microwave region, wideband amplifiers may be designed by the use of directional couplers. The directional couplers have symmetrical (balance) properties which offer a number of advantages over amplifiers previously considered. These include low input and output reflection factors, linear phase characteristics and low intermodulation distortions. No impedance transformation is effected by the directional coupler, so the input and output impedances of the applied transistors should not deviate substantially from the characteristic impedance Z_0 of the directional coupler. If this condition is not met, the transistor impedance should be modified by lumped circuit elements.

Figure 18.27 (a) shows the equivalent circuit of the 3-dB coupler (hybrid). The transmission line of length $\lambda/4$ between points *1* and *4* is coupled to a similar transmission line between points *2* and *3*. Figure 18.27 (b) shows details of a microstrip design of a 3-dB coupler, i.e. the top view of the inner conductor and the cross-sectional view of the coupled section [18.4]. The output connections serve as capacitive compensations of the transmission lines.

The input signal arriving at port *1* is equally divided to ports *2* and *4*, with 3-dB attenuation of the level and 90° phase difference, provided that these ports are terminated by the characteristic impedance to eliminate reflections. In this case, no power is transmitted to resistance Z_0 terminating port *3*. If the terminations at ports *2* and *4* are not free of reflection, but their deviations from the characteristic impedance are approximately the same, then the reflection at input port *1* is not increased substantially, although a power loss is introduced at port *3*. In other words, the input of the directional coupler is highly insensitive to the terminations of the output

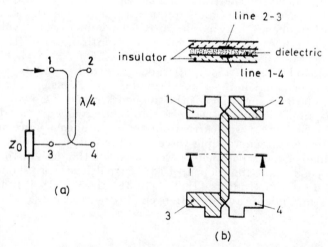

Fig. 18.27. 3-dB coupler. (a) Equivalent circuit, (b) practical layout

18.5. Directional couplers

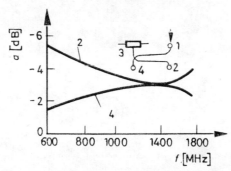

Fig. 18.28. Transmission loss at the two output ports of a 3-dB coupler versus frequency

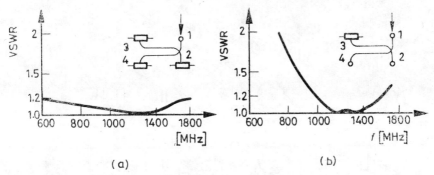

Fig. 18.29. Input VSWR of 3-dB coupler versus frequency. (a) Output ports terminated by characteristic impedance, (b) output ports open-circuited

ports. A further advantage of the directional coupler is the relatively wide transmission band.

Figure 18.28 shows the ratios of powers at output ports *2* and *4* to the input power at port *1*, as a function of frequency. At the band centre, the attenuation for both outputs is 3 dB, and this value changes rather slowly with frequency. Figure 18.29 shows the VSWR measured at port *1* as a function of frequency. If ports *2* and *4* are terminated by the characteristic impedance then the input VSWR is less than 1.2 over a wide frequency range (see Fig. 18.29(a)). Figure 18.29 (b) shows the input VSWR with output ports *2* and *4* open-circuited. It is seen that in a small frequency range, this VSWR is low even in this case.

Figure 18.30 shows the equivalent circuit of a symmetrical (balance) amplifier where directional couplers are used. The input power is equally divided to the two transistor base electrodes by the input directional coupler, with 90° phase difference. The output directional coupler operates in an inverse mode, i.e. the two collector signals are added, and no output power reaches the terminating resistance Z_0.

18. UHF and microwave amplifiers

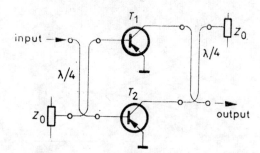

Fig. 18.30. Equivalent circuit of symmetrical wideband amplifier with directional couplers

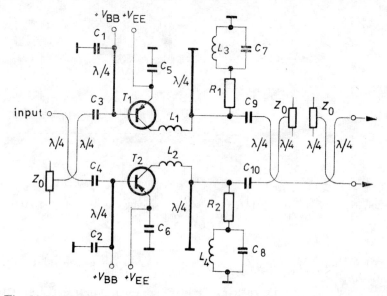

Fig. 18.31. Detailed circuit diagram of the amplifier shown in Fig. 18.30

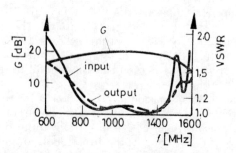

Fig. 18.32. Power gain, input and output VSWR of wideband microwave amplifier shown in Fig. 18.31

18.5. Directional couplers

Figure 18.31 shows a detailed circuit diagram of this amplifier. The microwave transistors applied have input impedances of approximately 50 ohms, so at the input side, the base electrodes are connected directly to the two outputs of the directional coupler. The input impedance may differ slightly from the characteristic impedance, but it is important that the two transistors should be selected by measurement, to be nearly identical. At the output side the situation is less favourable inasmuch as the output impedance differs considerably from the characteristic impedance of the directional coupler. By application of the lumped inductances L_1 and L_2 the characteristic impedance is approximated and thus the reflection due to the output directional coupler is rendered acceptable. Inductances L_1 and L_2 are produced in thin-film technology directly on the special transistor cases. Here identical output impedances are also prime objectives, so the transistors should be selected for this parameter.

Further compensating circuit elements are needed to guarantee flat transmission. These are R_1, L_3, C_7, and their symmetrical counterparts. The resonant frequency of the parallel tuned circuit is approximately at the upper cut-off frequency, so the loading effect of the compensating circuit may be neglected at this frequency. At lower frequencies, the transistor output is shunted by resistance R_1, thus compensating for the increasing transistor gain. By correct choice of compensating circuit elements, flat transmission over a wide frequency range may be achieved. DC connections are also realized via transmission lines. By applying a high-frequency short-circuit at the supply-voltage end and at the grounded end of the 150-ohm quarter-wave transmission lines, the open-circuit appearing at the other end does not load the circuit. This makes the connection of supply voltage V_{BB} and the DC grounding of the collectors possible. A further directional coupler is applied at the output for distributing the signal at the input of the next symmetrical stage.

Figure 18.32 shows the frequency response, and the input and output VSWR of a four-stage amplifier made up of stages as shown in Fig. 18.31. The gain is 18 $\pm$ 2 dB in the frequency range 600–1600 MHz, and the VSWR is below 1.8 over the whole range.

In Ref. [18.22] an amplifier is presented for the frequency range 1–2 GHz, using 3-dB hybrids and common-base transistors. The series LC network tuning the collector circuit is built into the stripline case in order to place it near the collector. Design was carried out with s parameters in several steps by iteration.

In Ref. [18.25] a 1.5 GHz power amplifier is presented, and an amplifier for the frequency range 4–8 GHz using GaAs MESFET's is considered in Ref. [18.23]. Both amplifiers use 3-dB hybrids as coupling elements.

19. HIGH-FREQUENCY POWER AMPLIFIERS. THEORY AND DESIGN PRINCIPLES

19.1. Factors limiting high-level operation

Several factors are responsible for high-level operation: those importance in a given circuit are determined by both the solid-state device itself and the circuit in which the device is operating. One group of limiting factors is presented in Fig. 19.1, showing the output characteristic of a bipolar transistor. In the following, the characteristics given in this figure will be considered and their effect on high-frequency operation will be investigated.

When the transistor is turned on, the collector–base voltage does not decrease to zero, but just to the saturation voltage V_{sat}. This residual voltage which cannot be used, depends primarily on the collector series resistance $r_{cc'}$, determined by the construction of the transistor, and obviously by the collector current at which saturation occurs. For the static case, the following approximate relation is valid:

$$V_{sat} \approx I_c r_{cc'}(I_c), \qquad (19.1.1)$$

where the resistance $r_{cc'}$ is itself dependent on the collector current. The situation is more complicated at high frequencies, since the collector current I_c in this expression is not only determined by the DC characteristic but is incremented by the current needed to discharge the collector–base capacitance. It will be shown later that in high-frequency tuned amplifiers it is not possible to define the dynamic load-line, and thus the peak collector current and saturation voltage are difficult to assess.

In computations which rely heavily on approximations, it is usual to use the static value of the saturation voltage which is easy to measure. The actual high-frequency saturation voltage definitely exceeds this value. Another method is based on the fact that the power gain of the transistor depends on the working point. From measurement of the power gain for different collector currents and collector voltages, and by connecting the identical values, we obtain the characteristic shown in Fig. 19.2. The variation of the small-signal power gain during the high-level operation is given by the intersections of the load-line with the constant-power-gain curves. The driving limit falls approximately to the point where the power gain is decreased to zero, so the point of intersection denoted by A approximately determines the saturation voltage. Finally, we should note that the following method may be used to reduce the saturation voltage limit of narrow-band tuned amplifiers. A parallel resonant circuit tuned to the third har-

19.1. Limiting factors

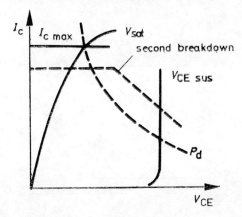

Fig. 19.1. Limiting factors of high-frequency operation

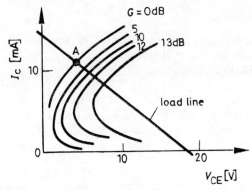

Fig. 19.2. Plot of the equal power gains on the output characteristic

monic is connected in series with the resonant circuit tuned to the operating frequency. In this case, a dip is produced at the peak of the collector voltage as a result of the two combined harmonics. Although it has a lower peak value, this waveform comprises an operating frequency component of higher amplitude, which allows better use of the available voltage range.

Another limiting factor in high-level operation is the *maximum collector voltage*, or more precisely the breakdown voltage shown in Fig. 19.3. The highest breakdown voltage may be attained on the collector-base diode with an open circuited emitter lead (V_{CB0}). The breakdown voltage is much lower in the common-emitter configuration, and depends on whether the base electrode is open-circuited (V_{CE0}), or terminated by a resistance (V_{CER}) or voltage source (V_{CEV}). All three voltage curves end up at the sustaining voltage called $V_{CE\,sus}$ which may be calculated approximately from the following relation:

$$V_{CE\,sus} = \frac{V_{CB0}}{\sqrt[n]{1 + h_{21e}}}, \qquad (19.1.2)$$

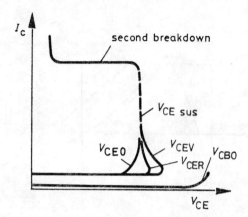

Fig. 19.3. Breakdown voltages of a bipolar transistor

where h_{21e} is the current gain factor and n is a material constant with a value falling in the range $n = 3\text{–}7$. The current gain factor is frequency-dependent, so the breakdown voltage will also be frequency-dependent according to eqn. (19.1.1). This may actually be observed, but the relation is more unambiguous because of the effect of other factors. The maximum collector voltage is exceeded mainly by *reflections* due to incorrect terminations [20.17].

Another phenomenon called *second breakdown* occurs at high collector currents as a result of a complex electrothermic process. During this process, the collector current is concentrated to a point where avalanche-breakdown begins, the high current density gives rise to locally heated "hot spot" [19.34] which could destroy the device. The voltage resulting from the second breakdown — as shown in Fig. 19.3 — is much smaller than the breakdown voltage $V_{CE\,sus}$ and the characteristic is not necessarily reversible. This means that in the best case, the device may be brought back into normal operating conditions after the second breakdown. This is only possible if the second breakdown does not last long enough to destroy the device. This has high-frequency significance, as the critical time interval, i.e. the simultaneous presence of high voltage and high collector current is extremely short, there is no time for completion of the electrothermic process, and the normal operating condition of the device is restored after the load is removed. The high-frequency high-power circuits are thus less susceptible to the second breakdown than are low-frequency circuits.

If the critical time interval is long enough to allow the completion of the current-concentration process, then the device is destroyed by the heating effect of the conducting path. Thus in this case the characteristic is not reversible, and this makes the investigation and measurement of the phenomenon more difficult.

The current concentration may be prevented by dividing the transistors into several elementary transistors, each having a small emitter resistance

19.1. Limiting factors

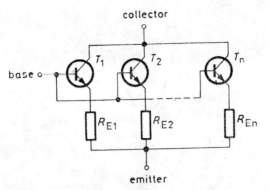

Fig. 19.4. Parallel connection of elementary transistors with separate emitter-feedback resistances for elimination of current crowding

R_E to introduce current feedback (see Fig. 19.4). This prevents the current concentration to a single elementary transistor, thus considerably reducing the danger of a second breakdown. In transistors protected against second breakdown, the low-valued metal-film resistances are contained within the same transistor package, and the elementary transistors are fabricated in monolithic form on a single substrate. Resistance R_E should be kept low to minimize power dissipation, and only high enough to prevent second breakdown in a certain output characteristic range (within the safe operating region).

The high-level operation is theoretically limited by the *maximum collector current* $I_{c\,max}$. In practical high-frequency applications, the saturation voltage limits the maximum current, and not the permissible peak current for the device.

The *dissipation power* is a basic limiting factor in high-power operation. As mentioned in previous chapters, good heat conduction is difficult to achieve for high-frequency transistors as the heat-sink must not introduce substantial parasitic elements (e.g. stray capacitances). In addition, the *transient temperature peak* [19.31], [19.33] is also important. However, the power-handling capability itself is not characteristic as the efficiency of the circuit is also important. For good efficiency (e.g. class C), much higher output power may be attained for a given dissipation power provided other limiting factors allow this.

The factors limiting the high-level operation were shown in Fig. 19.1. Operating conditions and parameters in a given circuit will determine which of the factors shown should be taken into account. General statements cannot be made as conditions may change over a wide range, as is shown by the following examples. For a relatively low supply voltage, the saturation voltage and the collector peak current will obviously be the limiting factors. For high supply voltages and especially for class C operation, the breakdown phenomena will be critical. Finally, for class A amplifiers, the dissipation hyperbola will be important as a result of the poor efficiency, which is not so critical in class C operation.

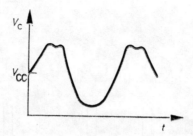

Fig. 19.5. Snap-back effect due to the collector–base capacitance

High-level operation is also limited by distortions introduced by nonlinear device properties such as production of harmonics, cross-modulation and intermodulation. If the requirements of these parameters are stringent, the transistor performance may not be used even though the limiting factors shown in Fig. 19.1 would allow to use. We shall return later to the different kinds of nonlinear distortions.

A typical waveform showing the so-called *snap-back effect* is illustrated in Fig. 19.5. Turning off the transistor the collector voltage will increase, which will have an effect through the collector–base capacitance C_c, modifying the base voltage. This effect may be observed when the emitter–base diode is off, i.e. it presents a high impedance. The base electrode is again driven slightly into conduction by the collector voltage, producing the "snap-back" on the collector voltage waveform. The effect obviously depends on the collector–base capacitance value and may be reduced by decreasing the impedance of the source driving the base electrode, thereby also decreasing the stage gain.

19.2. Instability of power amplifiers

Stability problems are worsened by nonlinear operation of high-power amplifiers. Stability problems of linear (small-signal) amplifiers and the stability factor have been treated extensively earlier. In these calculations, linear operation was always assumed, i.e. the possibility of exceeding the region of linear operation has been excluded. The situation is basically different for high-power amplifiers which normally operate in the nonlinear region. The high-power amplifier may therefore be considered theoretically to have two parts, a linear class A amplifier and a nonlinear (e.g. class C) amplifier. This theoretical aspect reveals the important role of the stability of the linear part which is a characteristic of a fictive, nonseparable circuit part. These considerations allow the following grouping of instabilities [19.10]:

(1) *Instabilities of class A linear amplifier:*
 — Instability due to internal feedback,
 — Instability due to external feedback (e.g. insufficient filtering),
 — Instability due to thermal feedback,

19.2. Instability

— Instability caused by negative resistance (transit-time effect, avalanche-breakdown phenomena).

(2) *Instabilities of class C nonlinear amplifier:*
— Parametric generation of harmonics,
— Parametric generation of subharmonics.

Of the linear instabilities, only those due to external feedbacks will be dealt with. As the operation of power amplifiers involves high collector currents, the mutual conductance of the transistor is extremely high, rendering the circuit sensitive to external positive feedback. In the circuit shown in Fig. 19.6 (a) which is typical for tuned power amplifiers, correct choice of filter capacities C_4 and C_5 provides sufficient supply voltage filtering both in the operating frequency range and at substantially lower frequencies: thus harmful feedback via the supply-voltage lead is eliminated.

In tuned power amplifiers, the danger of instability arises at frequencies which are well below the operating frequency range, since the transistor gain at these frequencies is high. Also, impedances likely to allow instability may be present because of the reactive filters and RF coils. Let us again consider Fig. 19.6 (a), which at both input and output shows matching networks tuned to the operating frequency range, and also shows the DC connections provided via high-frequency chokes. At frequencies well below the operating frequency range, the high-frequency chokes represent low-loss inductances, providing the actual termination of the transistor base and collector, as the external terminations are blocked by the series capacitances C_2 and C_6. This leads to instability in the case of high mutual conductance, and to avoid this, a series LR network has to be included at both the base and the collector side ($R_1 L_3$, $R_2 L_4$) to provide appropriate loading at low frequencies.

Resistive loads have the effect of increasing stability, because they reduce the loop gain. This means that the condition for stable operation is not the choice of a low inductance L_3, which might even worsen the situation, but the insertion of a suitable series loss which has no significance at the operating frequency but provides stability at the critical lower frequencies.

How does the stability depend on the values of inductances parallel with the base and collector, if the loss of these elements may be neglected? The

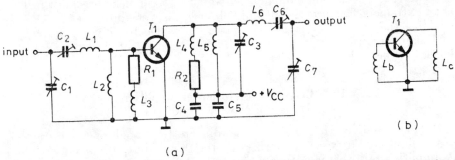

Fig. 19.6. Circuit of tuned power amplifier. (a) Practical circuits with lossy chokes, (b) low-frequency equivalent circuit

situation is investigated in Fig. 19.6 (b). Further circuit elements are not shown because of the low-frequency blocking of the series capacitances. Considering the feedback capacity C_c between collector and base, we have a *Hartley*-type oscillator, for which the condition of stability is:

$$L_b/L_c > h_{21e} C_c/C_e, \qquad (19.2.1)$$

where h_{21e} is the common-emitter current-gain factor and C_e is the emitter capacitance. It may be seen that the stability can be increased by a decrease of L_c. This approximate stability relation is frequently applied in the design of bias circuits of power amplifiers. It should be noted that the tendency for low-frequency oscillation to occur may go unobserved since this appears as a modulation of the operating frequency or may only be present during drive by the operating frequency signal, and cannot be observed in the absence of drive.

Finally, we present a short survey on linear instabilities in connection with avalanche effects. As shown in Section 1.6, certain diodes present a negative resistance during the avalanche effect occurring at high reverse voltage, and are thus suitable for generating oscillations. A similar phenomenon occurs in the collector–base junction of power transistors during the avalanche breakdown. This kind of parasitic oscillation occurring at typically high reverse voltages has been observed for a long time but the essence of the phenomenon is not yet clear, despite publication of theoretical explanations and calculations. This problem is connected with the other group of instabilities, the nonlinear instabilities characterized by the parametric generation of oscillations.

Parametric oscillations result from the fact that when a time-varying impedance (e.g. the voltage-dependent capacitance of a collector–base junction) is driven by a signal of frequency f_0, a number of signals with frequencies different from f_0 are generated. (This is the operating principle of parametric amplifiers.) If the variation of impedance with time is also affected by the signal of frequency f_0, then harmonic and subharmonic signals may be generated, depending on the impedance characteristic. These parasitic oscillations can often be observed on an oscilloscope. The situation is more dangerous if the device is destroyed by the increase in reverse voltage, since the parametric oscillations due to the voltage dependence of the collector-base diode bring about an avalanche breakdown. This is an extremely fast process, making observation very difficult [19.7]. This explains the trend to increase the maximum collector reverse voltage of high-power transistors, in order to minimize the risk of destructive instabilities in the vicinity of avalanche breakdown.

19.3. Classification of power amplifiers. Design of class A amplifiers

Power amplifiers are classified according to flow angle, i.e. the fraction of the sinusoidal drive characterized by the flow of output current. In class A amplifiers, transistor output current flows during the whole period, i.e. the flow angle is $2\alpha = 360°$. In class B amplifiers, current flow is present in alternate half periods, i.e. the flow angle is $2\alpha = 180°$. Class C amplifiers are characterized by a current flow of less than a half period. These are frequently found in high-frequency tuned amplifiers, because the distortion of the output current is eliminated by the selective load impedance, filtering out only the fundamental component.

The design of different types of power amplifiers varies significantly, warranting separate treatment. Class A amplifiers will be dealt with first. Of the design objectives, the realization of maximum output power is generally selected, with the indirect task of ensuring small nonlinear distortion at lower output levels. These two objectives are rarely contradictory, i.e. higher output power results in a more linear operation at smaller power levels. This, however, is not valid for small drive levels of class B push-pull amplifiers.

Another design objective may be maximum power gain. This has been amply treated in earlier chapters, so we need only to state that a design of this type will probably result in an amplifier with insufficient available output power.

The process is based on a load line corresponding to the load resistance chosen to obtain the circuit data. For high-frequency operation this will not correspond to the true situation, since the instantaneous output voltage and the instantaneous working point relating to the output current, will not move along the load line, but along a curve which is difficult to define (along an ellipse in the simplest case). This method of calculation will be dealt with later. The dynamic load line is thus an approximation used to carry out an approximate circuit design based on relatively simple relations. An acceptable approximation is expedient even if some circuit aspects, e.g. voltage– and current–time functions, are not revealed. Should the approximation lead to worthless results, as shown by control measurements, then a more profound investigation of the circuit is necessary, although this may not be possible by elementary methods. We shall demonstrate this in a specific example later.

We present here approximate design methods for class A high-frequency power amplifiers. Several methods, which depend on the requirements and transistor characteristics, are known. The difference between these methods is the limiting factor, as given in Section 19.1, which prevails under given circumstances, while others have no particular role. A few examples can illustrate this point. At low supply voltage, the peak current or saturation voltage will have a critical effect on optimum circuit adjustment, and the breakdown effect will not be significant. The reverse is true for high supply voltage, the saturation voltage being negligible compared with the high voltage amplitude, but this will enhance the breakdown effect.

In the following, basic design methods for single-ended class A power amplifiers will be surveyed, using the load line of approximate validity.

(a) *Design for maximum output power, for a given supply voltage V_{cc}^{-} and for a saturation resistance r_{sat} considered constant*

We shall assume that neither the permitted collector peak current nor the breakdown voltage are exceeded. Parameters are illustrated in Fig.

19.7 (a). The following relations will be used to calculate the more important circuit data such as load resistance R_L, output power P_{out}, quiescent collector current I_0, efficiency η and maximum voltage $V_{CE\,max}$ appearing at the collector:

$$R_L = 2r_{sat};$$
$$P_{out} = V_{CC}^2/16r_{sat};$$
$$I_0 = V_{CC}/4r_{sat}; \qquad (19.3.1)$$
$$\eta = 25\%;$$
$$V_{CE\,max} = 1.5 V_{CC}.$$

Saturation resistance r_{sat} may be determined from the voltage measurement at the maximum collector current. This design method may be used primarily at low supply voltages.

(b) *Design for maximum output power, for a given supply voltage V_{CC}, maximum collector current $I_{c\,max}$ and for a saturation resistance r_{sat} considered constant*

This design method is based on the requirement that the maximum collector current during operation may not exceed the peak current permitted for the transistor. Parameters are illustrated in Fig. 19.7 (b). The more important circuit parameters are given by the relations:

$$R_L = 2V_{CC}/I_{c\,max} - 2r_{sat};$$
$$P_{out} = (V_{CC} - V_{sat})I_{c\,max}/4;$$
$$I_0 = I_{c\,max}/2; \qquad (19.3.2)$$
$$\eta = (V_{CC} - V_{sat})/2V_{CC};$$
$$V_{CE\,max} = 2V_{CC} - V_{sat}.$$

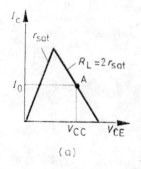
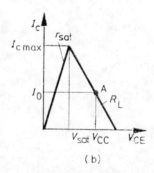
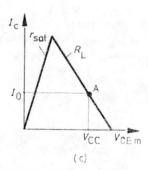

Fig. 19.7. Design of class A amplifiers

19.3. Class A amplifiers

The saturation voltage V_{sat} may be calculated from the saturation resistance at current $I_{c\,max}$. It has been assumed that the maximum collector voltage thus determined is less than the breakdown voltage specified for the transistor.

(c) *Design for maximum output power, for a given supply voltage V_{CC}, maximum collector voltage V_{CEm} and for a saturation resistance r_{sat} considered constant*

This design method is frequently applied in the case of high supply voltages, since it is based on the fact that the highest collector voltage during operation will be just equal to the voltage V_{CEm} assumed with a safety margin. This highest voltage may correspond to the normal breakdown, but can also take into account the second breakdown. In the latter case, it is often given as a series of plots in data sheets, including the effect of frequency and obviously the collector current. The collector current should not be determined by operational conditions alone; in the case of detuning or unloaded output (total reflection), the assumed load-line may not be valid, and a second breakdown may take place due to the high collector peak current and the simultaneous high reverse voltage.

The design process is illustrated by Fig. 19.7 (c). The more important circuit parameters are given by the relations:

$$R_L = 2r_{sat}(V_{CEm} - V_{CC})/(2V_{CC} - V_{CEm});$$
$$P_{out} = (V_{CEm} - V_{CC})(2V_{CC} - V_{CEm})/4r_{sat};$$
$$I_0 = (2V_{CC} - V_{CEm})/2r_{sat}; \qquad (19.3.3)$$
$$\eta = (V_{CEm} - V_{CC})/2V_{CC};$$
$$V_{CE\,max} = V_{CEm}.$$

(d) *Design for maximum output power, for a maximum allowable collector voltage V_{CEm} and for a saturation resistance r_{sat} considered constant*

This design process should determine the optimum supply voltage V_{CC} at which the highest collector voltage during operation equals the allowable voltage V_{CEm}. According to Fig. 19.7 (a), this condition is met by the supply voltage of

$$V_{CC} = 2V_{CEm}/3. \qquad (19.3.4)$$

The other circuit parameters are given by relations (19.3.1), by substituting in the above supply voltage.

(e) *Design for maximum output power, for a given supply voltage V_{CC} and for a saturation resistance r_{sat} increasing with collector current*

At high collector currents, the saturation resistance r_{sat} is not constant, but the saturation voltage increases with collector current as shown by Fig. 19.8. For the calculation, the curve $I_{c\,max}(V_{sat})$ will be approximated by the following function:

$$I_{c\,max} = I_{sat}(1 - e^{-V_{sat}/V_0}), \qquad (19.3.5)$$

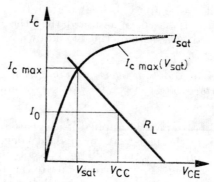

Fig. 19.8. The dependence of saturation voltage on collector current

where the saturation current I_{sat} and the voltage V_0 are constants and determine the shape of the characteristic. The maximum output power according to Fig. 19.8 is given by

$$P_{out} = (V_{CC} - V_{sat})I_{c\,max}/8 = (V_{CC} - V_{sat})I_{sat}(1 - e^{-V_{sat}/V_0})/8. \qquad (19.3.6)$$

The output power is a function of the saturation voltage V_{sat} and is expected to have a maximum value at the optimum saturation voltage. Calculating the extremum value from the equation

$$\partial P_{out}/\partial V_{sat} = 0, \qquad (19.3.7)$$

the characteristic shown in Fig. 19.9 is determined. This gives the optimum saturation voltage V^*_{sat} pertaining to various V_{CC}/V_0 values, normalized to the voltage V_0. Determining from this characteristic the value V^*_{sat}, the maximum collector current $I_{c\,max}$ may be calculated from relation (19.3.5). Similarly, the output power may be calculated from (19.3.6), and the slope of the load line gives the load resistance:

$$R_L = 2(V_{CC} - V^*_{sat})/I_{c\,max}. \qquad (19.3.8)$$

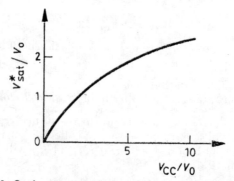

Fig. 19.9. Optimum saturation voltage versus supply voltage

19.4. Design of class B and class AB amplifiers

In single-ended class B power amplifiers the flow angle of the collector current is just $2\alpha = 180°$, i.e. the transistors only conducts in one half period. In Fig. 19.10, the current and voltage relations are indicated in connection with the output characteristic. The half-sinusoidal current pulses of peak value $I_{c\,max}$ generate an alternating voltage, having a peak value approximately equal to the supply voltage. The operation of high-frequency tuned amplifiers is substantially different from the operation of low-frequency amplifiers terminated by a resistance. This is because a reverse voltage much higher than the supply voltage may appear at the collector which is significant from the viewpoint of breakdown.

Let us restrict our investigations to the conditions shown in Fig. 19.10, applicable for class B amplifiers. Here the transistor is loaded by a selective impedance having a value of R_L at the fundamental frequency and a short-circuit at harmonics. The amplitude of the fundamental collector AC voltage component is determined by multiplying the amplitude of the fundamental component in the half sinusoidal current pulses by R_L. From Fourier analysis, this latter amplitude is given by

$$I_{c1} = I_{c\,max}/2, \tag{19.4.1}$$

i.e. half the peak current. The amplitude of the AC voltage at the collector is thus given by

$$V_1 = I_{c\,max} R_L/2. \tag{19.4.2}$$

This explains an important feature of high-frequency tuned power amplifiers: for flow angles of less than 360°, the dynamic load line is not determined by the load resistance. In Fig. 19.10, the load line of slope $R_L/2$ indicates the related instantaneous conditions, but only during the conducting phase. During the turned-off phase, the working point is shifted along

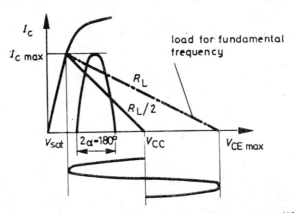

Fig. 19.10. Current–voltage relations of class B power amplifiers

the voltage axis ($I_c = 0$). The fictive load line indicated by a dashed line in Fig. 19.10, represents those states which would result from the actual collector voltage and the fundamental current component.

The situation is similar in the case of class C amplifiers. A smaller flow angle results in a smaller ratio of fundamental component to the peak current $I_{c\,max}$, so the slope of the actual load line will be higher than the slope of the fictive load line, related to the fundamental frequency component and computed from the load resistance.

In the design process of class AB amplifiers, the first step is the determination of the collector current–time function (flow angle). In the above, a fraction of a sinusoidal voltage was assumed (a half sine wave in class B operation), but other current waveforms may also be possible (we shall see examples presently). The amplitude of the fundamental frequency component or, in the case of frequency multipliers, of the harmonic frequency component, should be determined from the current pulses by Fourier analysis. The collector voltage–time function is determined by multiplying the amplitudes of the harmonics, calculated by Fourier analysis, by the impedance values at the appropriate frequencies, and by adding these voltages in correct phase. This is not in fact too complicated as the load impedance is highly selective, showing a finite impedance at a single frequency only (at the fundamental frequency or, in case of multipliers, at a given harmonic frequency), and represents a short-circuit at other frequencies, thus eliminating voltage components at these frequencies.

Let us now return to the waveform of the collector-current pulse. In the general case, the pulse sequence can be given by the sum of a series given by

$$f(t) = \frac{a_0}{2} + \sum_{n=1}^{\infty} A_n \sin(n\omega t + \varphi_n), \qquad (19.4.3)$$

where the DC component is given by

$$a_0 = \frac{1}{\pi} \int_0^{2\pi} f(t)\,d\omega t, \qquad (19.4.4)$$

and the harmonic coefficients are given by

$$a_n = \frac{1}{\pi} \int_0^{2\pi} f(t) \cos n\omega t\,d\omega t; \qquad (19.4.5)$$

$$b_n = \frac{1}{\pi} \int_0^{2\pi} f(t) \sin n\omega t\,d\omega t; \qquad (19.4.6)$$

$$A_n = \sqrt{a_n^2 + b_n^2}, \qquad (19.4.7)$$

the phase angle being

$$\tan \varphi_n = a_n/b_n. \qquad (19.4.8)$$

19.4. Class B and class AB amplifiers

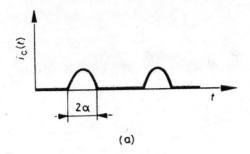

(a)

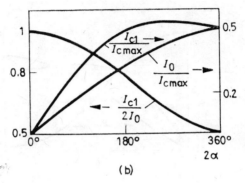

(b)

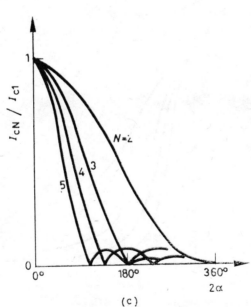

(c)

Fig. 19.11. Current components in the case of truncated sine-wave. (a) Definition of flow angle, (b) DC and fundamental frequency component versus flow angle, (c) harmonics versus flow angle

19. Power amplifiers. Theory

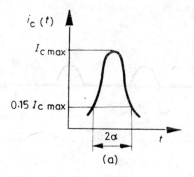

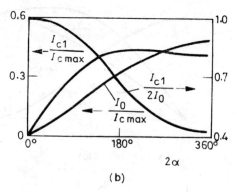

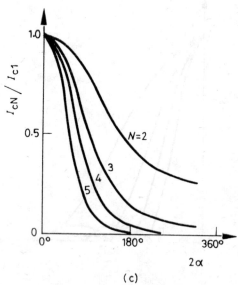

Fig. 19.12. Current components in the case of *expsin* waveform. (a) Definition of the flow angle, (b) DC and fundamental frequency component versus flow angle, (c) harmonics versus flow angle

19.4. Class B and class AB amplifiers

The more important parameters for a truncated sine-wave are plotted in Fig. 19.11. The flow angle is illustrated in Fig. (a) and in Fig. (b), the DC component I_0 and the fundamental frequency component I_{c1} related to the peak current $I_{c\,\text{max}}$, and the quantity $I_{c1}/2I_0$, characterizing the efficiency, are plotted as a function of flow angle 2α. Figure (c) shows the values of the harmonics I_{cN} for $N = 2\text{–}5$, related to the fundamental frequency component, again as a function of flow angle; these are useful for calculating the nonlinear distortion.

The collector-current-pulse waveform, approximated by a truncated sine-wave, is not advantageous since the collector current is an exponential function of the emitter–base voltage:

$$i_c(t) = i_0\, e^{qv_{eb}/kT}. \tag{19.4.9}$$

If the emitter–base voltage has a sinusoidal waveform, i.e. $v_{eb} = V_0 + V_1 \sin \omega t$, then the collector current time function is given by

$$i_c(t) = i_s\, e^{qV_0/kT}\, e^{qV_1 \sin \omega t/kT}, \tag{19.4.10}$$

and this is fairly different from the waveform of a truncated sine-wave. This is why the *expsin* function, defined by eqn. (19.4.10), is also used to approximate the collector current. Series expansion of this function results in the following expression:

$$i_c(t) = i_s\, e^{qV_0/kT} \left[J_0\!\left(\frac{qV_1}{kT}\right) + 2 \sum_{n=1}^{\infty} J_n\!\left(\frac{qV_1}{kT}\right) \cos n\omega t \right], \tag{19.4.11}$$

where J_n is a modified Bessel function of nth order, having an argument which is a normalized value of the driving voltage amplitude V_1. The values of the Bessel function are tabulated. The *expsin*-function has no zero point, and so the flow angle 2α cannot be defined by zero points. Let us define, the flow angle to be taken as the time interval between points $0.15 I_{c\,\text{max}}$ (Fig. 19.12), taking into account the similarity with the truncated sine-wave. The flow angle is thus given by

$$\alpha_{\text{expsin}} = \arccos\left(1 - \frac{kT}{qV_1} \ln \frac{1}{0.15}\right). \tag{19.4.12}$$

The DC component I_0 and the fundamental frequency component I_{c1} of this function, related to the peak current $I_{c\,\text{max}}$, and the quantity $I_{c1}/2I_0$, are plotted in Fig. 19.12 (b) as a function of flow angle. Figure (c) shows the values of the harmonics I_{cN} for $N = 2\text{–}5$, related to the fundamental frequency component, again as a function of flow angle.

Let us now express the output power and the efficiency, using Fig. 19.10. The DC power consumption is the product of the supply voltage and the DC current I_0:

$$P_{\text{DC}} = V_{CC} I_0, \tag{19.4.13}$$

and the output power is proportional with the fundamental frequency component I_{c1}:

$$P_{\text{out}} = (V_{CC} - V_{\text{sat}}) I_{c1}/2. \tag{19.4.14}$$

The efficiency follows from these two expressions:

$$\eta = \frac{V_{cc} - V_{sat}}{V_{cc}} \frac{I_{c1}}{2I_0} = \eta_v \eta_i, \qquad (19.4.15)$$

which is the product of the voltage efficiency, which depends on the saturation voltage, and the current efficiency which depends on the shape of the collector-pulse shape (see Figs. 19.11(b) and 19.12(b)). The load resistance is given by

$$R_L = (V_{cc} - V_{sat})/I_{c1}, \qquad (19.4.16)$$

and the maximum collector voltage is given by

$$V_{CE\,max} = 2V_{cc} - V_{sat}.$$

The design process of class B power amplifiers has the following fundamental variants.

(a) *Design for maximum output power, with an allowable highest collector voltage V_{CEm} and collector peak current $I_{c\,max}$.*

The first step is to determine the saturation voltage V_{sat} pertaining to the given collector peak current $I_{c\,max}$, using the saturation characteristic, or the saturation resistance r_{sat} assumed constant. The more important parameters of the class B amplifier are then:

$$\begin{aligned} V_{cc} &= (V_{CEm} + V_{sat})/2; \\ I_{c1} &= I_{c\,max}/2; \\ R_L &= (V_{cc} - V_{sat})/I_{c1}; \\ P_{out} &= (V_{cc} - V_{sat})I_{c\,max}/4; \\ \eta &= 0.785(1 - V_{sat}/V_{cc}). \end{aligned} \qquad (19.4.17)$$

(b) *Design for maximum output power, for a given supply voltage V_{cc} and saturation resistance r_{sat} considered constant*

The design is carried out as for class A amplifiers, point (a). The optimum load resistance is given by eqn. (19.3.1), the slope of the dynamic load-line is half this value, according to Fig. 19.10. The output power is similarly given by (19.3.1), and the efficiency is $\eta = 39\%$.

(c) *Design for maximum output power, for a given supply voltage V_{cc} and saturation resistance r_{sat} increasing with collector current*

The design process of the class B amplifier is similar to that of the class A amplifier treated earlier (Fig. 19.13). The calculation is similarly based on the fact that the function $P_{out}(V_{sat})$ has a maximum, obtainable by differentiation. The output power is given by

$$P_{out} = (V_{cc} - V_{sat})I_{c1}/2 = (V_{cc} - V_{sat})I_{c\,max}(V_{sat})/4. \qquad (19.4.18)$$

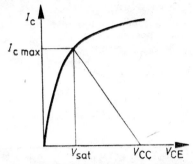

Fig. 19.13. Current–voltage relation of class B amplifiers, assuming a saturation resistance increasing with collector current

Here the function $I_{c\,max}(V_{sat})$ is approximated by the relation (19.3.5) derived earlier. Solving eqn. (19.3.7) yields the position of V_{sat}^* as shown by Fig. 19.9, normalized to the voltage V_0. It should be noted that V_0 should not be determined from the steady-state characteristics but from the curve presented in Fig. 19.2, especially in the UHF range. Thus the main circuit parameters are:

$$R_L = 2(V_{CC} - V_{sat}^*)/I_{c\,max};$$
$$P_{out} = (V_{CC} - V_{sat}^*)I_{c\,max}/4; \qquad (19.4.19)$$
$$\eta = 0.785(1 - V_{sat}^*/V_{CC}).$$

(d) *Design for maximum output power, for a given collector peak current $I_{c\,max}$ and negligible saturation voltage ($V_{sat} = 0$), by optimal selection of flow angle*

This design process will lead to a class AB operation, however, close to class B. It is seen from Figs. 19.11 (b) and 19.12 (b) that the ratio $I_{c1}/I_{c\,max}$ has a maximum value in the range $2\alpha > 180°$. Assuming a truncated sine-wave type collector-current pulse waveform, and using Fig. 19.11 (b), the maximum output power is at a flow angle of 240° with an efficiency of 0.64, and has a value of

$$P_{out\,max} = 0.134 I_{c\,max} V_{CE\,max}. \qquad (19.4.20)$$

Assuming an *expsin* shaped collector-current pulse and using Fig. 19.12 (b), the optimum flow angle is 207° and the efficiency is 0.52, and the value of the maximum output power is

$$P_{out\,max} = 0.11 I_{c\,max} V_{CE\,max}. \qquad (19.4.21)$$

19.5. Design of class C amplifiers

The principles outlined in the previous sections are also applicable for high-frequency class C power amplifiers: the collector AC voltage is calculated by multiplying the appropriate harmonic of the collector current pulse by the value of the terminating impedance at that frequency. Figures

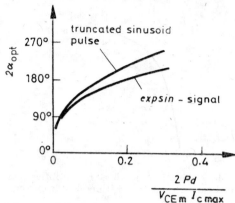

Fig. 19.14. Optimum flow angle as a function of the maximum ratings of transistors, for various waveforms

19.11 (a) and 19.12 (a), which give the fundamental frequency component for flow angles of less than 180°, are valid. It is seen that the fundamental frequency component is decreased by decrease in flow angle. In order to maintain the level of output power high, the load resistance should be increased, but this is limited by the maximum allowable collector voltage. This means that for class C amplifiers, the flow angle has an optimum which will be determined in the following. The initial condition is that the dissipation P_d, the maximum collector voltage V_{CEm} and the maximum collector current $I_{c\,max}$ given for the transistor may not be exceeded. To simplify the calculation, it will be assumed that the saturation voltage V_{sat} is negligible ($V_{sat} = 0$).

The power P_d dissipated by the transistor is the difference between the DC power consumption and the output power:

$$P_d = P_{DC} - P_{out} = V_{CC} I_0 - V_{CC} I_{c1}/2. \qquad (19.5.1)$$

Rearranging and dividing by the peak current $I_{c\,max}$, we have

$$\frac{I_0}{I_{c\,max}} - \frac{I_{c1}}{2I_{c\,max}} = f(2\alpha_{opt}) = \frac{P_d}{V_{CC} I_{c\,max}}. \qquad (19.5.2)$$

If the voltage appearing at the output is sinusoidal, i.e. it does not contain harmonics (which is approximately the case for tuned amplifiers), then the voltage peak at the output is twice the supply voltage V_{CC}. If the design is correct, the complete voltage range of the transistor will be utilized, i.e. the voltage peak equals the maximum allowable collector voltage: $2V_{CC} = V_{CEm}$. Taking this into account, eqn. (19.5.2) may be written as

$$f(2\alpha_{opt}) = 2P_d/V_{CEm} I_{c\,max}. \qquad (19.5.3)$$

On the right-hand side of this expression there are limiting parameters of the chosen transistor which thus allow the determination of the optimum flow angle, through Fig. 19.14. This figure shows eqn. (19.5.3) for current pulses having a truncated sinusoidal and an *expsin* shape.

19.5. Class C amplifiers

The optimum flow angles for the two pulse shapes will differ increasingly for high flow angles, but the actual optimum flow angle will always be situated between the two curves.

The design is thus carried out as follows: by choosing $I_{c\,max}$, I_{c1} is determined from the selected flow angle $2\alpha_{opt}$, and the load resistance is calculated from the expression $R_L = V_{CC}/I_{c1}$ as the ratio of the collector AC voltage and fundamental frequency component of the collector current (assuming $V_{sat} = 0$).

According to another design procedure, the saturation voltage V_{sat} is considered, but the requirements on dissipation and peak current are not taken into account. Let us first write the output power:

$$P_{out} = (V_{CC} - I_{c\,max} r_{sat}) I_{c1}/2, \quad (19.5.4)$$

which is half the product of the voltage and current amplitudes. The voltage amplitude is proportional to the load resistance:

$$V_{CC} - I_{c\,max} r_{sat} = I_{c1} R_L. \quad (19.5.5)$$

By eliminating I_{c1} from the above two relations, we obtain

$$P_{out} = \frac{V_{CC}^2}{2R_L} \left(\frac{I_{c1}}{I_{c\,max}}\right)^2 \bigg/ \left(\frac{I_{c1}}{I_{c\,max}} + \frac{r_{sat}}{R_L}\right)^2, \quad (19.5.6)$$

where $I_{c1}/I_{c\,max}$ is a function of flow angle. By investigating the dependence of the output power on the load resistance R_L with constant flow angle, it may be shown that the output power has a maximum value, reached at optimum load resistance

$$R_{L\,opt} = \frac{r_{sat}}{I_{c1}/I_{c\,max}}. \quad (19.5.7)$$

At this load resistance, the maximum output power is given by

$$P_{out\,max} = \frac{V_{CC}^2}{8 r_{sat}} \frac{I_{c1}}{I_{c\,max}}. \quad (19.5.8)$$

Investigation of the dependence on flow angle shows that the ratio $I_{c1}/I_{c\,max}$ has a maximum value, assuming a truncated sinusoidal current pulse, at a flow angle of $2\alpha = 240°$ (from Fig. 19.11b), and this maximum value is approximately 0.53. This results in an output power of

$$P_{out\,max} = 0.067 V_{CC}^2/r_{sat}, \quad (19.5.9)$$

and the load resistance is $R_{L\,opt} = 1.9 r_{sat}$, i.e. it is nearly equal to the value for class B operation.

However, this design method for the class C amplifier is not necessarily viable since no restrictions on the peak current or dissipation have been assumed, so there is a risk of exceeding these values. For this reason both data should be controlled through the equations:

$$I_{c\,max} = 1.9 V_{CC}/R_{L\,opt}, \quad (19.5.10)$$

and $P_d = V_{CC} I_0 \approx 0.4 V_{CC} I_{c\,max}$. Should this value exceed the dissipation power allowed for the transistor, then the flow angle should be decreased as shown in Fig. 19.14.

19.6. Push-pull amplifiers

A class A push-pull amplifier may be considered as a simultaneous operation of two class A power amplifiers, supplying simultaneously half-waves of opposite polarities through opposite drive. The two signals are added by the output transformer (Fig. 19.15). Power relations correspond to those of the normal class A amplifiers, so no advantages are offered. (This is in contrast with the class B push-pull amplifiers treated later.) However, push-pull operation is advantageous as regards even-order (primarily second-order) distortion. This is cancelled in the output transformer, so the overall second-order distortion of the push-pull amplifier will be smaller than that of the single transistors.

Let us now investigate the class B push-pull amplifiers in Fig. 19.10. The two half periods are supplied by the two transistors of the output stage, and a complete sine-wave is formed by the output transformer. For a resistive load, a wideband amplifier is obtained since disturbing harmonics are not generated.

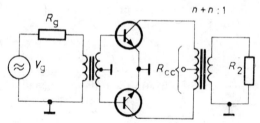

Fig. 19.15. Simplified circuit diagram of push-pull amplifier

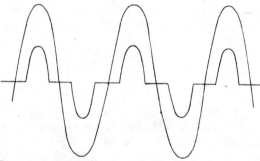

Fig. 19.16. Characteristic low-level nonlinear distortion in class B operation

19.6. Push-pull amplifiers

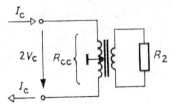

Fig. 19.17. Circuit for calculating the collector current

The distortion of a single-ended class B amplifier allows only selective applications. However, the push-pull class B amplifier, by addition of two half-waves, may be used over a wide frequency range, by simultaneously offering advantages in power relations and efficiency characteristic of class B operation. Further, even-order (primarily second-order) distortion may be substantially reduced through suitable circuit adjustment. At the same time, typical class B distortion at low output powers will be present. This is due to the fact that perfect matching of the two half periods may not be assured simultaneously at both maximum power and low power level. This is because of the nonlinear transistor characteristic and results in a limited output power range (limited "dynamical" operation). The waveforms for high and low output levels are shown in Fig. 19.16. As the harmonic output power at low levels is less than the power at full output, despite the increased distortion factor, this phenomenon may be disregarded in transmitter-design practice.

Class C push-pull amplifiers may generally be treated as class B amplifiers, since the flow angle is not much lower than 180°. This is because at lower flow angles, the above distortion is substantially increased and the amplification is substantially decreased.

For class A push-pull amplifiers, using the notation of Fig. 19.15, the value of the collector-to-collector resistance is given by

$$R_{cc} = 4n^2 R_2, \qquad (19.6.1)$$

where R_2 is the load resistance at the secondary side, and n is the turns ratio.

In the two primary half coils, the current flow is reversed, and the voltage across the whole primary coil is twice the collector voltage (Fig. 19.17). The collector current is given by

$$I_c = 2V_c/R_{cc}, \qquad (19.6.2)$$

and the resistance seen by one of the collectors may be calculated from the relation $R_L = V_c/I_c$. Assuming a symmetrical drive in class A operation, the load resistance of the single output transistor is given by

$$R_L = R_{cc}/2. \qquad (19.6.3)$$

This simultaneously gives the slope of the load line. It follows that the two primary half windings and the secondary winding of the output transformer may not be regarded as two separate transformers transforming the load to the appropriate collectors, since in this case, the load would be $R_{cc}/4$.

In class B push-pull amplifiers, the slope of the load line corresponds to the load resistance; this follows from the fact that during the passive half period of the transistor under test, the common load is supplied by the other transistor, and thus a common load line may be drawn.

When calculating the load resistance, it should be borne in mind that the current opposite to I_c (Fig. 19.17) does not appear in the other half winding at the same instant. This means that one half primary winding is excluded from operation during alternate half periods.

Thus the load resistance may be calculated by assuming that the load R_2 is transformed by only one half primary winding:

$$R_L = R_{cc}/4. \qquad (19.6.4)$$

Knowledge of the load resistances permits calculation of the voltage and current relations of push-pull amplifiers. The design of the class A push-pull amplifier is carried out using Section 19.3, according to the different basic cases given there, taking into account eqn. (19.6.3). The output power of the push-pull amplifier is the sum of the two transistor output powers, i.e. twice the simple amplifier output power:

$$P'_{out} = 2P_{out}. \qquad (19.6.5)$$

Class B push-pull amplifiers are designed according to Section 19.4 which deals with normal class B amplifiers, taking into account eqn. (19.6.4) giving the load resistance. Special attention should be given to the design procedure for a given supply voltage V_{CC} and constant saturation resistance r_{sat} as given in point (b). The overall output power of the two transistors, assuming an arbitrary load resistance R_L, is given by

$$P'_{out} = \frac{V_{CC}^2 R_L}{2(R_L + r_{sat})^2}, \qquad (19.6.6)$$

and the efficiency is

$$\eta = 0.78 R_L/(R_L + r_{sat}). \qquad (19.6.7)$$

At a saturation resistance of $R_L = r_{sat}$, the output power according to (19.6.6) has a maximum value

$$P'_{out} = V_{CC}^2/8r_{sat}. \qquad (19.6.8)$$

The efficiency is 39%, and the dissipation of one transistor is approximately

$$P_d \approx 0.1 V_{CC}^2/r_{sat}. \qquad (19.6.9)$$

Finally, according to eqn. (19.6.4), the collector-to-collector resistance pertaining to the maximum output power is given by $R_{cc} = 4r_{sat}$.

During the previous investigations, no distinction between selective and wideband terminations were made. Relations derived are valid for both amplifier types. Expressions given for the fundamental-frequency component are valid for both selective terminations implemented by reactive elements and for wideband terminations using transmission-line transformers. Obviously, the situation differs for the harmonic components.

The method of combining high-frequency power amplifiers by hybrids will be dealt with in a later chapter.

19.7. Nonlinear equivalent circuit of power amplifiers

In design procedures previously investigated, the steady-state output characteristic of the transistor has been used. A dynamic load-line has been assumed, along which the instantaneous electrical state of the device is changed. This approximation will not be valid at frequencies commensurable with the transistor cut-off frequency. In the following, the frequency response of the gain and the time dependence of the collector current will be investigated through the transistor high-frequency equivalent circuit [19.5].

Let us use the nonlinear high-frequency equivalent circuit shown in Fig. 2.27 as a starting point, and let us modify this circuit by replacing the parallel RC circuit representing the emitter impedance by a battery of voltage V_k (Fig. 19.18(a)). This is valid since the emitter–base diode only conducts above a certain threshold voltage V_k, and thus the input characteristic for high levels may be replaced by the straight-line characteristic shown in Fig. 19.18 (b). The input impedance at the base is infinite for no current flow, and has a value of r_b in the case of base current.

From the relation between base and collector currents we have

$$i_b(t) \approx \frac{1}{\omega_T} \frac{di_c}{dt}, \tag{19.7.1}$$

and thus the time dependence of the base voltage is given by

$$v_b(t) = \frac{r_b}{\omega_T} \frac{di_c}{dt} + V_k. \tag{19.7.2}$$

If the transistor is driven by a voltage $v_b(t) = V_1 \sin \omega t$, the collector current may be determined from eqn. (19.7.2) by integration:

$$i_c(t) = \frac{\omega_T V_1}{r_b \omega} \left[\cos \omega t_0 - \cos \omega t + \frac{V_k}{V_1} (\omega t_0 - \omega t) \right], \tag{19.7.3}$$

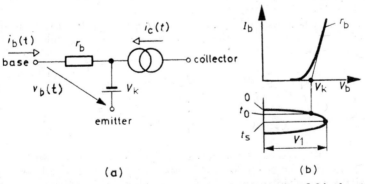

Fig. 19.18. (a) High-frequency nonlinear equivalent circuit of bipolar transistor, (b) broken-line approximation of the input characteristic

where t_0 is the time instant of transistor turn-on for which the relation $V_1 \sin \omega t_0 = V_k$ holds. The peak value of the collector current may be determined from the condition where the collector current has an extremum value at maximum drive voltage ($t = t_s$):

$$\frac{di_c}{dt}(t = t_s) = 0. \qquad (19.7.4)$$

From the value $i_{\max}$ at time instant $t = t_s$, the collector current decreases and reaches zero at $t = t_e$. The time t_e denoting the end of the current flow may be determined from the equation $i_e(t = t_e) = 0$.

Figure 19.19 shows the collector current time functions for various V_k/V_1 values, in normalized form. It is seen that the operation corresponds to class AB operation, but the waveform is asymmetrical, with a forward slope. At low values of V_k/V_1, the current peak value occurs in the turned-off state of the input. For gain determination, the fundamental frequency component I_{c1} of the collector current is important. Figure 19.20 shows the amplitude of the fundamental frequency component and of the DC component I_0 referred to the maximum current $I_{c\max}$. From the fundamental frequency component, the useful output power is given by

$$P_{\text{out}} = I_{c1}^2 R_L/2, \qquad (19.7.5)$$

where R_L is the load resistance for the fundamental frequency component. The input power is calculated from the input resistance which is given by the following relation:

$$R_{\text{in}} = \frac{4\pi r_b}{2\omega(t_e - t_0) - \sin 2\omega t_e - \sin 2\omega t_0 + 4 \sin \omega t_0 \cos \omega t_e}. \qquad (19.7.6)$$

The design is aided by Fig. 19.21, giving the normalized value of the fundamental frequency component as a function of ratio V_k/V_1. Knowing the transistor parameters r_b, ω_T and V_k, the operating frequency ω and the fundamental frequency amplitude I_{c1} pertaining to the required output power, the driving voltage V_1 may be read off, and this allows the remaining circuit parameters to be found. In Ref. [22.15] a design method also based on eqn. (19.7.1) is investigated.

In Ref. [19.7], the design of explicitly class C high-frequency amplifiers is dealt with. The initial equations for the emitter and collector currents (see Fig. 19.22) are:

$$i_e(t) = C_{Te}\frac{dv_{b'e}(t)}{dt} + \left[\tau \frac{qI_s}{kT}\frac{dv_{b'e}(t)}{dt} + I_s\right] e^{\frac{qv_{b'e}(t)}{kT}}; \qquad (19.7.7)$$

$$v_c(t) = I_s e^{\frac{qv_{b'e}(t)}{kT}}, \qquad (19.7.8)$$

where $v_{b'e}(t) = V_{b'e} \cos \omega t$ is the emitter–base voltage for the intrinsic transistor, τ is the time constant of the change of charge in the base, and C_{Te} is the emitter transition capacitance. The collector transition capacitance is embedded in the external tuning capacitance.

19.7. Nonlinear equivalent circuit

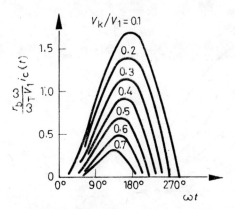

Fig. 19.19. Time function of the collector current for various V_k/V_1 values

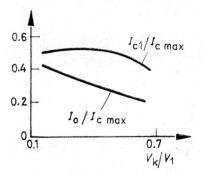

Fig. 19.20. Dependence of the fundamental frequency and DC components of the collector current on the ratio V_k/V_1

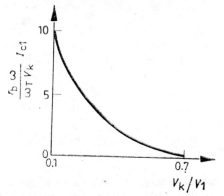

Fig. 19.21. Dependence of the normalized fundamental frequency value of the collector current on the ratio V_k/V_1

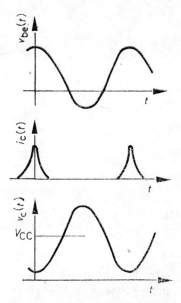

Fig. 19.22. Voltage and current time functions of class C tuned amplifier

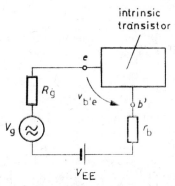

Fig. 19.23. Input circuit of class C amplifier

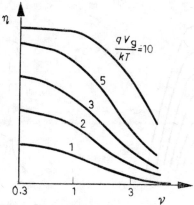

Fig. 19.24. Dependence of class C amplifier efficiency on the relative frequency, for various drive voltage amplitudes

Driving the input circuit by a generator of resistance R_g, the following equation based on Fig. 19.23 may be written:

$$V_{EE} + V_g = i_e R_g + (i_e - i_c) r_b + V_{b'e}, \qquad (19.7.9)$$

where V_{EE} is the emitter reverse bias voltage. With a few approximations, the differential equation may be simplified as follows:

$$\nu \frac{dz}{d\omega t} + z = \frac{qV_g}{kT} \cos \omega t. \qquad (19.7.10)$$

From the periodic solution of this equation, the collector current to be determined is given by

$$i_c(t) = I_s[e^{z(t)} - 1], \qquad (19.7.11)$$

where $z(t) = \Phi \cos(\omega t - \theta)$, the amplitude of this $\Phi = qV_g/kT \sqrt{1+\nu^2}$, and $\nu = \omega C_{Te}(R_g + r_b) = \tan \theta$. From the above, the efficiency of the amplifier for the fundamental frequency component, driven up to saturation ($V_{cb} = V_{CC}$) is given by

$$\eta = \frac{J_1(\Phi)}{J_0(\Phi)}, \qquad (19.7.12)$$

where J_0 and J_1 are Bessel functions of zero- and first-order. Figure 19.24 shows the efficiency as a function of factor ν for various generator voltages. It is seen that the efficiency reduces with increasing frequency, and an abrupt decrease takes place at high generator voltages. Substituting higher-order Bessel functions into the numerator of eqn. (19.7.12), the efficiency for higher-order harmonics, essential for frequency multipliers, may be determined.

19.7. Nonlinear equivalent circuit

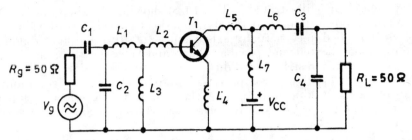

Fig. 19.25. Equivalent circuit of a class B amplifier used for computer analysis

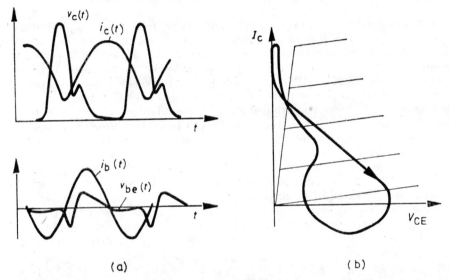

Fig. 19.26. Computed responses of class B amplifier. (a) Current and voltage time functions, (b) dynamic load curve

In the above, the behaviour of the high-level amplifier was investigated using the approximating nonlinear equivalent circuit. Since the calculation involved elementary methods, the equivalent circuit of the transistor and of the external passive elements had to be chosen simply to avoid mathematical difficulties. A computer-aided design allows the analysis of more complicated circuits.

Figure 19.25 shows a 30-MHz class B power amplifier investigated by the help of a computer [19.25]. The transistor has been replaced by a large-signal equivalent circuit based on the symmetrical equations given in Section 2.8. The circuit analysis involved several steps, by successive correction of external tuning elements (reactances), corresponding to the tuning procedure of the circuit. The results of the computer-aided design showed very good agreement with the waveforms measured by an oscilloscope. The high-frequency behaviour is illustrated in Fig. 19.26 showing the time functions of the voltages and currents and the dynamic load curve. According to Fig. 19.26 (b), no "load line" actually exists, since the working point follows a curve of a complicated and peculiar shape and thus the theorems on the load line have only restricted validity.

Reference [19.36] gives a computerized study of the class C power amplifier at 160 MHz, taking into account the base-widening effect of bipolar transistors. The power losses in the class E tuned power amplifier are considered in Ref. [19.38].

19.8. Approximate relations of power amplifiers

For the practical design of power amplifiers, relatively simple relations are needed, suitable for estimating the order of magnitude of the expected gain, input impedance, etc. Acceptable results are offered by applying the equivalent circuit with linear four-pole parameters (up to about 1 dB gain compression) [18.26], [19.32]. At high signal levels, these data become inaccurate, but can be corrected during the circuit measurement.

Figure 19.27 shows the applied transistor equivalent circuit, also taking into account the emitter inductance L_E [19.16]. The collector capacitance is made up of two parts: capacitance C'_c connected to the external base, and capacitance C_c connected to the internal base b'. If just the sum of these two feedback capacitances is given in the transistor data sheet, then only this latter capacitance C_c should be considered, and the external capacitance C'_c should be taken as zero.

From this equivalent circuit, both the input resistance of the transistor and the power gain pertaining to an assumed terminating resistance R_L may be calculated. The approximate value of the input resistance is given by

$$R_{in} \approx \frac{\omega_T L_E + (1 + \omega_T R_L C_c) r_b}{1 + \omega_T R_L (C_c + C'_c)}, \quad (19.8.1)$$

and the approximate value of the power gain is given by

$$G \approx \left(\frac{\omega_T}{\omega}\right)^2 \frac{R_L}{[1 + \omega_T R_L (C_c + C'_c)][\omega_T L_E + r_b(1 + \omega_T R_L C_c)]}. \quad (19.8.2)$$

If R_L is small, $\omega_T R_L (C_c + C'_c) \ll 1$ and $R_{in} \approx r_b + \omega_T L_E$. The power gain is then given by

$$G \approx \left(\frac{\omega_T}{\omega}\right)^2 \frac{R_L}{r_b + \omega_T L_E}. \quad (19.8.3)$$

In the case of matched termination, $R_L = 1/\omega_T C_c$ and by further substitution of $L_E = 0$, the unity power-gain frequency (maximum oscillation frequency) is determined:

$$f_{max} = \frac{1}{5}\sqrt{\frac{f_T}{r_b C_c}}. \quad (19.8.4)$$

19.8. Approximate relations

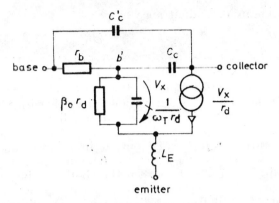

Fig. 19.27. Simplified equivalent circuit of high-frequency power transistors

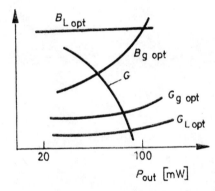

Fig. 19.28. Generator admittance and load admittance providing conjugated matching and power gain, as functions of output level

Assuming zero saturation voltage, the approximate value of the collector load resistance is given by

$$R_L \approx V_{CC}^2/2P_{out}, \qquad (19.8.5)$$

where P_{out} is the output power. The output and input matching four-poles may be designed by using the load resistance R_L and the input resistance R_{in}, respectively. These design relations are presented in the next chapters.

The above procedure is obviously inaccurate as a linear transistor equivalent circuit is applied to characterize a nonlinear circuit. The input resistance given by eqn. (19.8.1) is not actually constant but a function of the input power, although the determination of this function is difficult, even by measurement. It is more feasible to measure the level dependences of the optimum generator admittance

$$Y_{g\,opt} = G_{g\,opt} + jB_{g\,opt}, \qquad (19.8.6)$$

and of the optimum load admittance

$$Y_{L\,opt} = G_{L\,opt} + jB_{L\,opt}, \qquad (19.8.7)$$

assuring a conjugated impedance match [19.4], [19.17]. This may be accomplished by placing impedance transformers (e.g. transmission-line sections with several tuning stubs) at the input and output ports, and tuning these for maximum power at various power levels. After reading the value of P_{out}, the impedance of the passive coupling four-pole (seen by the transistor) is measured, and the conjugated values of these impedances will give the transistor input and output impedances at a given power P_{out}. The results of such a measurement (measurement frequency 2 GHz) are presented in Fig. 19.28. It is seen that both the impedances and the power gain G are highly dependent on the output power. This also proves that the results derived from the linear equivalent circuit are extremely approximate. Another method of measuring large-signal s parameters of transistors is given in Ref. [19.37].

20. HIGH-FREQUENCY POWER AMPLIFIERS. PRACTICAL DESIGN

20.1. Power amplifier matching four-poles

A matching four-pole is used to connect the load resistance of a high-frequency power amplifier to the amplifier output, to provide the necessary impedance transformation and frequency response. The situation is similar at the input of the power amplifier, where either the generator or the previous (driving) stage must be matched to the amplifier input.

Wideband matching. The most up-to-date solution is the application of a transmission-line transformer, particularly in the case of high relative bandwidths (several octave), assuming that no DC coupling is needed. Transmission-line transformers were dealt with extensively in Section 12.1; all transformers investigated there are suitable for wideband power amplifiers.

Selective matching. Band-pass and low-pass type four-poles should be distinguished. Both are characterized by substantial harmonics suppression, which is particularly important in class B and class C selective amplifiers. In the following, a few typical matching four-poles are presented, together with design relations.

In Fig. 20.1, a four-pole comprising a resonant circuit with inductive tapping is shown. The resonant circuit is matched by the series capacitance C_2 to the load resistance R_L. If the load R_c on the transistor collector is known the four-pole elements may be calculated using the following relations:

$$X_{L1} = \frac{n^2 R_c}{Q};$$

$$X_{C1} = \frac{n^2 R_c}{Q\left(1 - \dfrac{X_{c2}}{QR_L}\right)};$$

$$X_{C2} = R_L \sqrt{\frac{n^2 R_c}{R_L} - 1},$$

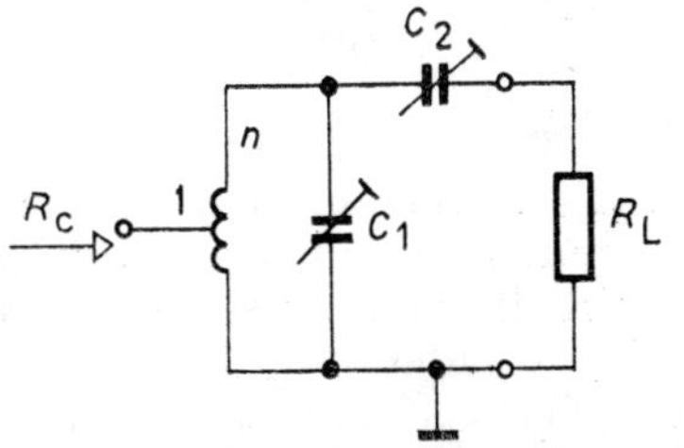

Fig. 20.1. Parallel resonant circuit with inductive tapping and series matching capacitance

where Q is the loaded Q factor of inductance L_1, n is the number of turns of the inductance, and the tapping is taken at the first turn. In Fig. 20.2, the load matching is given by the capacitive tapping of the resonant circuit, and thus the design relations are:

$$X_{L2} = \frac{n^2 R_c}{Q};$$

$$X_{C1} = \frac{n^2 R_c Q}{Q^2 + 1}\left(1 - \frac{R_L}{Q X_{C2}}\right);$$

$$X_{C2} = \frac{R_L}{\sqrt{\dfrac{(Q^2 + 1) R_L}{n^2 R_c} - 1}}.$$

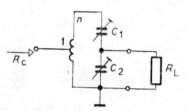

Fig. 20.2. Parallel resonant circuit with inductive tapping with capacitive divider

Figure 20.3 shows a matching four-pole consisting of a π network. The series inductance has a series-connected capacitance for trimming the inductance value. Capacitance C_0 is the output capacitance of the transistor which has to be embedded into the matching four-pole. Inductance L_1 is used for supply-voltage connection. Assuming $R_c < R_L$, then the design relations are:

$$X_{C1} = Q R_c;$$

$$X_{C2} = \sqrt{\dfrac{R_L}{\dfrac{(Q^2 + 1) R_L}{R_c Q^2} - 1}};$$

$$X_{L1} = \frac{Q R_c}{\dfrac{Q R_c}{X_{C0}} + 1};$$

$$X_{L2} = Q R_c \left(1 + \frac{R_L}{Q X_{C2}}\right),$$

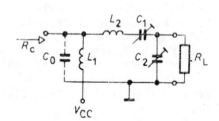

Fig. 20.3. π network matching four-pole

where X_{C0} is the output reactance of the transistor, and Q is the effective Q factor of the circuit from which the bandwidth is determined. This refers to the circuit and not to any of its elements, for example, inductance. In the following calculations, it is assumed that the applied reactive elements are lossless. The selectivity of the four-pole is proportional to the square of the effective Q factor. A finite Q factor introduces insertion loss. The choice of Q factor is a result of compromise, since it should also be taken into account that the reactive element values should be realizable.

The T network matching four-pole shown in Fig. 20.4 comprises the output capacitance of the transistor. DC connection is through a RF choke applied directly to the transistor collector and having an impedance which

20.1. Matching four-poles

can be neglected. If the following condition is met:

$$\frac{Q X_{C0}}{\sqrt{R_c R_L}} > 1, \qquad (20.1.1)$$

where Q is the selected effective Q factor and X_{C0} is the output reactance of the transistor, then the design equations for the matching four-pole circuit elements are as follows:

$$X_{L1} = \frac{Q X_{C0}^2}{R_c}\left(1 - \frac{\sqrt{R_c R_L}}{Q X_{C0}}\right);$$

$$X_{L2} = X_{C0}\sqrt{\frac{R_L}{R_c}};$$

$$X_{C1} = \frac{Q X_{C0}^2}{R_c}\left(1 - \frac{R_c}{Q X_{C0}}\right);$$

$$X_{C2} = \frac{R_L}{Q}\left(\frac{Q X_{C0}}{\sqrt{R_c R_L}} - 1\right).$$

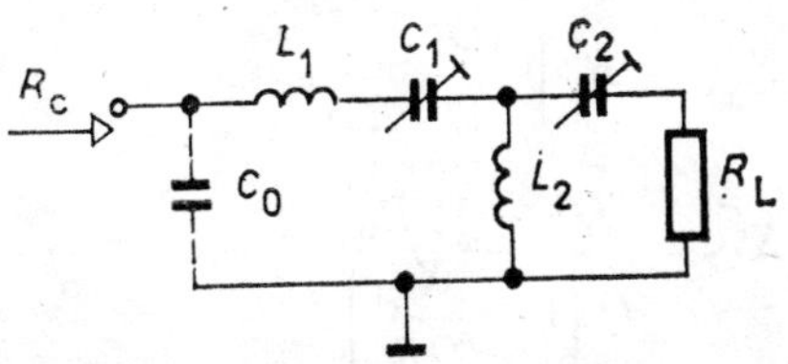

Fig. 20.4. T network matching four-pole

Figure 20.5 shows a further π network matching four-pole. It differs from that shown in Fig. 20.3 in the biasing which in this case is through a RF choke, since it has an impedance which can be neglected. Design equations for the circuit elements are given in the following.

$$X_{C1} = \frac{Q X_{C0}^2}{R_c}\left(1 - \frac{R_c}{Q X_{C0}}\right);$$

$$X_{C2} = \frac{R_L}{\sqrt{\frac{(Q^2 + 1) R_c R_L}{Q^2 X_{C0}^2} - 1}};$$

$$X_{L1} = \frac{Q X_{C0}^2}{R_c}\left(1 + \frac{R_L}{Q X_{C2}}\right).$$

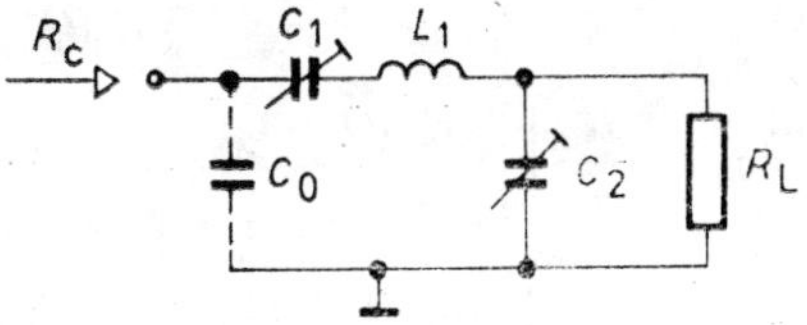

Fig. 20.5. π network matching four-pole

Because of the high signal level, the value of the output capacitance C_0 depends on the working point. Calculation shows that the average output capacitance C_0 is 1.5–2 times the collector–base capacitance corresponding to the working point. The exact value depends on the parameters of the collector–base junction, but is not critical since it is shunted by the parallel load resistance R_c.

The transistor input is similarly matched to the generator by reactive four-poles; the generator may be the previous (driving) stage. Input matching networks are somewhat different from output networks because of the type of required impedance. The output impedance is normally capacitive and may be replaced by a parallel RC network, while the input impedance is inductive and may be replaced by a series RL network. For this reason, other kinds of matching four-poles are used for input matching. A few are presented in the following.

Figure 20.6 shows a T network matching four-pole; the inductive component of the transistor input impedance is embedded into the series inductance. The real part of the input impedance is R_{in}, to which the generator resistance R_g has to be matched. Assuming $R_g > R_{in}$, the circuit-element design equations are:

$$X_{L1} = Q R_{in};$$

$$X_{C1} = R_g \sqrt{\frac{(Q^2+1)R_{in}}{R_g} - 1};$$

$$X_{C2} = \frac{(Q^2+1) R_{in}}{Q\left(1 - \dfrac{X_{C1}}{QR_g}\right)},$$

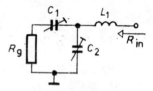

Fig. 20.6. T network used for input matching of transistor power amplifiers

where Q is the selected effective Q factor of the circuit. For the equations applying to Fig. 20.7, it is also assumed that $R_g > R_{in}$, so the design relations are:

$$X_{L1} = Q R_g;$$

$$X_{C1} = \frac{R_g(Q^2+1)}{Q}\left[1 - \sqrt{\frac{R_{in}}{R_g(Q^2+1)}}\right];$$

$$X_{L2} = \frac{R_{in}}{Q}\left[\sqrt{\frac{R_g(Q^2+1)}{R_{in}}} - 1\right];$$

$$X_{C2} = \frac{R_g}{Q}\sqrt{\frac{R_{in}(Q^2+1)}{R_g}}.$$

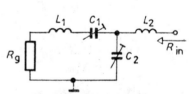

Fig. 20.7. T network used for input matching of transistor power amplifiers

Figure 20.8 shows an input matching network driven by a previous transistor stage. The specified load resistance of the driving stage is R_c, and the output capacitance of the driving stage is C_0, included in the matching four-pole. The input resistance of the transistor is assumed to be given by the base resistance r_b. The impedance of the series reactive element is much lower than that of inductance L_1, i.e. $X_{L1} > \mathrm{Im}(Z_{in})$. With this condition and also that $R_c > r_b$, the circuit elements of the matching four-pole are given as follows:

$$X_{L1} = Q\, r_b;$$

$$X_{C1} = X_{C0}\left[\sqrt{\frac{(Q^2+1)r_b}{R_c}} - 1\right];$$

$$X_{C2} = \frac{(Q^2+1)\,r_b}{Q\left[1 - \sqrt{\frac{(Q^2+1)\,R_c\, r_b}{X_{C0}^2\, Q^2}}\right]}.$$

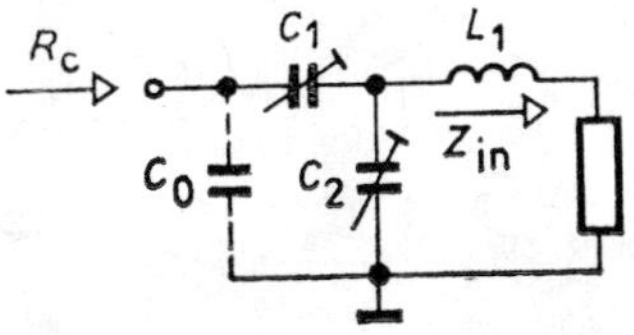

Fig. 20.8. Four-pole for matching the driving and output stages

Figure 20.9 also shows a matching four-pole between the driving and output stages. The output capacitance of the driving stage is C_0, and the real part of the common-emitter transistor input impedance is the base resistance r_b. Let us assume that the imaginary part of the input impedance is much lower than the impedance of the series capacitor C_2, i.e. $X_{C2} \gg \mathrm{Im}(Z_{in})$, and may thus be neglected. Assuming the inequality $R_c > r_b$, the circuit-element design equations are:

$$X_{L1} = \frac{r_b(Q^2+1)}{Q\left[1 + \sqrt{\frac{(Q^2+1)\,R_c\, r_b}{Q^2 X_{C0}^2}}\right]};$$

$$X_{C1} = X_{C0}\left[\sqrt{\frac{(Q^2+1)\,r_b}{R_c}} - 1\right];$$

$$X_{C2} = Q\, r_b.$$

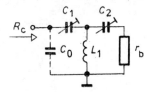

Fig. 20.9. High-pass four-pole for matching the driving and output stages

The circuit shown in Fig. 20.9 is a T network with an additional capacitance C_0. The DC connection of the driving transistor is applied through a choke having an impedance which can be neglected. If the impedance of the RF choke is commensurable with the impedances of the matching network elements then the circuit in Fig. 20.10 applies. If $R_c > r_b$, the element values of the matching network are:

$$X_{L2} = \frac{R_c}{Q};$$

$$X_{L1} = \frac{r_b \left(\sqrt{\frac{R_c}{r_b}} - 1 \right)}{Q \left(1 - \frac{R_c}{Q X_{C0}} \right)};$$

$$X_{C1} = \frac{R_c \left(1 - \sqrt{\frac{r_b}{R_c}} \right)}{Q \left(1 - \frac{R_c}{Q X_{C0}} \right)};$$

$$X_{C2} = \frac{R_c \sqrt{\frac{r_b}{R_c}}}{Q \left(1 - \frac{R_c}{Q X_{C0}} \right)}.$$

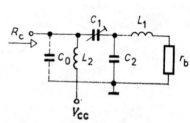

Fig. 20.10. Four-pole for matching the driving and output stages

Figure 20.11 shows a low-pass π network matching four-pole which can be applied at both the input and the output; therefore, the load resistances are denoted by the general symbols R_1 and R_2. Assuming $R_1 > R_2$, the elements are given as follows:

$$X_{C1} = \frac{R_1}{Q};$$

$$X_{C2} = \frac{R_2}{\sqrt{\frac{R_2(Q^2 + 1)}{R_1} - 1}};$$

$$X_{L1} = \frac{R_1 Q}{Q^2 + 1} \left(1 + \frac{R_2}{Q X_{C2}} \right).$$

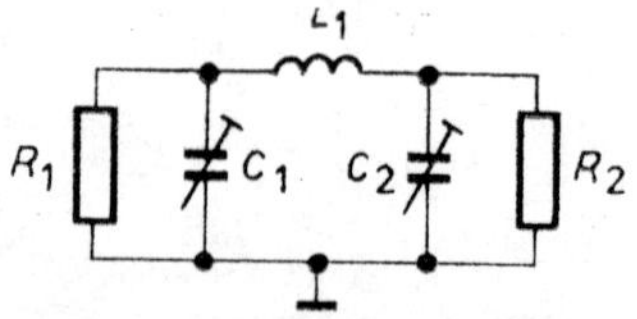

Fig. 20.11. Low-pass π network

Figure 20.12 shows a high-pass π network. Assuming $R_1 < R_2$, the elements are given as follows:

$$X_{C1} = \frac{R_1}{Q};$$

$$X_{C2} = \frac{R_1 Q}{Q^2 + 1}\left(\frac{R_2}{QX_{L1}} - 1\right);$$

$$X_{L1} = \frac{R_2}{\sqrt{\dfrac{R_2(Q^2+1)}{R_1} - 1}}.$$

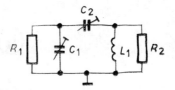

Fig. 20.12. High-pass π network

The high-pass T network shown in Fig. 20.13 is also frequently applied in high-frequency power amplifiers. The element values are the following:

$$X_{C1} = R_1\sqrt{\frac{R_2(Q^2+1)}{R_1} - 1};$$

$$X_{C2} = QR_2;$$

$$X_{L1} = \frac{R_2(Q^2+1)}{Q\left(1 + \dfrac{X_{C1}}{QR_1}\right)}.$$

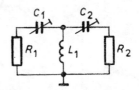

Fig. 20.13. High-pass T network

Finally, a short comment on the choice of effective Q factor and its effect on the loss of the matching network. Throughout the calculations, lossless reactances have been assumed, in which case the four-pole is also lossless. If the applied reactance elements are not lossless then the power loss of the circuit is governed mainly by the effective Q factor.

For example, let us investigate the four-pole shown in Fig. 20.5, used primarily for input matching of transistors. Omitting the calculation, only the ratio expressing the relative power loss (or attenuation) is given:

$$\frac{P_l}{P_0} \approx \frac{Q}{Q_{L1}} + \frac{Q}{\omega^2 C_1 L_1 Q_{C1}} + \frac{\omega R_2 C_2}{Q_{C2}}, \qquad (20.1.2)$$

where P_l is the power loss, and P_0 is the power transmitted. Q_{L1}, Q_{C1}, Q_{C2} are the unloaded Q factors of the reactances. It can be seen that by increasing the effective Q factor characteristic for the circuit, the loss is also increased for given L and C elements.

20.2. High-frequency power amplifiers with hybrid coupling

The output power of an output transistor may not be increased beyond a given dissipation limit. Better cooling, using effective heat-sinks, may increase the dissipation limits substantially but this is limited because of the need for reliable operation. Output power may only be increased by

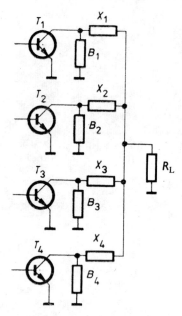

Fig. 20.14. Block diagram of parallel connected output stages

parallel operation of several output transistors, requiring a suitable network to combine the AC power supplied by the transistors. Further problems are then raised, particularly as regards operation reliability.

Let us investigate the relations using Fig. 20.14, showing four parallel connected transistors feeding a common load resistance R_L. To provide the required transistor load reactances, parallel elements B_1–B_4 and series elements X_1–X_4 are used. If one of the four transistors becomes faulty, e.g. a short-circuit appears at its output, then the complete output circuit will be detuned by this short-circuit, also changing the matching conditions of the three remaining output transistors. Insufficient matching results in less output power delivered to the load, i.e. higher dissipation. This leads to earlier failure of the other three transistors. In other words, a fault in one of the transistors destroys the others. This is obviously harmful, so this kind of interconnection should be avoided.

The interconnection of separate output stages should be such that the failure of a single stage only results in an output power loss, with the electrical conditions of the other stages remaining unchanged. This may be arranged by passive circuits isolating the single output stages, thus minimizing interaction. Such circuits are 3-dB *hybrids* (or quadrative hybrids), which may be used for both power division and power combination. Hybrids have already been investigated in previous chapters, stripline hybrids in Chapter 18 and transmission-line transformer hybrids in Chapter 12. In the following, we shall deal chiefly with lumped-element hybrids.

20.2. Hybrid coupling

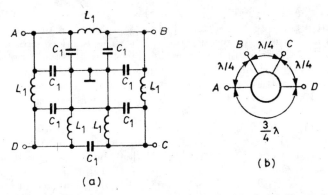

Fig. 20.15. (a) Lumped-element hybrid, (b) transmission-line hybrid

Figure 20.15 (a) shows a lumped-element hybrid [20.1], and Fig. 20.15 (b) shows the transmission-line equivalent. The principle of operation is best explained through the transmission-line variant. Power supplied to input port A is divided equally between loads connected to output ports D and B. There is no power at port C since the signals arriving via transmission-line sections $\lambda/2$ and λ respectively are out of phase, and thus cancel each other. Ports A and C are thus conjugated ports, isolated from each other. An equivalent description is that these two ports (e.g. generator at port A and load at port C) do not see each other. The situation is similar if generators are connected to both ports; mutual loading is eliminated. Obviously, B and D are also conjugated ports. The transmission-line characteristic impedance is $\sqrt{2 Z_0}$ where Z_0 is the impedance of the generator and load.

The lumped-element hybrid connection shown in Fig. 20.15 (a) may be derived from the transmission-line connection of Fig. 20.15 (b) by replacing the line sections of lengths $\lambda/4$ and $3\lambda/4$, by their π equivalent. The replacement will only be valid at the nominal frequency (band centre) where resonance of the transmission-line sections is present. The lumped-element values are calculated from:

$$L_1 = Z_0 \sqrt{2}/\omega_0 ; \qquad (20.2.1)$$

$$C_1 = 1/\omega_0 Z_0 \sqrt{2} , \qquad (20.2.2)$$

where Z_0 is the impedance of the loads connected to the output ports, and ω_0 is the band centre (nominal) frequency. Practically, the two capacitors C_1 connected to ports A and B may be combined. The isolation of the hybrid, i.e. the attenuation at the conjugated output port, is a function of the inductance Q factors. The isolation at the nominal frequency ω_0, expressed in attenuation, may be calculated from:

$$a \approx 20 \lg (4Q_0) . \qquad (20.2.3)$$

where Q_0 is the unloaded Q factor of the inductances, and it is assumed that all output ports are terminated by impedances of Z_0. Isolation is diminished at frequencies more remote from the band centre, as shown in Fig. 20.16. The hybrid under investigation is thus rather selective. The attenuation response shown in the figure refers to a hybrid operated at 6 MHz and terminated by 50 ohm, made up of inductances having Q factors of 100. Inductances should be shielded, and accurate element values are needed.

Figure 20.17 shows a simplified circuit of the lumped-element hybrid of Fig. 20.15 [20.2]. Element values are calculated from relations (20.2.1) and (20.2.2). Using the circuit as a power divider, the excess loss due to circuit element losses (over the 3 dB loss, i.e. power halving) is about 0.35 dB, the asymmetry is ± 0.25 dB, and the isolation between the two output ports is 20 dB. This means that the output powers are 3.35 ± 0.25 dB down relative to the input power.

Figure 20.18 shows the block diagram of an amplifier interconnected by hybrids [20.2]. The operating frequency is 132 MHz and the relative bandwidth is about $\pm 6\%$. The amplifier is made up of identical amplifying modules and hybrids. The output power from the input preamplifier is divided into two by a hybrid. Both output branches consist of a driving stage, a hybrid and two output stages. The output powers of the four output stages are combined by three hybrids in two steps. Power levels are given in the figure, and allow the determination of hybrid losses.

20.3. Biasing of power amplifiers

The biasing of power amplifiers is much more critical than that of low-level amplifiers. This is because a bias change of high-power transistors may have severe consequences (e.g. thermal runaway due to excess heating). The tendency to instability due to the high emitter currents also demands that perfect bias circuits are used — these must not contain parasitic elements which would lead to parasitic resonances and thus instability.

High-frequency power amplifiers may be divided according to the following parameters, when bias is considered: class of operation (A, B, C), configuration (common-emitter, common-base, etc.), and the electrode which is grounded. At high frequencies, "ground" is primarily the large heat-sink and metal chassis connected to it.

Figure 20.19 shows a few methods for biasing in class A and class AB amplifiers. In the general case shown in Fig. 20.19 (a), the base voltage is derived from a resistance divider connected to the supply voltage V_{CC}. If resistance R_1 has a low value, the transistor will be sensitive to thermal runaway, as the increase in base current is not sufficiently limited by a series resistance. At the same time, low-valued divider resistances result in high DC power consumption, thus resulting in poor efficiency. The feedback emitter resistance used for stabilizing the class A working point should have a low value since it will introduce a DC power loss. The AC return of the emitter, especially over a wide frequency range, is difficult, since the

20.3. Biasing

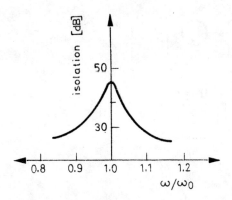

Fig. 20.16. Isolation of the lumped-element hybrid versus frequency

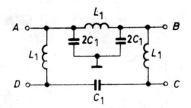

Fig. 20.17. Simplified circuit of the hybrid shown in Fig. 20.15(a)

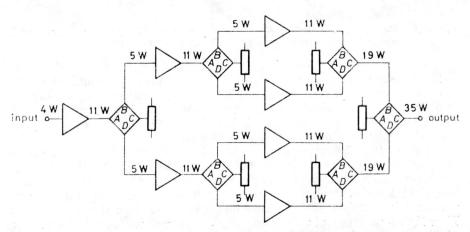

Fig. 20.18. Block diagram of power amplifier using hybrids

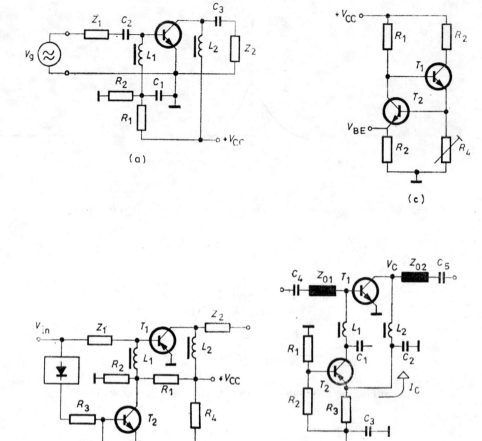

Fig. 20.19. Bias circuits. (a) Resistance voltage divider, (b) with guard circuit, (c) with temperature compensation, (d) with complementer transistor

series resistance and inductance of the bypass capacitor would be of the same order as the emitter-feedback resistance (less than one ohm). In single-frequency or narrow-band amplifiers, series resonant circuits may be used for emitter-return purposes, and this has the great advantage of tuning out the series emitter inductance of the transistor.

Figure 20.19 (b) shows a bias circuit with guard function. In the absence of input drive, the power dissipated in transistor T_1, which is a class A amplifier, would increase and have harmful effects. The guarding function is provided by switching transistor T_2. During normal operation, this transistor is switched off by a DC voltage rectified from the input signal. Thus T_2 has no effect of the biasing of T_1, and the bias is given by voltage divider $R_1 - R_2$. The decrease in input signal and thus in the rectified DC signal will turn on transistor T_2. Resistance R_2 is thus shunted, shifting the output transistor working point towards turn-off, thus protecting it from overheating.

Figure 20.19 (c) shows a bias circuit with temperature compensation. For the case of a power amplifier in class A or class AB operation, the base voltage should be adjusted for quiescent current. Since the transistor characteristic is highly dependent on temperature, the quiescent current increases at higher temperatures with constant V_{BE} bias, and the dissipation power becomes higher. The bias circuit shown in the figure has the effect of decreasing V_{BE} with increasing temperature, thus maintaining an approximately constant quiescent current. The optimum value of the quiescent current, which is particularly critical in class AB amplifiers with a low flow-angle, can be adjusted using potentiometer R_4.

Figure 20.19 (d) shows a bias circuit using complementer transistor T_2. This has the effect of maintaining constant the collector voltage V_C and collector current I_C of microwave amplifier transistor T_1, independent of temperature [18.26].

In Ref. [20.17], a guard circuit which senses the output VSWR is presented, with a PIN diode input power control.

Figure 20.20 shows the principle of biasing in a class B power amplifier. The effect of choke L_1 is to keep the base at earth potential. The transistor is turned on by the positive half cycles of the driving voltage.

Figure 20.21 shows different bias circuits of class C amplifiers. The reverse base voltage may be adjusted in several ways. In Fig. 20.21 (a), the reverse voltage, given on the base via choke L_1, is provided by an external voltage source V_{BB}. In Fig. 20.21 (b), the reverse voltage is provided by the driving signal itself. The emitter–base junction of the transistor acts as a diode, rectifying the high-amplitude input signal by which capacitor C_B is charged, as shown in the figure. This solution has the drawback of decreasing the collector–emitter breakdown voltage at high values of R_B.

Figure 20.21 (c) is similar to the latter, except that the rectified signal has the effect of charging the emitter capacitor C_E, and this DC voltage turns off the transistor. In order to make the negative feedback due to resistance R_E negligible, capacitor C_E should present a very low impedance at the operating frequency.

Figure 20.21 (d) is similar to (c), but choke L_1 is shunted by resistance R_1. As mentioned during the treatment of stability, a low-loss coil L_1 could lead to parasitic oscillations, so a suitable loading of the base should be provided over the whole frequency range: this is the role of resistance R_1. Theoretically, a series resistance could also be used as a series loss, but this would interfere with the DC conditions (Fig. 20.21e). This is eliminated by a further choke L_3 giving a DC short-circuit to R_2; this choke has to be negligible at high frequencies.

In this circuit, a further network comprising C_3 and R_3 is applied to increase stability. The normal value for R_3 is about 10 ohm. High-frequency and low-frequency grounds are provided by C_4 and C_3, respectively. At low frequencies, R_3 serves as a series loss for choke L_2 in the collector circuit, and thus avoids a high Q factor. This eliminates low-frequency instability,

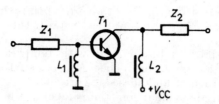

Fig. 20.20. Biasing of class B power amplifier

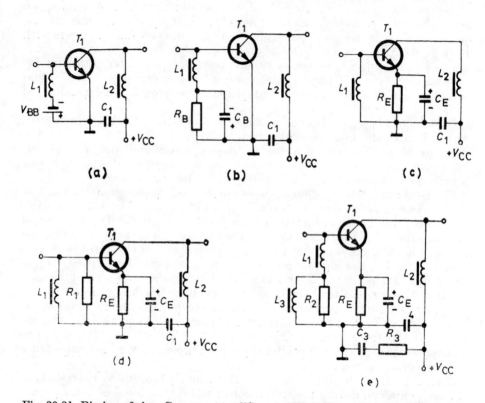

Fig. 20.21. Biasing of class C power amplifier. (a) With V_{BB}, (b) with RC network in the base circuit, (c) with RC network in the emitter circuit, (d) with damped choke, (e) with damped choke and series RC network for stability improvement

particularly when the internal resistance of the power supply is not negligible. At higher frequencies, the series RC network is shunted by C_4.

In the following, power amplifier aspects of various transistor-circuit configurations are considered. For power amplification, common-emitter circuits are primarily used because of the higher available power gain. Common-base and common-collector configurations are used, but less frequently.

20.3. Biasing

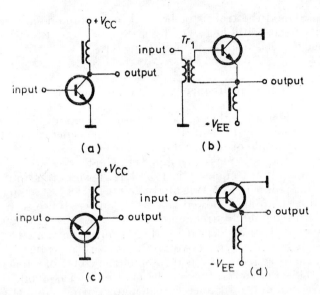

Fig. 20.22. (a) Common-emitter circuit, (b) common-emitter circuit with grounded collector, (c) common-base circuit, (d) common-collector circuit

Figure 20.22 shows the four most popular connections: (a) shows the simplified diagram of a common-emitter configuration. The cooling of the transistor in this circuit at high frequency is difficult, because the heat-sink itself produces a high value capacity in parallel with the collector, impairing the wideband matching. Therefore in high-power, high-frequency transistors the collector is isolated from the case (e.g. by a beryllium-oxide layer), allowing the case to be grounded, thus improving the power-handling capability of the device.

Figure 20.22 (b) shows a common-emitter configuration in which the collector is grounded. The driving voltage is connected between base and emitter by transformer Tr_1. This circuit has the advantage that the collector which must be cooled can be grounded directly, but the disadvantage of the transformer coupling. Another drawback is the fact that the base-to-earth stray capacitance acts as a feedback capacitance since the collector is at earth potential. Therefore, these circuits must usually be neutralized.

Figure 20.22 (c) shows a common-base amplifier usually applied at extremely high frequencies. Its cooling problems are similar to those of the common-emitter configuration. Finally, Fig. 20.22 (d) shows a common-collector circuit which, though advantageous as regards cooling, but is less frequently used because of the low power gain and impractical impedance values.

In conclusion, we mention a type of UHF power amplifier in which the output transistor is connected metallically to the inner conductor of a transmission line. Should the transistor be used in a common-emitter configuration, the collector can still be cooled by metallic contact to an inner conductor, provided this has a large enough diameter.

20.4. High-frequency power amplifier circuits

We shall consider a number of power amplifier circuits in this section to demonstrate the practical use of the theoretical considerations, and to show the order of magnitude of technical data in each case.

Figure 20.23 shows a class B amplifier which uses a single transistor [19.16]. The operating frequency is 175 MHz, terminating impedances are 50 ohm, output power is 13.5 W, supply voltage is 28 V. Calculations based on simplified assumptions showed that the load resistance seen by the collector at the fundamental frequency is $R_c = V_{CC}^2/2P_{out} = 29$ ohm, and this value should be transformed by the output π network to the 50-ohm load resistance. For the applied overlay transistor, the transition frequency is 300 MHz, the base resistance is 0.75 ohm, the collector capacity is about 16 pF, and the series inductance of the emitter lead is about 3 nH. From these data, the input resistance is calculated using eqn. (19.8.1) to be 2.7 ohm, the power gain is 8 dB.

Figure 20.24 shows a class A power amplifier circuit using a common-emitter connection [20.2]. Output transistor T_1 is driven via the high-frequency transformer made up of inductances L_1 and L_2, resulting in a driving voltage between emitter and base electrodes. The collector of T_1 is grounded directly, giving efficient heat transfer. The supply voltage is connected via series resistance R_5 and choke L_5. The output signal is taken from the emitter which is connected to the external load via the π section made up of elements C_5, C_6, C_7 and L_6.

The circuit uses class A operation, with protection as given in Fig. 20.19 (b) against excess dissipation in the absence of the driving signal. Transistor T_2 together with resistances R_2, R_3 and R_4 form a voltage divider which supplies the base voltage of the output stage. The base voltage V_{B2} needed for turning on transistor T_2 is rectified by diode D_1 from the incoming driving signal. Should this decrease, a decreased DC voltage across resistance R_1 results in turning off transistor T_2 and thus representing a higher resistance. This also has the effect of decreasing the base voltage and collector current of transistor T_1 and thus the power dissipated in the transistor. Inductance L_4 has a neutralizing effect by tuning out the collector-base capacitance of transistor T_1 which has a value of about 14 pF for the output transistor used. L_4 is returned to earth via capacitance C_4. Output power is 10 W at the operating frequency of 132 MHz.

Figure 20.25 shows a two-stage wideband power amplifier [12.10]. T network matching four-poles are applied at both the input and the output; L section matching is used between driving transistor T_1 and output transistor T_2. The Q factors of the chokes L_3 and L_6 used for supply-voltage connection are decreased by damping resistances $R_1 = R_2 = 2$ kohm, thus improving stability.

20.4. Power amplifier circuits

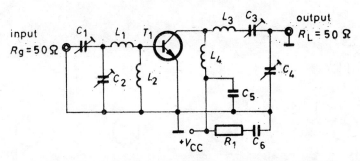

Fig. 20.23. Class B transistor power amplifier for 175 MHz

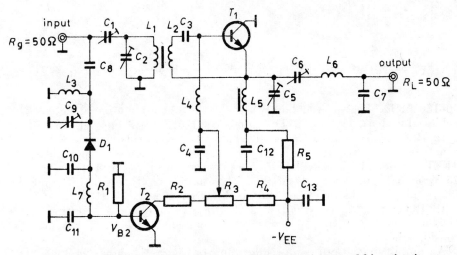

Fig. 20.24. Class A power amplifier for 132 MHz with guard bias circuit

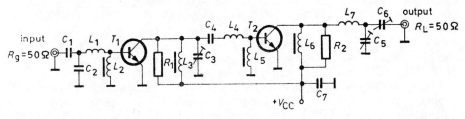

Fig. 20.25. Two-stage wideband power amplifier for the frequency range 80–160 MHz

L_2 and L_5 chokes are chosen to meet the practical requirement of $5Z_{in} < Z_c < 50Z_{in}$, where Z_c is the choke impedance and Z_{in} is the input impedance of the parallel-connected transistor. With a supply voltage of 27 V, the amplifier delivers an output power of 10 W in the frequency range

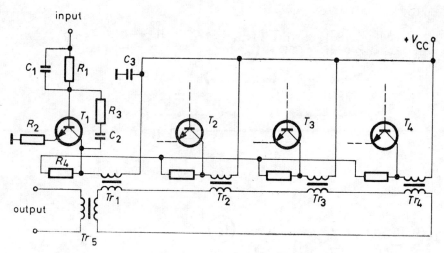

Fig. 20.26. Part of a 1 kW power amplifier for the frequency range 2–32 MHz

80–160 MHz. The desired bandwidth of the output matching network is achieved with a Q factor of 2. The gain is 10 dB at an efficiency of about 55%.

In Ref. [20.9], a high-frequency amplifier using a JFET transistor and π network matching at both input and output is presented. The high-power JFET developed specifically for this purpose, supplies an output power of 25 W at a frequency of 30 MHz, with a supply voltage of 28 V, and the intermodulation distortion product is down by at least 37 dB.

In Ref. [29.16], an amplifier with three parallel-connected transistors delivering 50 W output power in the frequency range 102–128 MHz is considered.

An amplifying group comprising four transistors is shown in Fig. 20.26 [20.4]. The transistor outputs are connected by transmission-line transformers to provide the necessary matching and the combining of the output powers. Several of these amplifying groups may be combined by transformers, thus achieving extremely high output powers. In the amplifier in question, 60 amplifying groups are combined (using partly series and partly parallel combination), thus achieving 1 kW output power in the frequency range 2–32 MHz, with less than -20 dB intermodulation product level.

Figure 20.27 shows a push-pull amplifier with transmission-line transformer coupling [20.10]. Driving transistors T_1 and T_2 are matched to the output stages T_3 and T_4 by transformers having ratios of 4 : 1. The push-pull output signals are combined by hybrids made up of transformers Tr_5 and Tr_6, also providing matching to the external load of 30 ohms. For achieving low distortion, a class AB output amplifier is used, supplying an output power of 60 W in the frequency range 2–30 MHz. Because of the push-pull operation, even-order and thus second-order harmonic levels are significantly reduced. Overall distortion is therefore governed by the level of the third harmonic which is at about -12 dB referred to the fundamental

20.4. Power amplifier circuits

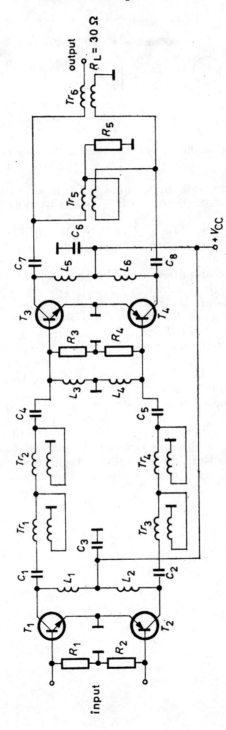

Fig. 20.27. Power amplifier using transmission-line transformers for the frequency range 2–30 MHz

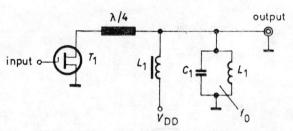

Fig. 20.28. Class F power amplifier using MOS transistor matched with $\lambda/4$ transformer

component at full drive (assuming equal transistors for each pair) [12.11]. The intermodulation distortion level is less than -30 dB, which is acceptable for transmitters.

Reference [20.19] describes a class A push-pull linear amplifier with coupled transmission lines arranged as striplines on the printed circuit board and delivering 50 W output power in the frequency range 160–240 MHz.

The class F power amplifier using MOS transistor matched with $\lambda/4$ transformer is shown in Fig. 20.28 [20.18]. Transistor T_1 operates as a switch. The harmonics are shorted with the parallel resonant circuit at the output tuned at frequency f_0. Due to the $\lambda/4$ transformer a short circuit appears on the drain of the T_1 transistor for the even-order harmonics, and an open circuit for the odd ones. Owing to the latter, the drain current does not contain odd harmonics. The even-order harmonics circulate in a short circuit, therefore, they do not consum power. This is the reason why this circuit has very good efficiency.

Reference [20.20] desribes a broadband amplifier using VMOS transistors and delivering 16 W output power in the frequency range 1.8–54 MHz. In Ref. [18.31] microwave power amplifiers are presented. In Ref. [20.7] an 800-W, 400-MHz amplifier is described, using moduls interconnected by hybrids. Finally in Ref. [20.14] a 50-W amplifier is studied for the frequency range 100–160 MHz, using Chebyshev-type filters.

21. HIGH-FREQUENCY CONVERTERS

21.1. Theoretical fundamentals of conversion

Converter circuits are used to generate combination frequency signals from the two input signals of different frequencies, which results from a nonlinear transfer characteristic. Modulators and, in a certain sense, detectors (conversion detectors) have similar functions. The input converter of a receiver or mixer has the function of converting the incoming RF signal to the IF signal. The properties of nonlinear transfer functions, detailed in the following, are applicable to modulators and also to any nonlinear device, e.g. amplifiers. The relations given are also suitable for the calculation of nonlinear distortion in power amplifiers.

In the low-frequency approximation of the nonlinear circuit, the transfer function is expressed by a power series:

$$i = a_0 + a_1 v + a_2 v^2 + \ldots, \tag{21.1.1}$$

where the current i is a nonlinear function of the voltage v given on the device with the nonlinear characteristic. The function is characterized by coefficients a_0, a_1 etc. which can be determined by DC or low-frequency AC methods. The approximation excludes any frequency dependence of the device.

In the power series (21.1.1), the higher-order terms are usually neglected, and only the quadratic and cubic terms are taken into account:

$$i = a_0 + a_1 v + a_2 v^2 + a_3 v^3. \tag{21.1.2}$$

We want to know what the device output current will be if the input voltage v consists of two sinusoidal voltages of different frequencies, in addition to the DC voltage V_0:

$$v(t) = V_0 + V_1 \sin \omega_1 t + V_2 \sin \omega_2 t. \tag{21.1.3}$$

Substituting this voltage into eqn. (21.1.2) and separating the different frequency terms yields the amplitudes given in Table 21.1. It can be seen that in addition to the DC component and the fundamental frequency, the ω_1 and ω_2, double- and triple-frequency components, and the so-called combination-frequency component will also be present. In receiver mixers, the combination-frequency components $\omega_1 - \omega_2$ is usually used as an intermediate frequency; in this case, ω_1 is the local oscillator frequency, and ω_2 is the modulated signal frequency. The component amplitudes depend on the coefficients of the transfer function and on the amplitudes V_0, V_1 and V_2.

Let us investigate the amplitude of the component with frequency $\omega_1 - \omega_2$. Assuming a quadratic transfer function, then by substituting $a_3 = 0$ and introducing the notations $\omega_1 = \omega_{osc}$, $\omega_2 = \omega_{RF}$ and $\omega_1 - \omega_2 = \omega_{IF}$, we obtain:

$$I_{IF} = a_2 V_{osc} V_{RF} = g_M V_{RF}, \qquad (21.1.4)$$

where $V_{osc} = V_1$ is the amplitude of the local oscillator signal, $V_{RF} = V_2$ is the amplitude of the incoming signal, and $g_M = a_2 V_{osc}$ is the mixer transconductance which is seen to be dependent on the oscillator voltage.

The power-series expression of the nonlinear transfer function together with its use will be demonstrated by a practical example. The collector current of a bipolar transistor, as a function of the emitter–base voltage, may be characterized by an exponential relation:

$$i_c = I_0^* \exp(q v_{eb}/kT). \qquad (21.1.5)$$

Let us assume an emitter–base voltage of the form:

$$v_{eb}(t) = V_{EB} + V_{RF} \sin \omega_{RF} t + V_{osc} \sin \omega_{osc} t = V_{EB} + v(t), \qquad (21.1.6)$$

where V_{EB} is the DC component. Substituting (21.1.6) into expression (21.1.5) of the collector current, and performing the power series expansion, we have

$$i_c(t) = I_0 \left\{ 1 + \frac{q v(t)}{kT} + \frac{1}{2}\left[\frac{q v(t)}{kT}\right]^2 \right\}, \qquad (21.1.7)$$

where $I_0 = I_0^* \exp(qV_{EB}/kT)$. It can be seen from eqn. (21.1.7) that the value of the quadratic coefficient, characteristic for the conversion, is $a_2 = I_0/2$, so the amplitude of the IF current, through (21.1.4), is given by

$$I_{IF} = \frac{I_0}{2} \frac{qV_{osc}}{kT} \frac{qV_{RF}}{kT}. \qquad (21.1.8)$$

The transconductance at the fundamental frequency is given by $g_0 = qI_0/kT$, so the conversion transconductance is given by

$$g_M = \frac{qI_0}{2kT} \frac{qV_{osc}}{kT} = \frac{g_0}{2} \frac{qV_{osc}}{kT}, \qquad (21.1.9)$$

and is proportional to the oscillator voltage.

In another calculation method of mixers, the time function of the transconductance is determined. Differantiating eqn. (21.1.2) with respect to v, the transconductance of the nonlinear device is determined:

$$g_m = \frac{di}{dv} = a_1 + 2a_2 v + 3a_3 v^2, \qquad (21.1.10)$$

21.1. Theoretical fundamentals

Table 21.1. Component amplitudes of different frequencies for third-order characteristic

frequency	amplitude
$\omega = 0$	$a_0 V_0 + a_2 V_0^2 + \dfrac{a_2}{2}(V_1^2 + V_2^2) + a_3 V_0^3 + \dfrac{3a_3}{2} V_0 (V_1^2 + V_2^2)$
ω_1	$a_1 V_1 + 2a_2 V_0 V_1 + \dfrac{3a_3}{2} V_1 \left[2V_0^2 + \dfrac{V_1^2}{2} + V_2^2 \right]$
ω_2	$a_1 V_2 + 2a_2 V_0 V_2 + \dfrac{3a_3}{2} V_2 \left[2V_0^2 + \dfrac{V_2^2}{2} + V_1^2 \right]$
$2\omega_1$	$-\dfrac{a_2}{2} V_1^2 + \dfrac{3a_3}{2} V_0 V_1^2$
$2\omega_2$	$-\dfrac{a_2}{2} V_2^2 + \dfrac{3a_3}{2} V_0 V_2^2$
$3\omega_1$	$-\dfrac{a_3}{4} V_1^3$
$3\omega_2$	$-\dfrac{a_3}{4} V_2^3$
$\omega_1 \pm \omega_2$	$\mp a_2 V_1 V_2 \mp 3a_3 V_0 V_1 V_2$
$2\omega_1 \pm \omega_2$	$\mp \dfrac{3a_3}{4} V_1^2 V_2$
$2\omega_2 \pm \omega_1$	$\mp \dfrac{3a_3}{4} V_2^2 V_1$

which is the function of voltage v. If this voltage is time-dependent and of the form $v = V_{osc} \cos \omega_{osc} t$, then the transconductance will also be time-dependent, $g_m(t) = g_{m0} + g_{m1} \cos \omega_{osc} t + g_{m2} \cos 2\omega_{osc} t$, where the coefficients are:

$$g_{m0} = a_1 + \frac{3}{2} a_3 V_{osc} ; \qquad (21.1.11)$$

$$g_{m1} = 2 a_2 V_{osc} ; \qquad (21.1.12)$$

$$g_{m2} = \frac{3}{2} a_3 V_{osc}^2 . \qquad (21.1.13)$$

If the time dependence of the transconductance is known, the current $i(t)$ due to a small amplitude AC voltage $V_{RF}(t) = V_{RF} \cos \omega t$ may be calculated from the relation $i(t) = g_m(t)\,u_{RF}(t)$. The component of frequency $\omega_{osc} - \omega_{RF}$ has the amplitude value of $I_{IF} = g_{m1} V_{RF}/2$. Transconductance g_{m1} is twice g_M: $g_{m1} = 2g_M$. This is valid for both additive and multiplicative conversion.

In the expression of the transconductance time function, it has been assumed that neither the current nor the transconductance ever reaches zero. In other words, the converter operates in class A. However, converter adjustments are possible, allowing the mixing device to conduct for just a brief time interval. In the remaining part of the cycle, no current flows, and thus transconductance is zero. These class B and class C mixers are suitable for achieving extremely high conversion transconductance and good efficiency.

In the case of class C biasing, the time dependence of the transconductance defined by eqn. (21.1.9) is shown in Fig. 21.1 (a). During the flow angle 2α, the transconductance is increased to the maximum value g_{max}. To express this time function by a Fourier-series, a knowledge of the exact waveform would be necessary; however, good approximation may be achieved by using a truncated sinusoidal approximation. This method was applied in Chapter 19. With this approximation and knowledge of flow angle 2α, the coefficients may be determined from Fig. 21.1 (b). This figure corresponds to Fig. 19.11 (b), only the notations have been altered, introducing transconductance components instead of collector-current components.

It can be seen from the figure that both the transconductance g_{m1} for the fundamental frequency and the conversion transconductance g_M have a maximum as a function of the flow angle; it is advisable to choose this working point. The flow angle providing the highest conversion transconductance depends on the waveform. For practical time functions of $g_m(t)$, the optimum flow angle is between 90° and 130°.

From a knowledge of the conversion transconductance, the voltage gain may be calculated from the relation:

$$V_{IF}/V_{RF} = A_v = g_M R_L, \qquad (21.1.14)$$

where V_{IF} is the output IF voltage, and R_L is the IF load resistance. The load is selective, i.e. it represents approximately a short circuit for the frequency of the input signal (ω_{RF}) and for the frequency of the local oscillator signal (ω_{osc}), and also for other combination frequencies.

The fictive equivalent circuit of mixers has admittances for two different frequencies (Fig. 21.2). The equivalent circuit is made up of an input admittance referred to the frequency ω_{RF} an output admittance referred to the frequency ω_{IF} and transfer parameters. Although all parameters are complex quantities, the transfer parameters are denoted as conductances in the figure since the phase lag is not definable because of the differing frequencies.

The inverse transfer parameter g_i has no real significance since for the customary solid-state devices, no inverse conversion takes place.

A knowledge of the high-frequency nonlinear behaviour of the transistor is of importance from the aspect of nonlinear distortion. Related calculations may be carried out by using the various nonlinear equivalent circuits.

21.1. Theoretical fundamentals

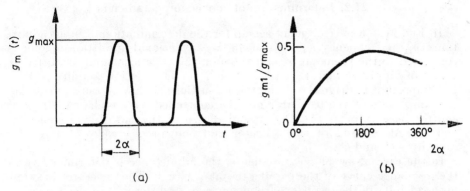

Fig. 21.1. (a) Transconductance time function, (b) fundamental frequency component of transconductance as a function of flow angle

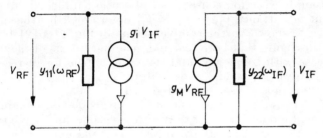

Fig. 21.2. Equivalent circuit of a mixer

From these, the best approximation is attained by the charge-controlled transistor models [21.15], [21.16], [21.17]. In Ref. [21.16], the third-order nonlinear distortion is expressed as a function of the load resistance, the transition frequency f_T and the ratio $\partial^2 f_T/\partial I_C^2$. In Ref. [21.14], the relation between high-frequency cross-modulation and third-order intermodulation is dealt with. In Ref. [10.20], nonlinear distortions are investigated using the relation

$$i_b(t) = [\tau(i_c) + R_L C_c(i_c)]\frac{\partial i_c}{\partial t}, \qquad (21.1.15)$$

where i_b is the base current, i_c is the collector current, τ is the delay time, and C_c is the collector capacity dependent on the collector current. According to calculation, it is theoretically possible to eliminate the second-order intermodulation distortion by suitable choice of quiescent collector current.

In Ref. [21.12], the distortion of JFET's, based on the nonlinear equivalent circuit, is considered. In Ref. [21.18], the high-frequency cross-modulation in JFET's is discussed.

21.2. Determination of conversion parameters

In Fig. 21.3, a measurement set-up for the determination of input admittance y_{11} at frequency ω_{RF} is shown. A conventional admittance-meter is connected to the input ports, and the admittance is measured for different values of supply voltage V_{CC}, DC current I_E and oscillation amplitude V_{osc}. It is important to drive the circuit with a sufficiently low measuring voltage. The amplitude of the oscillator may be controlled by the change ΔV_{BE} of emitter–base voltage. The DC ground return of the base is via choke L_1, and the AC ground of the collector and emitter electrodes is given by capacitors C_2 and C_3.

In the case of swept measurements, the admittance is determined from the resonance curve of the parallel resonant circuit. The generator is swept in the vicinity of the resonance frequency ω_{RF}, and the shift of the resonance frequency allows the determination of the imaginary part of the admittance, while the flattening of the curve yields the real part. At higher frequencies, a transmission-line section is used instead of the parallel circuit in the base.

The input admittance $y_{11} = g_{11} + jb_{11}$ determined by the measurement set-up above is shown on the complex plane in Fig. 21.4 [21.11]. It can be seen from this figure that both real and imaginary parts of the input admittance increase with frequency, and the same is true for increasing emitter current I_E. However, at higher oscillator voltages (higher ΔV_{BE} voltages) the input admittance is somewhat reduced.

Figure 21.5 shows the measurement of conversion transconductance. The DC adjustment circuit is similar to the previous one, and again the measurement parameters are the supply voltage V_{CC}, the emitter DC current I_E, and the change ΔV_{BE} in emitter–base voltage due to the oscillator voltage. The input voltage V_{RF} is supplied by a generator of resistance R_{g2}. The converted intermediate-frequency signal, appearing at the collector, is

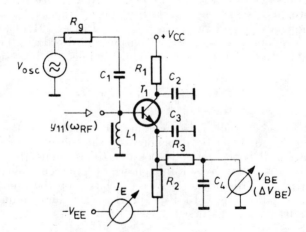

Fig. 21.3. Set-up for measuring the input admittance of a transistor mixer using an admittance-meter

21.2. Conversion parameters

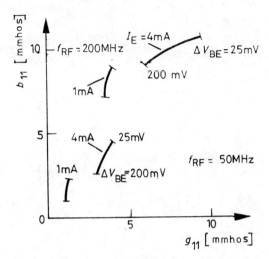

Fig. 21.4. Input admittance of a bipolar transistor mixer on the complex plane for various frequencies, emitter currents and oscillation amplitudes

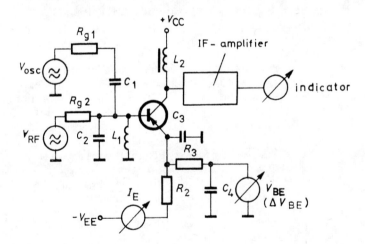

Fig. 21.5. Determination of transistor conversion transconductance

transmitted to the indicator through an amplifier with a low input resistance. The indicator may be calibrated to show the absolute value of transconductance g_M directly. Calibration is achieved by short-circuiting the base and collector electrodes and suitable adjustment of the intermediate frequency amplification.

In Fig. 21.6, the transconductance measured as above is presented as a function of voltage change ΔV_{BE} (i.e. oscillation amplitude), for various values of emitter current I_E and two input frequencies ω_{RF}. In the figure,

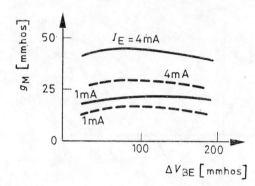

Fig. 21.6. Dependence of conversion transconductance on the oscillator amplitude for various emitter currents. Solid line: $f = 50$ MHz, dashed line: $f = 200$ MHz

the measured parameters of a modern converter transistor are shown, and it can be seen that the conversion transconductance is barely dependent on the oscillation amplitude.

The output admittance of the transistor mixer is measured by the circuit shown in Fig. 21.7. The parameters are: collector voltage V_{CC}, emitter DC current I_E and voltage ΔV_{BE} used indirectly for the determination of the oscillator voltage. The output admittance may be measured either by a conventional admittance-meter or by a sweep-frequency method.

An important property of converters is the *signal-handling capability*, i.e. the sensitivity to high-amplitude interfering signals. If a cubic term is contained in the nonlinear characteristic, an interfering modulated signal reaching the input may cause cross-modulation above a certain level. This means that its modulation is transferred to the required signal, despite the fact that the signal converted from the interfering signal is completely attenuated. In Fig. 21.8, data of two different converter transistors are presented. For a given level of cross-modulation distortion, substantially higher interfering input signal level V_d (given in mV) may be allowed for the better transistor denoted by 2. The interfering signal is shown as a function of the emitter DC current I_E, and it is seen that there is an optimum (maximum value of V_d) at some value of I_E which, however, does not necessarily yield the highest conversion transconductance.

The signal-handling capability is much greater with field-effect transistors. This may be understood by considering that the voltage range needed to drive field-effect transistors is of the order of volts, while this range is a few tens of mV for bipolar transistors.

Figure 21.9 (a) shows the dependence of the source current I_s on the gate voltage V_G for a depletion-type MOS field-effect transistor. For low values of negative V_G voltages, the transistor is highly turned on, the current is approximately a linear function of voltage V_G, and the transconductance is high. For amplifier operation, the working point should be chosen somewhere in this range. However, the linear characteristic is not suitable for conversion, so the working point has to be chosen around point M which

21.2. Conversion parameters

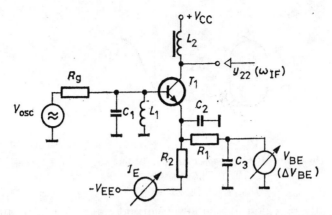

Fig. 21.7. Determination of transistor mixer output impedance

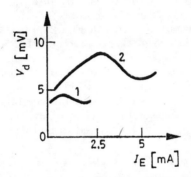

Fig. 21.8. Level of interfering voltage introducing 1% cross-modulation as a function of emitter current

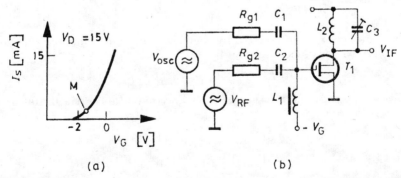

Fig. 21.9. (a) Transfer characteristic of MOS transistor, (b) application of MOS transistor in a mixer

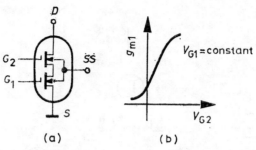

Fig. 21.10. (a) Dual-gate MOS transistor circuit, (b) dependence of the transconductance on the opposite gate voltage

is in the nonlinear region, in the lower current range. With suitable choice of working point, both the nonlinearity and the transconductance are acceptable, so a favourable conversion gain may be attained. At very low currents, the situation is again worse: despite higher nonlinearity, the transconductance will be extremely small.

An additive mixer circuit is shown in Fig. 21.9 (b), where both the radio-frequency and the oscillator signals are given on the gate, and the converted IF signal is selected out by the resonant circuit connected to the drain.

Figure 21.10 (a) shows the circuit of a *dual-gate* MOSFET which may be applied both as a gain-controlled amplifier and a high-frequency mixer [15.15]. Figure 21.10 (b) shows the dependence of the lower FET transconductance g_{m1} on the voltage V_{G2} given on gate G_2, at a constant voltage V_{G1}. The characteristic may be considered to be linear in the middle region) so that

$$g_{m1}(V_{G2}) = g_{m1}(V_{G20}) + a_1(V_{G2} - V_{G20}), \qquad (21.2.1)$$

which is a linear relation, characterized by coefficient a_1. A similar expression may be written to express the dependence of the second (upper) FET transconductance g_{m2} on the voltage V_{G1} given on gate G_1, at a constant voltage V_{G2}:

$$g_{m2}(V_{G1}) = g_{m2}(V_{G10}) + a_1'(V_{G1} - V_{G10}), \qquad (21.2.2)$$

where the slope is characterized by coefficient a_1'. The common current of this two-transistor device is given by

$$I_D = g_{m1}V_{G1} + g_{m2}V_{G2}. \qquad (21.2.3)$$

Substituting the equations giving the transconductances, the drain current will be given as:

$$I_D = g_{m1}(V_{G20})V_{G1} + g_{m2}(V_{G10})V_{G2} + \\ + a_1(V_{G2} - V_{G20})V_{G1} + a_1'(V_{G1} - V_{G10})V_{G2}. \qquad (21.2.4)$$

From the above terms, only those containing the products of the two gate voltage will be important as these yield the conversion products. Let us

21.3. Mixer circuits

denote this current component containing the gate voltage products by i_d, and let us separate it out in eqn. (21.2.4):

$$i_d = (a_1 + a_1')v_{G1}\,v_{G2}\,. \qquad (21.2.5)$$

If G_1 is driven by the RF input signal $v_{G1} = V_{RF} \sin \omega_{RF} t$, and G_2 is driven by $v_{G2} = V_{osc} \sin \omega_{osc} t$, then the current to be determined will be

$$i_d = \frac{1}{2}(a_1 + a_1')V_{RF}\,V_{osc}\,[\cos(\omega_{osc} + \omega_{RF})t + \cos(\omega_{osc} - \omega_{RF})t]\,. \quad (21.2.6)$$

Since the second term in the above expression is the converted IF component ($\omega_{IF} = \omega_{osc} - \omega_{RF}$), the transconductance of the device will be

$$g_M = \frac{di_d}{dV_{RF}} = \frac{a_1 + a_1'}{2} V_{osc}\,. \qquad (21.2.7)$$

According to measurements, the conversion transconductance of a dual-FET is a linear function of the oscillator voltage over a wide voltage range, and its maximum value can be attained by correct choice of coefficients a_1 and a_1', i.e. of the DC bias voltages V_{G10} and V_{G20}.

21.3. High-frequency mixer circuits

In this section, we shall present a few specific high-frequency converter circuits, made up partly by discrete circuit elements such as diodes and transistors, and partly by integrated circuits.

Figure 21.11 shows a diode mixer circuit. The working point along the diode characteristic may be chosen by correct selection of elements R_1 and C_1. This is possible because the oscillator voltage is rectified by the diode, and thus capacitor C_1 is charged to turn off the diode. Diode current flows only during positive peaks of the oscillator voltage, and this determines the time function of the transconductance. A disadvantage of the diode converter is the lack of isolation between IF output and RF input. These

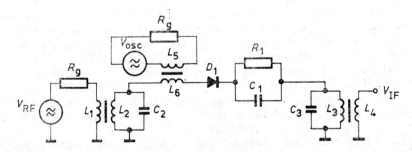

Fig. 21.11. Diode mixer

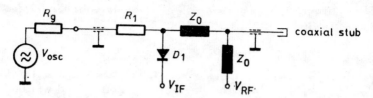

Fig. 21.12. High-frequency diode mixer with coaxial mount

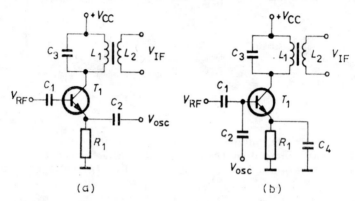

Fig. 21.13. Oscillator signal coupling to the mixer transistor. (a) To the emitter electrode, (b) to the base electrode

are strongly interdependent, as shown by the set of mixer equations:

$$I_{RF} = g_{mo} V_{RF} + g_M V_{IF} ; \qquad (21.3.1)$$

$$I_{IF} = g_M V_{RF} + g_{mo} V_{IF} . \qquad (21.3.2)$$

Here g_{mo} is the DC component of the time-dependent transconductance, and g_M is the conversion transconductance. The coefficients of the transconductance time function may be calculated from the diode characteristic, the bias voltage (i.e. the elements $R_1 C_1$), and also from the amplitude of the applied oscillator voltage.

Figure 21.12 shows the simplified circuit of a high-frequency mixer using coaxial transmission-line sections. Mixer diode D_1 and the resistors are contained in a coaxial mount of characteristic impedance Z_0. The coaxial stub prevents the oscillator voltage from reaching the RF input by acting as a short-circuit. At the IF output port, the DC current flow of the mixer diode should be made possible (e.g. by applying a choke). Resistance R_1 and a coaxial voltage divider following the generator of voltage V_{osc} provide an optimum voltage level at the diode. This configuration is used primarily in the frequency range above 100 MHz.

In Fig. 21.13, transistor mixer circuits operating in the common-emitter configuration are shown. In Fig. 21.13 (a), the RF signal is given on the

21.3. Mixer circuits

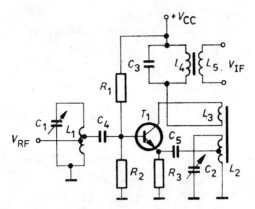

Fig. 21.14. Self-oscillating transistor mixer

base, while the emitter is driven by the oscillator signal. For correct operation the base should be at earth potential at the oscillator frequency, and the emitter should be at earth potential at the radio frequency. This may be acomplished by parallel resonant circuits, assuming suitable Q factors. In Fig. 21.13 (b), signals of both frequencies are given on the base. Interaction between the two voltage sources may be decreased by applying small coupling capacitors C_1 and C_2. This latter type of mixer is mainly applied above 100 MHz, because of the low-Q resonant circuits, and because of the relatively close spacing of the radio frequency and oscillator frequency.

A typical self-oscillating mixer circuit for short-wave applications is shown in Fig. 21.14. Resonant circuits of frequencies ω_{RF} and ω_{osc} are tuned by ganged capacitors C_1 and C_2. Tappings of these resonant circuits are connected to the base and emitter of the mixer transistor. The oscillator operates in a common-base configuration, inductive feedback is supplied by inductance L_3. At short-wave frequencies, the oscillator frequency is much higher than the intermediate frequency, so the impedance transformed by inductance L_3 in series with the collector may be neglected in comparison with the impedance of the IF resonant circuit. The reverse is true for the impedances appearing at the oscillator frequency.

A common-base mixer circuit is shown in Fig. 21.15. The base of mixer transistor T_1 is grounded by C_4, and bias is set by the voltage divider $R_1 - R_2$. The input signal reaches the emitter via C_1. The collector IF resonant circuit is matched to the low-resistance load by capacitor C_9. The oscillator signal is generated by transistor T_2 which has a feedback given by capacitance C_6. This oscillator circuit is widely used in the UHF band up to the GHz frequency range because of its simplicity and advantageous properties. The bias circuit is similar to that of the mixer transistor. The oscillator signal is coupled to the mixer emitter electrode from a tapping of the oscillator-frequency resonant circuit through a small series capacitor C_3.

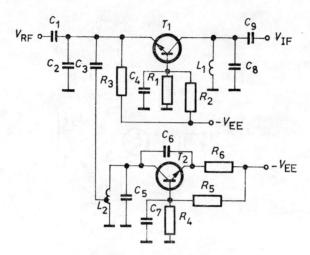

Fig. 21.15. Mixer circuit for the UHF range

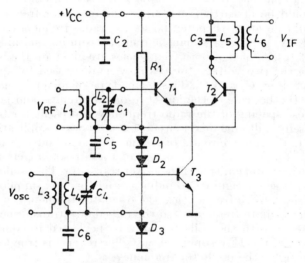

Fig. 21.16. Monolithic integrated-circuit mixer

Figure 21.16 shows a monolithic integrated-circuit mixer. The RF signal is coupled via inductances $L_1 - L_2$ to the base of transistor T_1 whose current is controlled by transistor T_3 according to the instantaneous value of the oscillator voltage. The time-varying emitter current introduces a time-varying transconductance, thus providing a conversion effect. The working point is set by resistance R_1 and diodes $D_1 - D_3$, and the resonant circuit tuned by C_3 serves to select the converted intermediate-frequency signal. Transistor T_2, as a common-base amplifier, prevents the IF voltage from reaching the input port.

21.3. Mixer circuits

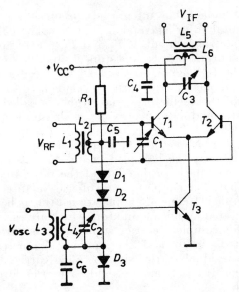

Fig. 21.17. Monolithic integrated-circuit balanced modulator

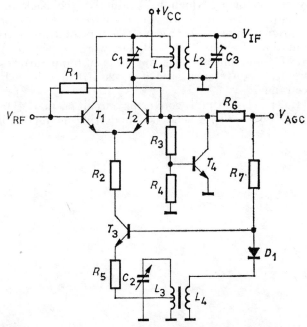

Fig. 21.18. Balanced mixer with local oscillator and bias circuit

In Fig. 21.17 a monolithic integrated circuit is shown as a balanced modulator, also suitable for conversion purposes. The balanced property is given by the symmetrical circuit of T_1 and T_2, preventing oscillator signal from reaching the output port. Transistors T_1 and T_2 are driven in a push-pull manner by the RF signal, so the load is placed symmetrically, with the aid of a high-frequency transformer, between collectors of T_1 and T_2. On the other hand, the oscillator signal (which is called the carrier signal in conjunction with modulators) provides a simultaneous turning on and off of transistors T_1 and T_2 through transistor T_3, and thus the oscillator signal does not appear between the two collectors, provided the symmetry is perfect. This means that the balanced modulator has carrier suppression, allowing only the modulation sidebands of the radio frequency voltage to appear at the output. The simplest form of the balanced modulator is the ring modulator; in high-frequency applications, this is being replaced more and more by integrated-circuit modulators which are easier to apply.

The condition for complete carrier suppression is perfect symmetry for both the integrated circuit and the applied symmetrical transformers. In practice, some residual carrier signal, called carrier leakage, is always present. Careful construction may provide a carrier level -30 dB below the sideband level.

Figure 21.18 shows an integrated-circuit balanced mixer. Operation is similar to the previous circuit, but here the base of T_2, which is one of the push-pull controlled transistors, has an AC ground potential via the low-impedance feedback circuit provided by transistor T_4. Oscillator voltage is generated by transistor T_3 which has inductive feedback in the emitter circuit, and this provides an identical phase drive to transistors T_1 and T_2. The oscillator signal appears between collectors of transistors T_1 and T_2 with an attenuation of about 25–30 dB. The working point of the oscillator is set by diode D_1, and this also has the effect of temperature stabilization. With this circuit, a conversion gain of 15–20 dB may be achieved in the medium wave range.

In Ref. [1.33], a push-pull converter circuit using MOS transistors is presented. In Ref. [1.30], a Schottky-diode ring-modulator is treated.

22. FREQUENCY MULTIPLIERS

22.1. Basics of operation

Frequency-multiplier circuits are intended to generate harmonic signals from the input signal as the fundamental-frequency signal. Second or third harmonics, or at a maximum fifth or sixth harmonics, are usually generated. For higher-order multiplication, two multiplier stages are required, assuming that the multiplication number may be expressed as the product of two integers. The level of harmonics higher than sixth-order is extremely low, making the multiplication process uneconomical.

The fundamental-frequency signal may be distorted by the use of nonlinear characteristic. To compensate for the attenuation thus introduced, it is possible to use an amplifying device, e.g. a bipolar transistor as a nonlinear element. The transistor input characteristic has high nonlinearity, especially in class C operation, and is thus suitable for distorting the fundamental frequency signal.

Figure 22.1 (b) shows the simplified circuit of the class C transistor multiplier, while the broken-line transfer characteristic and the signal waveforms are shown in Fig. 22.1 (a). The transistor working point is set to class C operation by the reverse voltage V_{BB}, so collector current only flows during positive peaks of the generator voltage waveform, during the time interval of the flow angle 2α. From the collector current pulses, the harmonic amplitudes are determined by Fourier analysis. The first step is the determina-

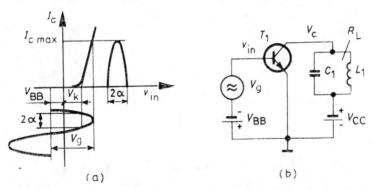

Fig. 22.1. Multiplier operating with a bipolar transistor. (a) Transfer characteristic, (b) simplified circuit

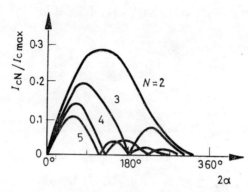

Fig. 22.2. Harmonic content relative to the maximum current value for the case of a truncated sine-wave collector current pulse as a function of flow angle

tion of the flow angle: from Fig. 22.1 (b),

$$\cos \alpha = (V_{BB} + V_k)/V_g, \qquad (22.1.1)$$

where V_k is the break-point voltage of the characteristic. The second important parameter is the shape of the collector-current pulse. According to our assumption, the transfer characteristic is a broken-line characteristic, i.e. it is linear in the turning-on region. This results in collector-current pulses with truncated sinusoidal waveform, allowing the harmonic content to be determined from Fig. 19.11. In this figure, the harmonic values are not referred to the maximum current $I_{c\,max}$, so the set of curves are repeated in Fig. 22.2. As a function of flow angle, the harmonic values show several local maxima, and third- and higher-order harmonics attain zero value for certain flow angles. The maximum amplitude as given in Fig. 22.2 should be approximated when adjusting the flow angle. The optimum flow angle may be determined from the relation $2\alpha_{opt} \approx 240°/N$, and the relative amplitude pertaining to this angle is given by

$$\frac{I_{cN}}{I_{c\,max}} (2\alpha_{opt}) \approx \frac{0.56}{N}. \qquad (22.1.2)$$

This allows the voltage amplitude of the Nth harmonic at the transistor collector to be calculated. If the impedance of the resonant circuit tuned to the investigated harmonic frequency is R_L, then the collector voltage is given by

$$V_{cN} = I_{c\,max} \left(\frac{I_{cN}}{I_{c\,max}}\right) R_L, \qquad (22.1.3)$$

where the ratio $I_{cN}/I_{c\,max}$ is a function of the flow angle 2α.

During the above investigation, the maximum attainable output voltage was considered, without taking into account the power relations of the multiplier. In the case of low-power multipliers, the limiting factor is only

22.1. Basics of operation

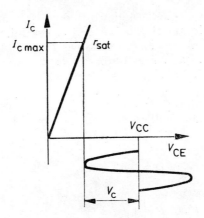

Fig. 22.3. Transfer characteristic for calculation of high-power multipliers

the maximum collector current $I_{c\,max}$. However, high-power circuits are frequently limited by the allowable dissipation and sometimes by the collector breakdown voltage.

The design procedure for high-power multipliers is similar to that for power amplifiers. The difference is that the selective network in the collector circuit is not tuned to the fundamental frequency but to some harmonic frequency. This means a more severe collector peak-current limitation, because the higher the harmonic number, the higher the collector peak current.

Let us now calculate the maximum output power as a function of the load resistance R_L for a given multiplication number N, a given supply voltage V_{CC} and a constant saturation resistance r_{sat}; this will also result in minimum dissipation. Calculation will be based on Fig. 22.3; the maximum collector current at the negative peak of the collector voltage of amplitude V_c will be given by

$$I_{c\,max} = \frac{V_{CC} - V_c}{r_{sat}} = \frac{V_{CC} - I_{cN} R_L}{r_{sat}}. \tag{22.1.4}$$

Utilizing this relation for the peak current, the output power appearing across resistance R_L may be expressed in the form:

$$P_{out} = P_0 \frac{4 \left(\dfrac{I_{cN}}{I_{c\,max}}\right)^2 \dfrac{R_L}{r_{sat}}}{\left[1 + \left(\dfrac{I_{cN}}{I_{c\,max}}\right) \dfrac{R_L}{r_{sat}}\right]^2}, \tag{22.1.5}$$

where the power coefficient is $P_0 = V_{CC}^2/8 r_{sat}$, and the ratio $I_{cN}/I_{c\,max}$ is function of the flow angle 2α. It is seen that power P_{out} is a function of resistance R_L and flow angle 2α and will have a maximum value at their given values. The optimum flow angle $2\alpha_{opt}$ pertaining to the maximum output power P_{max} and the load resistance R_{opt} may be determined by extremum-value calculation. The optimum flow angle is again given by $2\alpha_{opt} \approx 240°/N$, and the optimum load resistance is given by

$$R_{opt} = r_{sat} N/0.56 \,. \tag{22.1.6}$$

With these values, the maximum output power will be given by

$$P_{max} = 0.56\, P_0/N \,, \tag{22.1.7}$$

and the dissipation power in this case is

$$P_d = P_0 \left\{ 4 \left(\frac{I_{c0}}{I_{c\,max}} \right) - \frac{0.56}{N} \right\}. \tag{22.1.8}$$

In this expression, the ratio $I_{c0}/I_{c\,max}$ which determines the DC component may be read off from Fig. 19.11 (b), for different flow angles. If the dissipation power is known, the multiplication efficiency may also be given:

$$\eta = \frac{P_{max}}{P_{DC}} = \frac{0.14}{N \left(\dfrac{I_{c0}}{I_{c\,max}} \right)} \,, \tag{22.1.9}$$

where P_{DC} is the total DC power consumption.

The parameters resulting in maximum output power are summarized in Table 22.1 for several values of multiplication factor N.

Table 22.1. Optimum flow angles and power levels for various multiplication numbers

N	$2\alpha_{opt}$	$\dfrac{I_{cN}}{I_{c\,max}}$	$\dfrac{P_{max}}{P_0}$	$\dfrac{R_{opt}}{r_{sat}}$	$\dfrac{P_d}{P_0}$	$\eta = \dfrac{P_{max}}{P_{DC}}$ [%]
2	120°	0.28	0.28	3.6	0.6	32
3	80°	0.186	0.186	5.4	0.375	33
4	60°	0.14	0.14	7.1	0.26	35
5	48°	0.11	0.11	8.7	0.2	44

What happens if the dissipation power P_d exceeds that allowed for the transistor, despite optimum adjustment? As P_d depends on the supply voltage V_{CC}, it seems logical to decrease the supply voltage. However, the supply voltage is frequently a given value which should be considered in the multiplier design procedure, and not vice versa. What adjustment is then needed to just achieve the allowable dissipation power P_d with the original supply voltage V_{CC}? There are no elementary relations which give flow angle and load resistance needed for such an adjustment. The following

22.1. Basics of operation

method will therefore be used. Firstly, the flow angle should be decreased and the load resistance should be increased for less power dissipation. The modified parameters will result in a power dissipation which can be calculated from the relation:

$$P_\mathrm{d} = 4P_0 \left\{ \frac{2\left(\dfrac{I_{c0}}{I_{c\,\mathrm{max}}}\right)}{1 + \left(\dfrac{I_{cN}}{I_{c\,\mathrm{max}}}\right)\dfrac{R_\mathrm{L}}{r_\mathrm{sat}}} - \frac{\left(\dfrac{I_{cN}}{I_{c\,\mathrm{max}}}\right)^2 \dfrac{R_\mathrm{L}}{r_\mathrm{sat}}}{\left[1 + \left(\dfrac{I_{cN}}{I_{c\,\mathrm{max}}}\right)\dfrac{R_\mathrm{L}}{r_\mathrm{sat}}\right]^2} \right\}. \quad (22.1.10)$$

During the next procedure, several related 2α and R_L values will be used to determine by interpolation the plotted values resulting in the desired dissipation according to eqn. (22.1.10), at the same time giving maximum output power. High accuracy is not needed for this calculation, since other factors not considered will impair the accuracy in any case. Such factors are the truncated sine-wave collector-current pulse assumption, the constant value of the saturation resistance r_sat, and last but not least the frequency-dependent behaviour of the transistor, which has not yet been taken into account. It should be noted in this connection that up to one tenth of the transition frequency f_T, this calculation will have sufficient accuracy for most purposes. After building the circuit, the actual parameters should be ascertained by measurements.

In Ref. [22.15], nonlinear transfer is taken into account in the calculation of a transistor multiplier.

Another multiplication method is the application of a nonlinear reactive element, in particular junction diodes having a junction capacity which is dependent on the reverse voltage applied. For high-efficiency multiplication, solid-state diodes having rather high capacity change in the available reverse voltage range are needed, such as varicap diodes or varactor diodes. The principle of operation with such diodes is the following. The diode shows a time-dependent impedance (capacitance) as a result of the fundamental-frequency drive. Expanding this into a Fourier series results in time-varying impedance terms corresponding to the different harmonics. As shown for the converter circuits, time-varying impedance results in output current harmonics which may be selected by suitable selective tuned circuits. The three main types of varactor diode and their multiplication properties were dealt with in Chapter 1.5.

If varying capacity diodes are applied as multipliers, one great disadvantage is the interaction between input and output signals, and further, the rather high attenuation in the absence of an amplifying element. The series loss resistance of the diode also has an adverse effect on the multiplication efficiency and influences the usable upper cut-off frequency. Bipolar transistors may also be used for frequency multiplication because of the voltage-dependent collector-base capacitance, and in some cases they are used in this role.

A suitable device for frequency multiplication is the step-recovery diode, especially for comb-generator uses. If the step-recovery diode is switched from forward to reverse direction, the charge accumulated in the diode is

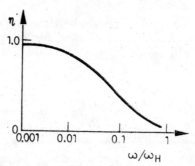

Fig. 22.4. Efficiency of a step-recovery diode frequency-doubler

stored for a while, and is then discharged during a very short time of about 100 ps. The steep voltage rise thus produced comprises a high number of harmonics useful for multiplication. The multiplication efficiency of a step-recovery diode is determined by the series loss resistance r_s. Figure 22.4 shows the frequency dependence of the efficiency, and the cut-off frequency of the diode is given by the relation

$$\omega_H = 1/r_s C \,, \tag{22.1.11}$$

where C is the capacity of the diode.

22.2. Frequency-multiplier circuits

In this section, a few typical frequency-multiplier circuits will be presented, together with the main technical data and details of operation. In Fig. 22.5, a common-base doubler circuit is shown. The transistor input is matched to the generator by trimmer capacitor C_1, and choke L_1 provides the DC ground for the emitter circuit.

This circuit is thus a class C multiplier, the reverse voltage equal the break-point voltage V_k of the input characteristic. The reactance network in the collector circuit made up of elements C_2, L_2 and C_3 form a series resonant circuit for the fundamental frequency and thus attenuates this component, simultaneously acting as a parallel resonant circuit for the required second harmonic, thus not loading this component. The latter condition is expressed by the equation:

$$2\omega_0 \sqrt{L_2 C_2} = 1 \,. \tag{22.2.1}$$

Substituting this into the series resonant-circuit expression for the fundamental frequency, we have

$$j\omega_0 L_2 \| \frac{1}{j\omega C_2} + \frac{1}{j\omega C_3} = 0, \tag{22.2.2}$$

and the relation between the two tuning capacitors is given by $C_3 = 3C_2$.

This method allows good suppression of the fundamental frequency. This not only provides a pure output signal waveform, but prevents the high-

22.2. Frequency-multiplier circuits

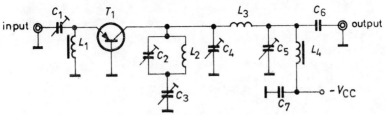

Fig. 22.5. Low-power frequency doubler

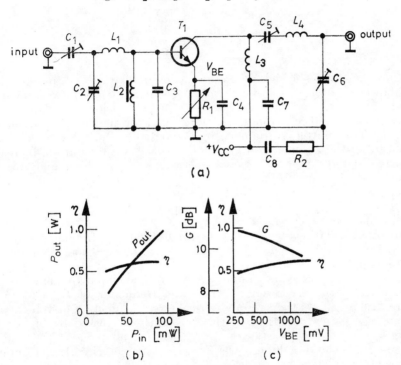

Fig. 22.6. High-power frequency doubler. (a) Circuit diagram, (b) output power and efficiency as a function of input power, (c) gain and efficiency as a function of reverse bias V_{BE}

level fundamental component from driving the transistor into saturation. The saturation may also be decreased by a lower collector impedance, but this would significantly decrease the gain and the available power.

The load is matched to the collector impedance by a low-pass π network made up of elements C_4, C_5 and L_3. This has no effect on the fundamental-frequency suppression. Capacitor C_6 is there for both impedance transformation and DC blocking; the collector DC current flows through choke L_4. The circuit is used for doubling the input frequency of 120 MHz at a relatively low power level, the fundamental frequency suppression is about 50 dB.

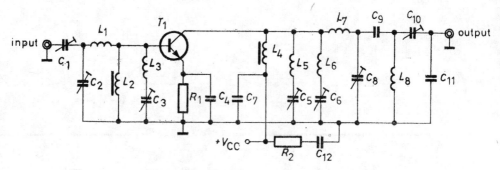

Fig. 22.7. High-power frequency tripler

Figure 22.6 shows a high-power frequency doubler. The transistor input impedance is matched to the generator resistance by a T network made up of elements C_1, C_2 and L_1. Capacitor C_3 represents a low impedance at the base for the second harmonic. The DC ground for the base is provided by choke L_2. The emitter is grounded by C_4, and the adjustable resistance R_1 is used to set the DC voltage V_{BE} required for class C operation. Increasing R_1 increases the voltage V_{BE}, and operation will be closer and closer to class C operation, which increases the efficiency but reduces the gain. The collector impedance is matched to the load by a π network made up of elements L_3, L_4, C_5 and C_6. Capacitor C_7 provides a ground for one end of the high-inductance coil L_3.

Parasitic oscillations which may be set up at lower frequencies are prevented by a shunt capacitor C_8, together with a series loss resistance R_2. Figure 22.6 (b) shows that this doubler circuit generates an output power of 0.5–1 W at an efficiency of about 50%, from an input frequency of 87.5 MHz, using a transistor with a transition frequency of $f_T \approx 300$ MHz. Figure 22.6 (c) shows the gain and efficiency as a function of voltage V_{BE}. The average gain is 10 dB, and this is slightly decreased by higher V_{BE} voltages. Terminating resistances are $R_g = R_L = 50$ ohms.

Figure 22.7 shows a high-power frequency-tripler circuit. The input is matched by the T network made up of elements C_1, C_2 and L_1. Series resonant circuit $L_3 - C_3$ provides a short circuit for the second-harmonic component. The base DC return is through choke L_2. The emitter is AC grounded by C_4, and the very low valued resistance R_1 is used to set the DC bias for class C operation. The series resonant circuits $L_5 C_5$ and $L_6 C_6$ are provided for shorting the fundamental- and second-harmonic frequency components in order to prevent saturation. The remaining reactive elements are used to select the third-harmonic component and for matching to the load.

The high-pass T network made up of elements C_9, C_{10} and L_8 is used for further suppression of the fundamental and second-harmonic component. Choke L_4 is used for DC-supply connection, and AC grounds are provided by C_7 and C_{12}, the latter in connection with series loss resistance R_2. The reactance elements in the collector circuit are adjusted experimentally, by tuning to maximum third-harmonic power. The circuit operates between

22.2. Frequency-multiplier circuits

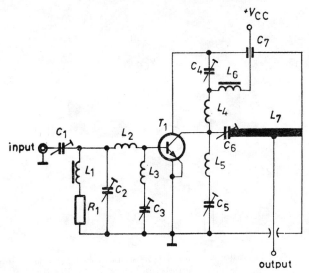

Fig. 22.8. Frequency tripler operating in the UHF range

50-ohm loads, and when driven by an input power of 0.3 W at 150 MHz, provides an output power of 1.5 W at 450 MHz. Efficiency is 45%, the supply voltage is +28 V, collector DC current is about 130 mA, and the transition frequency is about 300 MHz.

Figure 22.8 shows a very high-frequency tripler which uses a microwave overlay transistor [22.2]. The input circuit is similar to the previous T network, the short circuit for the second harmonic is provided by the series resonant circuit L_3C_3. The base DC return is via choke L_1, and a series resistance R_1 is applied to prevent low-frequency oscillations (see the corresponding part in connection with power amplifiers). Because of the very high frequency, the inductance in the collector circuit is implemented by a shorted transmission-line section. The load is connected to a low-impedance point along the transmission line, near the short-circuit. Resonant circuits L_4C_4 and L_5C_5 are used for filtering out the fundamental- and second-harmonic components. Tuning is for maximum output power. The collector DC connection is through choke L_6. This circuit generates a 1.02 GHz frequency signal at a power of 1–2.5 W from an input signal of frequency 340 MHz, between 50-ohm terminations.

Figure 22.9 shows a microwave frequency doubler using microstrip technique and resonant transmission-line sections [22.18]. The input signal frequency for the class C multiplier is 1.15 GHz, input power is 140 mW, gain is 7 dB, and efficiency is about 25%. Components of undesirable frequencies are shorted at the input and at the output by transmission-line sections of lengths $\lambda/4$.

An integrated-circuit multiplier is shown in Fig. 22.10 (a), and uses the previously met three-transistor monolithic amplifier. This is a class C amplifier, and transistors T_1 and T_2 turn on only above a certain input voltage

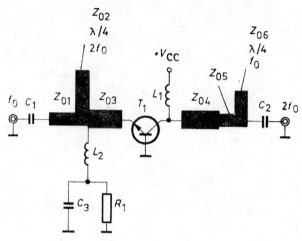

Fig. 22.9. Microwave frequency-doubler with transistor using transmission-line-secti matching

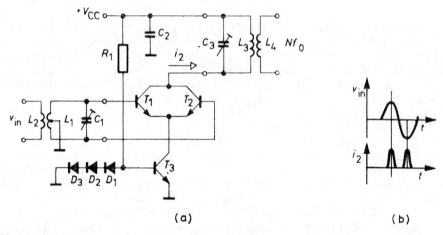

Fig. 22.10. Integrated-circuit frequency multiplier. (a) Circuit diagram, (b) time function of output current

level v_{in} although resembling a push-pull circuit, the connection of the output resonant circuit $C_3 L_3$ causes the circuit to operate in a "push-push" manner.

Circuit operation is illustrated in Fig. 22.10 (b). At positive and negative peak values of the voltage v_{in}, the two transistors become alternately conducting, producing current pulses in the common-collector lead, with a frequency equal to the second harmonic of the driving waveform. Resonant circuit $L_3 C_3$ may be used to select the even-order harmonics from these current pulses.

22.2. Frequency-multiplier circuits

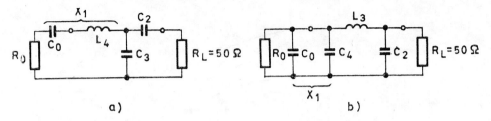

Fig. 22.11. Varactor diode high-power frequency doubler. (a) Utilizing L-section matching networks, (b) utilizing tapped parallel-tuned circuits

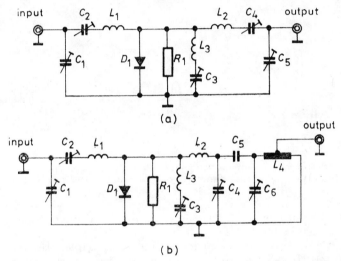

Fig. 22.12. Varactor diode high-power frequency tripler. (a) Utilizing L-section matching network, (b) utilizing an output transmission-line matching circuit

A diode frequency doubler is shown in Fig. 22.11. The applied diodes are varicap or step-recovery diodes at lower power levels [22.12], or varactor diodes at power levels above 0.1 W [22.17]. If the efficiency is not critical, conventional silicon epitaxial switching diodes may also give good results. Input and output reactance elements are used for impedance matching and for filtering out the second harmonic at the output. Class C operation is set by the rectified bias voltage across resistance $R_1 > 30$ kohm.

The filters are L networks in Fig. 22.11 (a), and L networks realized by parallel-tuned circuit in Fig. 22.11 (b). The low-impedance generator and load are connected to tappings on the parallel tuned circuits. At higher frequencies, the inductance of the parallel circuit is replaced by a shorted transmission-line section; in this case, the generator and load are connected to a low-impedance point of this section. These circuits may be used to achieve, by use of modern varactor diodes, extremely high power levels

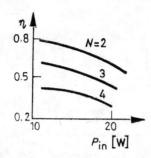

Fig. 22.13. Efficiency of a varactor-diode frequency multiplier in the frequency range 0.5–2 GHz as a function of input power, for different orders of multiplication

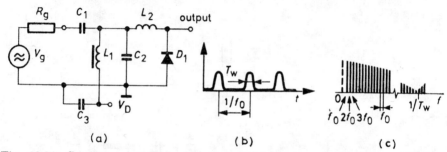

Fig. 22.14. Step-recovery diode frequency-comb generator. (a) Circuit diagram, (b) output signal time function, (c) spectrum of the output signal

above 100 W with 50–60% efficiency multiplication, particularly in the frequency range below 1 GHz.

Figure 22.12 shows a varactor diode frequency tripler. Input and output reactive filters are similar to the previous solution. Series resonant circuit L_3C_3 serves as a short for the second harmonic. In the circuit shown in Fig. 22.12 (b), a shorted output transmission-line section is applied, with a tapping connected to the load. Similarly to the doubler circuit, class C operation is again provided by the DC voltage across high resistance R_1. When applying a high-power varactor with a cut-off frequency of $f_H = 25$ GHz, an input driving power of 25 W at a frequency of 150 MHz will result in an efficiency of about 60%. The cut-off frequency of the varactor is calculated from the relation $f_H = 1/2\pi r_s C$, where r_s is the series loss resistance, usually a few ohms, and C is the varactor capacitance, generally at the highest reverse voltage.

The multiplication efficiency is lower at higher power levels, and also depends on the order of multiplication. Figure 22.13 shows the efficiency of multipliers, utilizing modern varactor diodes, in the GHz frequency range as a function of the input power for doubler, tripler and quadrupler circuits.

The *frequency-comb* generators constitute a separate class of multipliers, utilizing the spectrum of an extremely narrow pulse train. In the circuit of Fig. 22.14 (a), the driving sine-wave voltage is coupled via capacitor C_1 to capacitor C_2 which acts as a low-value generator impedance, driving the circuit comprised of inductance L_2 and step-recovery diode D_1 [22.7]. Choke L_1 is used to connect the reverse bias voltage to the diode. At the amplitude peaks of the sinusoidal driving voltage, the diode is switched on for a short time, and the switching off of the diode current flow occurs in an extremely short time interval. Thus high amplitude, narrow voltage spikes are produced cross inductance L_2 as shown in Fig. 22.14 (b). The efficiency, i.e. the ratio of the pulse power to the consumed input power, may be as high as 80% with modern diodes, making the circuit very economical. The pulse width may be very small, e.g. 70 ps, resulting in a spectrum extremely rich in harmonics, as shown in Fig. 22.13 (c).

The spacing between components of the line spectrum is f_0, the driving voltage frequency, and the amplitude of the components is nearly constant over a wide frequency range, which may be taken as approximately $1/\pi T_w$. The amplitude of component of frequency $1/T_w$ is zero, and then there is a further amplitude increase. It is seen that up to the frequency $1/\pi T_w$, a frequency comb is generated, comprising a high number of components of nearly equal amplitude with a given raster. With a 10-MHz input drive and $T_w = 100$-ps pulse width, the frequency comb is generated in the frequency range 10 MHz–3 GHz with nearly constant amplitude and with a 10 MHz raster [22.8].

23. HIGH-FREQUENCY OSCILLATORS

23.1. Operation and design methods

The condition for operation, i.e. the condition for generating sustained oscillations is the presence of *positive feedback*. Oscillations are generated when a particular condition is met: this condition can be formulated using the small-signal parameters of four-pole theory. For providing constant amplitude oscillations, a limiter is needed to limit the oscillation amplitude which would otherwise continue to grow because of the positive feedback. Further oscillator parameters are the frequency stability, the harmonic distortion, noise and, in particular for power oscillators, good efficiency. For tuned oscillators, these parameters should be appropriate over the whole frequency range to be covered, so the frequency dependence of the parameters is also important.

Regarding the oscillator as a feedback amplifier (Fig. 23.1), the gain is given by the common formula $A_f = A/(1 - \beta A)$ where A is the open-loop gain and β is the feedback factor. For the case where

$$\beta A = 1, \qquad (23.1.1)$$

the feedback amplifier gain is infinite, which means that there will be an output voltage without an input voltage, i.e. sustained oscillations will be produced. The condition for oscillation as given in eqn. (23.1.1) may be expressed for all kinds of oscillators from the knowledge of the open-loop gain and the feedback factor.

Applying the pole-zero method, the first step is to express the transfer function of the network. Depending on the position of the poles on the complex plane, either sustained or damped oscillations are produced. The

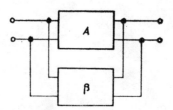

Fig. 23.1. Circuit principle of oscillators as positive feedback amplifiers

same condition of oscillation is determined either from the poles of the transfer function or from the relation (23.1.1).

The condition for oscillation may also be determined from the four-pole parameters of the circuit, as in the stability investigation of small-signal amplifiers. Taking an arbitrary point of the circuit as input or output, the currents flowing in or out should be zero in the set of admittance-parameter equations. The condition for oscillation is given by

$$y_{11}y_{22} - y_{12}y_{21} = 0, \qquad (23.1.2)$$

where the parameters in question are the overall admittance parameters, also including the terminations.

Finally, the condition for oscillation may also be determined from the equality of positive and negative real resistances in the circuit. This is because a positive feedback circuit may be represented by a negative real resistance, compensated by the load as a positive resistance. The negative resistance is directly applicable for tunnel diodes, IMPATT diodes etc., so this method illustrates rather well the case where oscillators use these types of devices.

It has been shown that the condition for oscillation may be determined in several ways, but is expressed by a single *complex equation*. Equating the real and imaginary parts, the complex equation may be separated into two scalar equations. The oscillation frequency is determined by equating the imaginary parts, while the terminations and feedback ratio are calculated from the equation containing the real parts.

The working point of an oscillator differs from the point at which oscillations start to build up and to which the condition for oscillations is given. If this is overlooked and the working point needed to start the oscillations is not provided, then no oscillations will build up, or a starting stimulus is needed after switching on.

23.2. Circuit principles

Several types of oscillators are possible, depending on arrangement of the positive-feedback circuit. In the following, we shall investigate these methods in detail. A transistor will always be used as an active element, but with suitable modifications, the relations are valid for other active elements, e.g. integrated circuits.

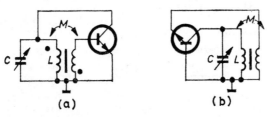

Fig. 23.2. Transformer-coupled oscillator. (a) Common-emitter circuit, (b) common-base circuit

Figure 23.2 (a) shows the simplified circuit diagram of an oscillator with a tuned collector circuit and transformer feedback to the base. The transformer windings are in opposite directions, since the transformer has to provide a phase reversal for positive feedback, because of the phase reversal of the common-emitter circuit.

Let us assume that the operating frequency is much lower than the transistor cut-off frequency and the transistor has no internal feedback. This means that the input impedance h_{11e} and the current gain h_{21e} are real, their phase shift being zero. In case the output admittance h_{22e} is not negligible it should be included in the collector circuit by considering its real value as a circuit loss. With these approximations, the condition for oscillation for this circuit is given by

$$h_{21e} > \frac{h_{11e}}{R_c n} + n, \qquad (23.2.1)$$

where R_c is the resistance of the collector circuit, at resonance, and the feedback ratio is $n = -M/L$. For the case of tight coupling, the feedback ratio is equal to the turns ratio $n = n_b/n_c$ where n_b and n_c are the number of turns of the base coil and collector coil, respectively. The oscillation frequency is determined by the tuning capacity C and the primary inductance L. If the input or output includes reactances, these should be included in the resonant circuit. For the case of loose coupling, the secondary inductance in the base circuit and the transistor input capacity may form a separate resonant circuit which interfere with the operation. Finally, if the phase shift of the current gain factor h_{21e} cannot be neglected, then the whole calculation may become invalid. In practice, low phase shifts of less than $-45°$ allow the condition for oscillation to be used as an approximation, with subsequent correction required.

Figure 23.2 (b) shows an oscillator with common-base circuit and transformer feedback to the emitter. The common-base circuit provides no phase reversal, so the transformer windings have identical directions. The condition for oscillation is similar to that shown in eqn. (23.2.1) but the common-base hybrid parameters have to be replaced:

$$h_{21b} > \frac{h_{11b}}{R_c n} + n. \qquad (23.2.2)$$

For the case of tight coupling, $n = n_e/n_c$, where n_e and n_c are the number of turns of the secondary winding and the collector winding, respectively.

The above statements are invalid for complex four-pole parameters. However, for a common-base configuration, the current-gain factor h_{21b} is real over a considerably wide frequency range, and the phase shift only increases slowly with frequency. Thus the relation can be used over a wider frequency range.

Figure 23.3 (a) shows a *Hartley* three-point oscillator with inductive feedback. In practice, the emitter is grounded instead of the base, and a suitable coil tapping is used for DC supply-voltage connection, with an AC

23.2. Circuit principles

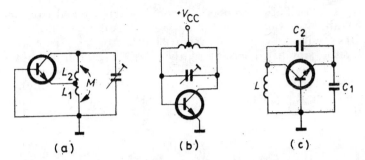

Fig. 23.3. Three-point-type oscillators. (a) Hartley oscillator, (b) common-emitter variant of the above, (c) Colpitts oscillator

ground as shown in Fig. 23.3 (b). Using the notations of the figure, the condition for oscillation for a Hartley circuit is given by

$$h_{21e} > \frac{h_{11e}(1+n)^2}{R_c n} + n, \qquad (23.2.3)$$

where R_c is the resistance of the resonant circuit at resonance, including the transistor output conductance, and n is the feedback ratio:

$$n = \frac{L_1 + M}{L_2 + M}. \qquad (23.2.4)$$

To determine the resonant frequency, capacitance C and overall inductance $L = L_1 + L_2 + 2M$ should be taken considered. The Hartley oscillator is extremely popular for its simplicity, but is not used extensively in the UHF range.

The *Colpitts* oscillator, the capacitive equivalent of the Hartley type, is shown in Fig. 23.3 (c). Feedback is realized here by a capacitive path for which the capacitive feedback within the transistor may be utilized. For this reason, this type of oscillator is used extensively in the UHF range. The condition for oscillation is identical to the condition for the Hartley oscillator as given in (23.2.3), with a modified feedback ratio given by

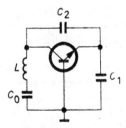

Fig. 23.4. Transistor Clapp oscillator

$n = C_2/C_1$. The oscillation frequency should be determined from inductance L and overall capacitance $C = C_1 C_2/(C_1 + C_2)$. A detailed design procedure for Colpitts oscillators is given in Ref. [23.29].

Finally, the simplified circuit of the *Clapp* oscillator is shown in Fig. 23.4. This is a modified Colpitts oscillator, having a series resonant circuit in the collector circuit. A feature of this oscillator is the high-frequency stability making it especially useful as a crystal oscillator. The condition for oscillation is given by

$$h_{21e} > \frac{h_{11e}}{R_c n} \left(\frac{C_2}{C}\right)^2 + n, \qquad (23.2.5)$$

where the feedback ratio is $n = C_2/C_1$, similarly to the Colpitts oscillator, and the capacitance responsible for the oscillation frequency is

$$C = \frac{C_0 C_1 C_2}{C_0 C_1 + C_1 C_2 + C_0 C_2}. \qquad (23.2.6)$$

If the values of C_1 and C_2 are high, then $C \approx C_0$, so the oscillation frequency is determined primarily by the series resonant circuit. The frequency is hardly influenced by other circuit elements, including transistor parameters.

In all previous relations for three-point oscillators, it was assumed that the parameters are real. However, the relations may also be used at high frequencies with limited accuracy, up to a phase angle φ_{21e} of $-45°$. Reactances appearing at the input and output should be included in the appropriate reactances of the external circuit. The resistance at resonance in the above relations may be calculated as follows:

$$R_c = \frac{L}{r_s C} \parallel \frac{1}{\mathrm{Re}(h_{22})} \parallel R_L, \qquad (23.2.7)$$

where r_s is the series loss resistance of inductance L, $\mathrm{Re}(h_{22})$ is the real part of the output admittance, and R_L is the external load resistance.

It has been assumed in the preceding analysis that the cut-off frequency of the active element is much higher than the oscillation frequency, i.e. the circuits have been designed by low-frequency approximation. At higher frequencies, there is an increase in the phase angle of the transfer parameters gradually invalidating all previous relations. We should therefore calculate the condition for oscillation for the common-emitter transformer-coupled oscillator shown in Fig. 23.2 (a), taking into account high-frequency properties.

The circuit diagram used for the calculation is shown in Fig. 23.5; here the transistor is defined by the h parameters. Oscillation may be considered as the effect of the output impedance Y_{out} the real part of which is made negative by the feedback. The criterion for oscillation is given by the value of this negative resistance. The higher the output load conductance g_L needed to compensate this negative resistance, the better the oscillator is loaded.

23.2. Circuit principles

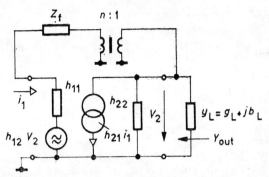

Fig. 23.5. Equivalent circuit of transformer-coupled oscillator

As a first step, let us express the output admittance by the transistor parameters:

$$Y_{out} = h_{22} + \frac{(n - h_{12})(h_{21} + n)}{h_{11} + Z_f}. \qquad (23.2.8)$$

Separating the second term into real and imaginary parts, we have

$$Y_{out} = h_{22} - (g_n + jb_n), \qquad (23.2.9)$$

where the quantities g_n and b_n are given by rather complicated expressions. Let us introduce the factor

$$\Theta = \frac{\text{Im}(h_{11}) + \text{Im}(Z_f)}{\text{Re}(h_{11}) + \text{Re}(Z_f)}, \qquad (23.2.10)$$

and solve the equations $\partial g_n/\partial n = 0$ and $\partial g_n/\partial \Theta = 0$. From these equations, the transformer ratio and the feedback impedance Z_f giving the highest negative conductance g_n at the output may be calculated. Solving the two equations needed for extremum value calculations, the optimum value of the factor Θ characterizing the transformer ratio and the feedback impedance may be determined:

$$n_{opt} = \frac{1}{2}\left[r_{12} - r_{21} + \frac{x_{12} + x_{21}}{r_{12} + r_{21}}(x_{12} - x_{21})\right]; \qquad (23.2.11)$$

$$\Theta_{opt} = \frac{x_{12} + x_{21}}{r_{12} + r_{21}}. \qquad (23.2.12)$$

For clarity, the transfer h parameters have been separated into real and imaginary parts in the above expression, i.e. $h_{12} = r_{12} + jx_{12}$ and $h_{21} = r_{21} + jx_{21}$. Substituting these values back, the maximum output-conductance negative value is determined:

$$\mathrm{Re}(Y_M) = \mathrm{Re}(h_{22}) - \frac{(r_{12}+r_{21})^2 + (x_{12}-x_{21})^2}{4[\mathrm{Re}(h_{11})+\mathrm{Re}(Z_f)]};\qquad (23.2.13)$$

$$\mathrm{Im}(Y_M) = \mathrm{Im}(h_{22}) + \Theta_{opt}[\mathrm{Re}(Y_M) - \mathrm{Re}(h_{22})].\qquad (23.2.14)$$

Let us first determine the low-frequency value of the output conductance. At low frequencies, $r_{12} = x_{12} = x_{21} = 0$, so the current-gain factor is given by $r_{21} = h_{21}$, and $\mathrm{Re}(h_{11}) = h_{11}$. Let us further assume that $\mathrm{Re}(h_{22}) = \mathrm{Re}(Z_f) = 0$, so the maximum output conductance negative value is given by

$$\mathrm{Re}(Y_M) = -h_{21}^2/4h_{11}.\qquad (23.2.15)$$

For the common-emitter circuit, $h_{21e} = \beta_0$ and $h_{11} \approx \beta_0/g_m$, and with these values, the negative output conductance maximum value is $\mathrm{Re}(Y_M) = -\beta_0 g_m/4$ where g_m is the transconductance, and the optimum transformer ratio is $n_{opt} = \beta_0/2$. For the common-base circuit, $\mathrm{Re}(Y_M) = -g_m/4$ and $n_{opt} = 0.5$.

At high frequencies, relation (23.2.13) is extremely difficult to evaluate, especially when the equivalent circuit of the applied transistor is given. To simplify calculations, an approximate expression for the maximum negative value of the output conductance is given using the elements of the equivalent circuit:

$$\mathrm{Re}(Y_M) \simeq g_m \frac{C_e}{C_e+C_c}\left(1 - \frac{2\pi f_T}{4r_b C_c \omega^2}\right),\qquad (23.2.16)$$

where ω is the oscillation frequency, and C_e and C_c are the emitter and collector capacitances, respectively. The above expression is only valid for frequencies high enough that $\omega \gg 1/r_b C_e$. The factor in parentheses in eqn. (23.2.16) is related to the highest oscillation frequency f_{max}, since at this frequency the value of $\mathrm{Re}(Y_M)$ is decreased to zero. This means that the second term in the parenthesis will be unity, and from this relation, the highest oscillator frequency can be calculated.

Finally, the operation of the *negative-impedance oscillator* will be investigated, using the equivalent circuit shown in Fig. 23.6 (a) [23.10], [23.12]. It will be shown that because of the phase shift of the current-gain factor for the case of a capacitive emitter impedance Z_E, the real part of the transistor input impedance will be negative. In the figure, the transistor is accounted for by a simplified equivalent circuit. The two important currents in question are related by the parameter h_{21e}, its frequency dependence being given in the frequency range $\omega \gg \omega_\beta$ by the following approximation:

$$\frac{i_c}{i_b} = h_{21e} \approx \frac{\beta_0}{1+j\omega/\omega_\beta} \approx \frac{\omega_T}{j\omega},\qquad (23.2.17)$$

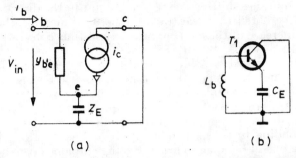

Fig. 23.6. Negative impedance oscillator. (a) Equivalent circuit, (b) circuit principle

where $\omega_T = 2\pi f_T$. The voltage appearing at the input, V_{in}, for a high conductance value $\mathrm{Re}(Y_{in})$, is approximately equal to the voltage drop across impedance Z_E:

$$V_{in} = (i_b + i_c) Z_E = i_b \left(1 + \frac{\omega_T}{j\omega}\right) \frac{1}{j\omega C_E}, \qquad (23.2.18)$$

and from this expression, the input impedance is

$$Z_{in} = -\frac{\omega_T}{\omega^2 C_E} + \frac{1}{j\omega C_E}.$$

It is seen that this has a negative real part and is thus suitable for producing oscillations. A simplified circuit of a high-frequency oscillator using this principle is shown in Fig. 23.6 (b) [23.12]. The condition for operation is a given phase shift of the current-gain factor, and this phase shift is obviously not present at low frequencies.

Further design aspects of high-frequency oscillators are given in Ref. [23.29] and [23.30], and a computer-aided transient analysis is presented in Ref. [23.19]. The design of MESFET oscillator is described in Refs. [23.34] and [23.35].

23.3. Amplitude and frequency stability

Amplitude and frequency stability are basic requirements of oscillators. Let us first investigate amplitude stability, i.e. the practical design of limiter circuits for transistor oscillators.

The function of limiting is to influence the condition for oscillation unfavourably, while the amplitude of oscillation increases; finally, an equilibrium state takes place, eliminating the further increase of amplitude. This function may occur as a decrease in transconductance, which, in turn, appears as an increase in impedance h_{11} in the equations giving the oscillation condition. Transistor transconductance may be decreased by decrease in emitter current, i.e. shifting the working point in the direction of turning off. It should be noted that the transconductance may also be reduced by a substantial increase of the emitter current and thus a decrease in collector voltage, in other words, by driving the transistor into saturation. This method is actually used in gain-controlled amplifiers, but the method is not utilized in high-frequency oscillators. In the following, two different types of limiters are presented, with the common property of shifting the working point towards cut-off, i.e. class C operation.

Figure 23.7 shows a common-emitter oscillator with transformer feedback, corresponding to the principle shown in Fig. 23.2 (a). The DC voltage on the base, is set by a high-current voltage divider given by resistances R_1 and R_2, so the bias V_B is independent of the currents flowing through the transistor. The turn-off voltage needed for limiting is provided by the emitter circuit elements R_E and C_E.

During the build-up of oscillations, oscillation amplitude steadily increases, so the high-frequency voltage reaching the base, V_{in}, also increased. Because of the nonlinear transistor characteristic, the collector-current waveform gradually becomes distorted, and the DC component increases, causing a voltage drop across resistance R_E, shifting the working point towards turn-off. The effect can also be attributed to the base–emitter diode, acting as a rectifier and thus charging the capacitor C_E, causing the shift of the working point. In the limiting case, class C operation will thus prevail. The gradual turn-off brings about a decreased transconductance which may be given approximately, for $V_{in} > 150$ mV, by the relation

$$g_m \approx \frac{2}{R_E}\left(1 + \frac{V_B}{V_{in}}\right). \qquad (23.3.1)$$

Knowing the resistance at resonance R_c, the collector voltage amplitude V_c to be determined may be calculated from the equation $V_c = V_{in} g_m R_c$, and thus

$$R_E = 2R_c(V_B + V_{in})/V_c. \qquad (23.3.2)$$

From the knowledge of base voltage V_B set by the resistive divider, the emitter resistance may be calculated if the input voltage is chosen suitably ($V_{in} > 150$ mV). This calculation is only approximately valid at frequencies approaching the transistor cut-off frequency.

Another kind of limiting is applied in Fig. 23.8 showing a common-base oscillator with transformer feedback. Here again, the limiting action is basically achieved by shifting the working point towards turn-off.

The turn-off voltage is generated as follows. During the increase in oscillation amplitude, the collector voltage increases above the supply voltage V_{CC}, and during the voltage peak intervals, the polarity of the collector-base diode voltage is reversed, and the diode turns on. During the brief period of conduction, a high current flows through the base circuit, charging the capacitor C_2 so as to decrease the positive voltage V_B supplied by the resistance divider. This process shifts the working point towards turn-off (class C operation), decreasing the transconductance until equilibrium is

23.3. Amplitude and frequency stability

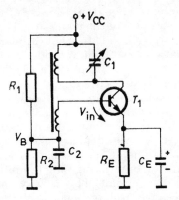

Fig. 23.7. Oscillator circuit with limiting elements in the emitter circuit

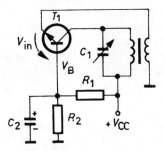

Fig. 23.8. Oscillator circuit with limiting elements in the base circuit

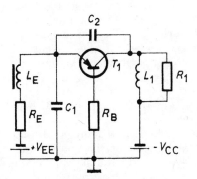

Fig. 23.9. Simplified circuit for the calculation of a high-frequency Colpitts oscillator

reached. In this state, the collector-voltage amplitude is equal to the supply voltage to good approximation.

The amplitude-stability problem and the dependence of amplitude stability on the DC working point are dealt with in Ref. [23.13] by considering the Colpitts oscillator shown in Fig. 23.9. In this case, the circuit equation becomes a nonlinear differential equation because of the reactance elements and because of the charge-storage of the transistor. The equation is evaluated by an approximate solution similar to that applied for oscillating motions, and this yields relations for the amplitudes and distortions.

The frequency stability of high-frequency oscillators is determined in most cases by both the resonant circuit and the active elements. If the overall phase shift of the amplifier and the following feedback network is not zero, then the oscillation frequency will not be exactly equal to the resonant frequency, so that the resonant circuit should be able to compensate for the phase error (the overall phase shift should be zero). It thus follows that the phase relations of the oscillation and thus the oscillation frequency will depend on the external circuit (amplifier and feedback network), not only the resonant circuit. This is why the amplifier plus feedback circuit should be designed to have a small phase shift.

A change in load impedance may result in a shift of frequency if the amplifier loads the resonant circuit, so a resonant circuit which is unloaded and has a high Q factor is needed for high-frequency stability. We shall not deal with the considerations for stable realization of the resonant circuit elements (temperature stability, aging, etc.) here, since this is a general problem, not related specifically to solid-state devices.

Reference [23. 32] deals with the change in the oscillation frequency due to the change in of the VSWR.

23.4. Circuit implementation

In this section, a few oscillator circuits actually built will be investigated, with emphasis of the circuit properties of different solutions. Figure 23.10 (a) shows a high-frequency Hartley oscillator, containing a trimmer capacitor C_3 for optimum feedback adjustment. A phase shift is introduced into the feedback path by this series capacitor, but this compensates the high-frequency phase shift of the transistor. The circuit is a common-emitter circuit, owing to the blocking capacitor C_2. The base voltage is set by a resistive voltage divider, via the choke L_3. Load matching is realized by trimmer capacitor C_4. The operating frequency is 30 MHz, efficiency is about 30%, and the transition frequency of the applied transistor is 250 MHz.

Figure 23.10 (b) shows the circuit diagram of a common-base Hartley oscillator. Feedback is implemented from a tapping of the inductance via trimmer capacitor C_3 to the emitter. The base ground return at the operating frequency is given by C_2, and the bias is provided by a resistive divider. C_1 tunes the resonant circuit, and load matching is given by trimmer capacitor C_4. This circuit provides about 10% efficiency at 60 MHz, with a transistor having a transition frequency of 250 MHz.

Figure 23.11 shows two oscillator circuits using field-effect transistors [23.16]. Figure 23.11 (a) shows a transformer-coupled oscillator, with the resonant circuit in the drain. C_3 gives a high-frequency ground for the source, and R_1 serves for biasing. In Fig. 23.11 (b) a Colpitts oscillator operating in the UHF range is presented, where limiting is accomplished

23.4. Circuit implementation

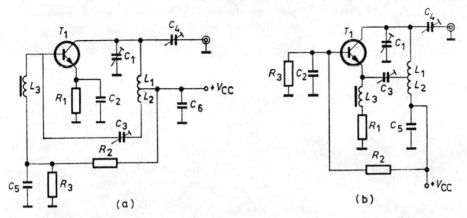

Fig. 23.10. High-frequency Hartley oscillators. (a) Common-emitter circuit, (b) common-base circuit

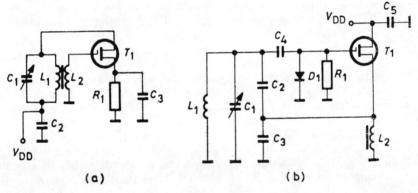

Fig. 23.11. Oscillators using MOS transistors. (a) Transformer-coupled oscillator, (b) Colpitts oscillator

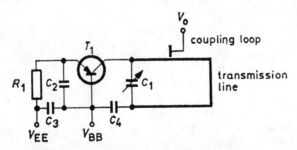

Fig. 23.12. UHF oscillator circuit with transmission-line tuning

by the rectifier circuit containing diode D_1, resistance R_1 and coupling capacitor C_4. In Ref. [1.30], another VHF oscillator utilizing a JFET is discussed.

Figure 23.12 shows a common-base oscillator circuit with a parallel resonant circuit in the collector circuit, the inductance being realized by a shorted transmission-line section. This oscillator may be tuned over a wide frequency range and operates at very high frequencies [23.17]. The circuit is essentially a Colpitts oscillator which has a feedback capacitor (C_2 in Fig. 23.3(c)) given by the collector–emitter capacitance of the transistor. The transmission line is formed by a microstrip, output coupling is provided by a coupling loop. The frequency range which may be covered is determined by the capacity change of capacitor C_1. According to results of measurement, oscillation may be maintained in the range 400–1000 MHz if the transition frequency of the transistor is 800 MHz and C_1 is adjustable between 2 and 40 pF. At the output coupling termination of 60 ohms, a high-frequency voltage in the range 80–500 mV is generated.

Figure 23.13 shows two oscillators operating in the GHz frequency range [18.14]. Figure 23.13 (b) shows the practical design of the common-collector Colpitts oscillator shown in Fig. 23.13 (a). Inductance L is given by a transmission-line section of suitable length and characteristic impedance Z_0; this is used to transform the impedance given by the 50-ohm load and by trimmer capacitor C_3 and C_4 to the collector. Capacitors C_1 and C_2 shown in the figure are in fact internal transistor capacities and stray capacities. The supply voltage to the base and emitter is connected via chokes L_1 and L_2; the base DC voltage and the limiting are provided by resistances R_1, R_2 and R_3. With a supply voltage of 24 V, the circuit gives 0.8 W output power at a frequency of 1.25 GHz. Utilizing a varactor diode as a tuning element, the oscillation frequency is variable between 1 and 1.5 GHz.

Figure 23.13 (c) shows the simplified circuit of a Hartley oscillator, with the collector grounded. Figure 23.13 (d) gives the practical design. Capacitor C_1 is the collector–base capacitance of T_1, and inductance L_1 is made up of two parts, L_1' and L_1''. Their common point is connected to the 50-ohm load via capacitors C_2 and C_3, intended for matching. Inductance L_1' is realized by the output leads of the capacitor C_1 used for DC decoupling. The base voltage, generated by divider $R_2 - R_3$, is connected via choke L_3. Another microwave oscillator is presented in [23.28].

Figure 23.14 (a) shows a 10-MHz oscillator which uses an integrated circuit. Inductances L_1 and L_2 are equal, so the feedback transformer ratio is 1 : 1. The condition for oscillation is given by $y_{21} > g_2$ where y_{21} is the transconductance of the integrated circuit and g_2 is the sum of all output conductances. This will also give the highest loading permitted. The lower limit is given by the voltage amplitude which will not damage the input of the integrated circuit (generally a few volts peak-to-peak). The load resistance should be chosen somewhere between these two limits. One advantage of this circuit is that the oscillation frequency is only slightly dependent on the supply voltage V_{CC}, as shown by Fig. 23.14 (b).

Figure 23.15 shows a push-pull oscillator made up of two Hartley circuits as given in Fig. 23.10 (b). Push-pull operation allows a lower second-order

23.4. Circuit implementation

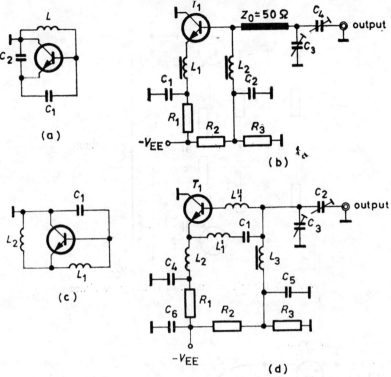

Fig. 23.13. High-frequency oscillators with transmission-line tuning. (a) Colpitts oscillator, circuit principle, (b) Colpitts oscillator, practical design, (c) Hartley oscillator, circuit principle, (d) Hartley oscillator, practical design

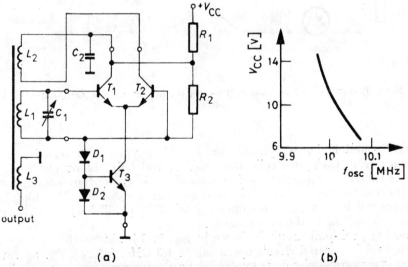

Fig. 23.14. Monolithic integrated-circuit oscillator. (a) Circuit diagram, (b) dependence of oscillation frequency on supply voltage

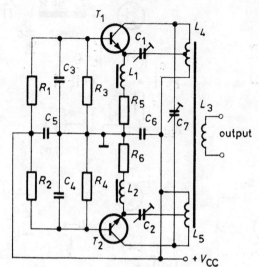

Fig. 23.15. Circuit of push-pull Hartley oscillator

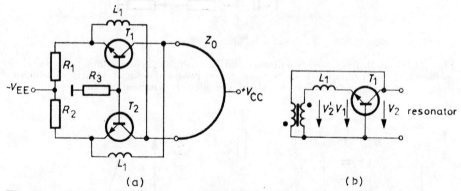

Fig. 23.16. Transmission-line push-pull oscillator. (a) Circuit diagram, (b) equivalent circuit

distortion and higher output power. In this circuit, both transistors are fed back from their own collector winding, and only the output resonant circuit, tuned by capacitor C_7, is common. L_3 is the output coupling coil, and the DC working point is adjusted as in earlier circuits. Matching may be improved by placing a trimmer capacitor in parallel with the output, tuning the coil L_3 to the operating frequency.

Another push-pull circuit is shown in Fig. 23.16. This oscillator is a thin-film hybrid integrated circuit operating at 4.3 GHz. [23.23]. The operation of the symmetrical circuit is explained by the asymmetrical circuit shown

23.4. Circuit implementation

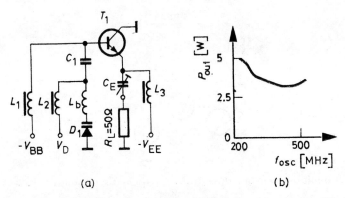

Fig. 23.17. High-power negative impedance oscillator tunable over a wide frequency range. (a) Circuit diagram, (b) frequency dependence of output power

in Fig. 23.16 (b), in which push-pull operation is replaced by the phase reversal of the ideal transformer. Feedback is supplied by the other transistor via phase-correcting element L_1. Frequency is determined by a U-shaped stripline resonator, whose centre point is connected to the collector DC supply. The stripline resonator is essentially a resonant transmission-line section of characteristic impedance Z_0 which appears at the collectors as a resonant circuit with high Q factor.

Assuming that the resonator impedance as seen by the collector is 50 ohms, the output voltage may be expressed by the following relation:

$$V_2 = V_1 \frac{s_{21}}{1 + s_{11}}, \qquad (23.4.1)$$

where V_1 is the emitter voltage, and s_{21} and s_{11} are the transistor scattering parameters. The phase of voltage V_2 is reversed by the ideal transformer, thus generating voltage V_2'. Finally, the phase of this voltage is shifted by inductance L_1 so that its direction coincides with the direction of V_1. Resistance R_3 in the common-base circuit is intended to suppress unwanted parasitic oscillations, and may be ignored regarding push-pull operation. High-frequency power is coupled out by a suitably placed stripline section. The thin-film version of the circuit has extremely small dimensions and favourable location of components. Output power of the oscillator is $+10-+13$ dBm at 4.3 GHz, harmonic level is 40 dB below the fundamental. Temperature coefficient is about -500 kHz/°C. Fine tuning is possible by the adjustment of the collector voltage.

Figure 23.17 shows a negative impedance oscillator delivering high output power, and tunable over a wide frequency range [23.12]. The requirements of wide tuning range and flat output power are best met by negative impedance oscillators with the load in the emitter circuit. According to the circuit shown in Fig. 23.17 (a), electronic tuning is realized by varactor diode

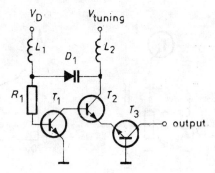

Fig. 23.18. Circuit diagram of a three-transistor oscillator tunable over a wide frequency range

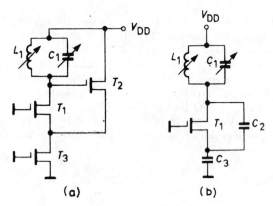

Fig. 23.19. Tunable oscillator suitable for MOS integrated circuits. (a) Circuit diagram, (b) equivalent circuit

D_1, its capacity being changed by voltage V_D via choke L_2. Base voltage V_{BB} is connected via choke L_1, emitter voltage V_{EE} is connected via choke L_3. According to the simplified circuit of Fig. 23.6 (b), oscillation is provided by the base inductance L_b (adjustable by the varactor diode) and the lossy emitter capacity C_E. The tuning range is 200–500 MHz, and output power variation within this range is shown in Fig. 23.17 (b). A microwave negative impedance oscillator is treated in Ref. [2.39], and a YIG-tuned oscillator is presented in Ref. [23.22].

Figure 23.18 shows a voltage tuned oscillator using a varactor diode; it is made up of two oscillator transistors and an output amplifier transistor [23.18]. Feedback is provided by a low-pass π network made up of inductances L_1 and L_2 and of the varactor diode D_1 having adjustable capacitance. The series resistance serves to stabilize the input impedance of transistor

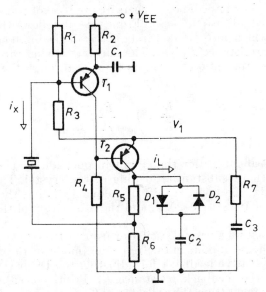

Fig. 23.20. Circuit diagram of crystal oscillator with limiting diodes of opposite polarities

T_1. The tuning voltage is connected via inductance L_2. Transistors T_1 and T_2 make up a two-stage common-emitter amplifier. Transistor T_3 which is in the emitter circuit of T_2 introduces only little negative feedback. A well separated oscillator signal appears at the collector of T_3. The oscillator may be tuned in the range 200–310 MHz, and is eminently suitable for voltage-controlled oscillator purposes in PLL loops or sweepers. Application of a high-Q resonant circuit results in good frequency stability and low phase noise. Another oscillator tunable over a wide frequency range is dealt with in Ref. [23.26].

A modified Colpitts oscillator utilizing field-effect transistors is shown in Fig. 23.19, which is especially suitable for application in MOS integrated circuits [23.15]. Only transistor T_1 acts as an active element, while T_2 and T_3 form a capacitive voltage divider. Looking from the parallel tuned circuit, the output impedance has a negative real component over a wide frequency range. Thus wideband operation can be attained by exchanging elements L_1 and C_1 of the external tuned circuit. This is in contrast with the conventional Colpitts oscillators which also require the change of the capacitive divider for providing oscillation over a wide frequency range. This method allows oscillations in the frequency range 0.7–65 MHz.

Figure 23.20 shows the circuit of a crystal oscillator [23.3]. Limiting is provided by the two opposite-polarity point-contact diodes D_1 and D_2 as nonlinear elements. The amplifier is made up of DC-coupled transistors T_1 and T_2, and the feedback including both stages is given by resistance R_3 which is also effective at DC and is thus suitable for adjusting the bias voltages. A very low impedance appears at the base of transistor T_1 making this point a "virtual" earth point. From this, it follows that current i_L of transistor T_2 (which also flows through the diode limiter) is the ratio of voltage V_1 and resistance $R_3 \parallel R_7$. The current flowing through the crystal is $i_x = V_1/R_3$, so the ratio of currents flowing through the limiter and the crystal is given by

$$\frac{i_L}{i_x} \approx 1 + \frac{R_3}{R_7}. \tag{23.4.2}$$

This value is about 10 for the circuit in question, so good limiting may be attained without substantial loading of the crystal. The operating frequency is 1 MHz, the relative shift of frequency is less than 2×10^{-8} for a supply voltage change of 10–20 V, and the dissipation of the quartz crystal is about 1 μW.

Figure 23.21 shows a crystal oscillator using a two-stage amplifier [23.5]. This operates in the fundamental frequency mode in the frequency range of 1–14 MHz. Negative feedback for adjusting the DC bias of the amplifier is provided by the feedback resistance R_4. Positive feedback, setting the oscillation frequency, from the collector of T_2 to the base of T_1, is provided by the crystal. The output voltage is limited by diodes D_1 and D_2. Resistance R_6 in series with the crystal is intended for adjusting the value of the positive feedback and the dissipation of the crystal. Too high a value of R_6 will prevent oscillations, whereas a low resistance will severely load transistor T_2. The relative shift of frequency is less than $\pm 1.5 \times 10^{-5}$ in the temperature range $-30 - +100$ °C.

Figure 23.22 shows a high-frequency crystal oscillator utilizing a single transistor [23.11]. According to the circuit principle illustrated in Fig. 23.22 (a), the circuit is essentially a Colpitts oscillator containing a series

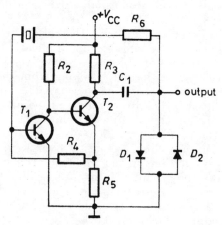

Fig. 23.21. Circuit diagram of two-transistor crystal oscillator

23.4. Circuit implementation

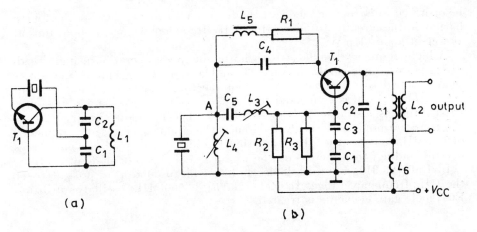

Fig. 23.22. High-frequency crystal oscillator. (a) Circuit principle, (b) practical arrangement

crystal in the feedback path. The operation is as follows. Capacitor C_3 provides a high-frequency short-circuit, so the resonant circuit made up of elements L_1, C_1 and C_2 can be clearly recognized between the collector and the base. The common point of voltage divider $C_1 - C_2$ is at earth potential; we have seen that this has no significance for oscillators, and any of the points may be earthed. From the voltage-division point (i.e. from the earth point) the crystal provides feedback to point A, from which point the signal is returned to the emitter via matching elements C_4 and L_3. These are provided for transforming up the emitter impedance and thus decreasing the load on the crystal. Capacitor C_5 provides a high-frequency short-circuit, while choke L_5 is used for biasing the emitter. The case capacity of the crystal is tuned out by inductance L_4; the remaining elements are used for DC bias adjustments.

The output signal appears across coil L_2. The circuit supplies an output power of 30 mW at 41 MHz with an efficiency of 40%. Exchanging capacitor C_2 and inductances $L_1 - L_2$ by a suitable band-pass filter, the oscillator may

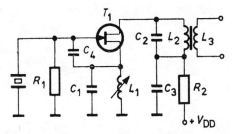

Fig. 23.23. Circuit diagram of crystal oscillator using a JFET

be operated in harmonic mode. If the band-pass filter is tuned to the fourth harmonic the circuit provides a 164 MHz frequency signal with an output power of about 1 mW.

A high-frequency oscillator with harmonic mode operation using a field-effect transistor is shown in Fig. 23.23. The resonant circuit in the source lead is tuned to a frequency less than the fundamental frequency (about $0.7 f_0$). The crystal is connected to the gate, and the transistor gate-source capacitance is also essential for oscillation. The resonant circuit tuned to the third harmonic is in the drain circuit. The circuit operates at a frequency of 120 MHz and has the advantage that the high gate impedance eliminates crystal loading, thus increasing the useful life of the crystal significantly.

In Ref. [23.31], a tunnel-diode subharmonic oscillator is presented for the GHz-frequency range. In Ref. [23.25], a 200-MHz oscillator utilizing ECL logic gate elements is treated.

24. HIGH-FREQUENCY NOISE OF SOLID-STATE DEVICES

24.1. The concept of noise figure and noise four-pole

The noise figure of an amplifying four-pole is the ratio of input to output signal-to-noise ratios:

$$F = \frac{(\text{signal/noise})_{\text{input}}}{(\text{signal/noise})_{\text{output}}}. \qquad (24.1.1)$$

As the input signal and noise are amplified by equal amounts, the physical significance of the difference between the two signal-to-noise ratios is the noise generated within the amplifier itself.

The noise figure is thus the measure of four-pole noisiness expressing how many times the input noise appears at the output; $F = 1$ would correspond to a noise-free amplifier which cannot be realized. The noise figure is usually expressed in decibels, and is an important parameter of solid-state devices, especially in the input stage.

In a multi-stage amplifier, the overall noise figure is influenced by the second, third, etc. amplifier stages but to an extent which is inversely proportional to the first stage gain. In other words, if the high gain of the first stage raises the signal above the noise level then the effect of the remaining stages is not critical. The overall noise figure of a multi-stage amplifier is given by

$$F = F_1 + \frac{F_2 - 1}{G_1} + \frac{F_3 - 1}{G_1 G_2} + \ldots, \qquad (24.1.2)$$

where F_1, F_2, etc. are the noise figures and G_1, G_2, etc. are the stage gains. It can be seen that if the noise figures do not differ too much and G_1 is high, then the second and subsequent terms may be neglected.

Figure 24.1 shows the equivalent circuit of a noisy four-pole; in this equivalent circuit, the noise generators are separated from the remaining four-pole elements. The circuit is thus separated in two parts: the noisy four-pole comprising only active elements (noise generators) and the noiseless four-pole. The four-pole input is terminated by the generator admittance y_g which generates noise in itself, taken into account by the noise generator of source current i_{ng}.

The noisy four-pole may be characterized by three noise sources, of which those having a source voltage e_n and a source current i_n are independent,

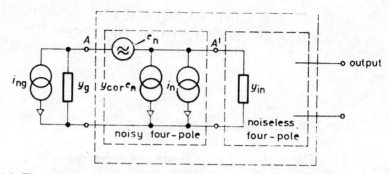

Fig. 24.1. Equivalent circuit of a noisy four-pole, with separation of the noise sources and the noiseless four-pole

while the source current of the third current generator is correlated to the voltage e_n; this is taken into account with the correlation admittance y_{cor}. The equivalent circuit of Fig. 24.1 can be adapted for all noisy four-poles which can be considered linear, independently of the components within the four-pole which may be bipolar or field-effect transistors, integrated circuits, etc.

According to the equivalent circuit, the noise figure of the noisy four-pole is determined by adding the noise powers at the internal point A', and relating this sum to the noise power available from the generator at the input point A. Since the part following point A' is noise-free, it has no effect on the noise figure; further, the effect of the gain is also eliminated by reducing the noise generator to the input. Expressing the power ratio in question, the noise figure is given by

$$F = 1 + \frac{\overline{|i_n|^2} + \overline{|e_n|^2} \, |y_g + y_{cor}|^2}{\overline{|i_{ng}|^2}}. \tag{24.1.3}$$

The noise power of the noise generator with source current i_{ng} (average of the square of the absolute value) is given by

$$\overline{|i_{ng}|^2} = 4 g_g k T \Delta f, \tag{24.1.4}$$

where g_g is the real part of the generator admittance, k is the Boltzmann constant, T is the absolute temperature, and Δf is the bandwidth.

Instead of the noise powers, let us now introduce the concepts of noise resistance and noise conductance as defined in the following:

$$\overline{|e_n|^2} = 4 k T \Delta f \, R_n; \tag{24.1.5}$$

$$\overline{|i_n|^2} = 4 k T \Delta f \, g_n. \tag{24.1.6}$$

24.1. Noise figure and noise four-pole

Expressing the noise-figure formula given in (24.1.3) by quantities R_n and g_n, we obtain

$$F = 1 + \frac{g_n + R_n |y_g + y_{cor}|^2}{g_g}. \qquad (24.1.7)$$

Here the generator admittance and correlation admittance are complex quantities characterized by their real and imaginary parts, i.e. $y_g = g_g + jb_g$ and $y_{cor} = g_{cor} + jb_{cor}$. It can be seen from the noise-figure expression given in (24.1.7) that a minimum occurs when the following condition is satisfied:

$$b_g + b_{cor} = 0, \qquad (24.1.8)$$

(see Fig. 24.2a). This condition is called "tuned-out noise" because of the cancelling of the imaginary parts. This condition does not necessarily coincide with the condition providing optimum conjugated matching for power gain; in fact, there is generally a considerable gap between these conditions.

It can be seen from Fig. 24.2 that in the case of generator susceptance $b_{g\,opt}$ giving conjugated matching the noise figure F' is higher than the minimum value which is unfavourable. For instance, in the case of bipolar transistors in the UHF range, there is a significant difference between noise figures F' and F_{min}. In the common-base configuration, the optimum noise figure requires a more inductive generator reactance than is required by the condition for conjugated matching. This means that a compromise should be reached either for the power gain or for the noise figure; sometimes both of these are chosen as a result of compromise.

By substitution of (24.1.8), the noise figure in the "tuned-out noise" condition is given by

$$F_{min} = 1 + \frac{g_n + R_n (g_g + g_{cor})^2}{g_g}. \qquad (24.1.9)$$

This expression may be considered to be a function of generator conductance g_g, and reaches a minimum value F^*_{min} at the conductance value

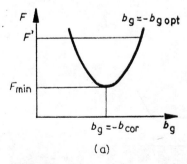
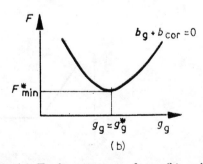

Fig. 24.2. Reaching minimum noise figure. (a) Tuning-out procedure, (b) noise-matching procedure

$g_g = g_g^*$ (see Fig. 24.2b). The generator conductance resulting in the lowest noise figure is determined from the equation

$$\frac{\partial F}{\partial g_g} = 0 , \qquad (24.1.10)$$

and from this equation,

$$g_g^* = \sqrt{\frac{g_g}{R_n} + g_{cor}^2} . \qquad (24.1.11)$$

With this value of *matched* generator conductance, the minimum noise figure will be

$$F_{min}^* = 1 + 2R_n \left(\sqrt{\frac{g_n}{R_n} + g_{cor}^2} + g_{cor} \right) . \qquad (24.1.12)$$

This noise figure is an absolute minimum value which is valid when the condition of "tuned-out noise" and noise matching are simultaneously met. It should be noted that the generator conductance resulting in optimum noise does not coincide with that resulting in maximum power gain. However, relations for the real parts are more favourable than for the imaginary parts, since neither the power gain nor the noise figure shows fast changes as a function of g_g. It is thus easier to satisfy both conditions simultaneously.

24.2. Noise figure of solid-state devices

Figure 24.3 shows the typical frequency dependence of bipolar transistor noise figures. The curve consists of a low-frequency part of 3-dB/octave slope, a constant part, and a high-frequency part of 6-dB/octave slope. We are primarily interested in the medium- and high-frequency ranges. The knee between these ranges, i.e. the start of noise-figure increase, is at the frequency given in the following:

$$f_A \approx f_T/\sqrt{\beta_0} . \qquad (24.2.1)$$

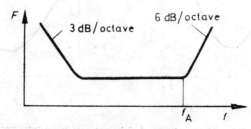

Fig. 24.3. Typical frequency dependence of bipolar transistor noise figures

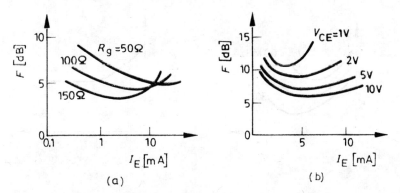

Fig. 24.4. Dependence of bipolar transistor noise figure on emitter current. (a) For various generator resistances, (b) for various collector voltages

In these two frequency ranges, the following relation can be used to express the noise-figure frequency function for bipolar transistors:

$$F \approx 1 + \frac{2r_b + r_e}{R_g} + \frac{(r_b + r_e + R_g)^2}{2\beta_0 r_e R_g}\left[1 + \beta_0\left(\frac{f}{f_T}\right)^2\right], \quad (24.2.2)$$

where R_g is the generator resistance, and r_e is the dynamic resistance of the emitter–base diode.

The transistor noise figure depends on both the generator resistance R_g and the transistor DC bias through the parameters given in (24.2.2). A few typical characteristics illustrating the dependence of noise figure on the DC bias are shown in Fig. 24.4. In Fig. 24.4 (a), the noise figure is plotted as a function of emitter current for a few generator resistances. The noise figure has a minimum at a given emitter current. This minimum is shifted towards higher emitter currents when the generator resistance is decreased, and simultaneously the minimum noise-figure value increases. In Fig. 24.4 (b), the noise figure dependence on emitter current is shown for various collector voltages. At low collector voltages, there is a strong increase in noise figure, with a simultaneous decrease in power gain.

Figure 24.5 shows noise-figure functions of a converter using a bipolar transistor. In Fig. 24.5 (a), the dependence on generator resistance is plotted for different input signal frequencies. With increasing frequencies, the minimum is located towards lower generator resistances and the noise figure minimum gradually increases. Figure 24.5 (b) shows the dependence on oscillator voltage. At low voltage amplitudes there is a substantial increase in noise figure. However, above a certain amplitude value, the noise figure remains constant.

Figure 24.6 shows an approximate equivalent circuit of a field-effect transistor showing all noise sources. At the input, the noise generators representing the internal transistor noise have been separated; the following

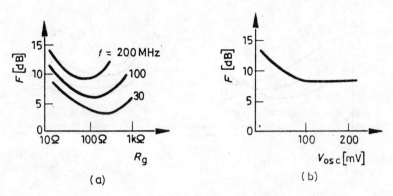

Fig. 24.5. Noise-figure function of a converter circuit using a bipolar transistor. (a) Dependence on generator resistance, (b) dependence on oscillator voltage

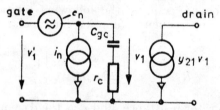

Fig. 24.6. Equivalent circuit of field-effect transistor showing noise sources

relations hold for the source parameters:

$$\overline{|e_n|^2} \approx 4kT\Delta f r_c ; \qquad (24.2.3)$$

$$\overline{|i_n|^2} \approx 4kT\Delta f g_m b , \qquad (24.2.4)$$

where g_m is the low-frequency transconductance, r_c is the channel resistance and b is an experimental constant having a value between 0.6 and 1 for JFET's and between 0.6 and 4 for MOS transistors.

An important parameter in the equivalent circuit is the complex transconductance y_{21} which is given by the following approximate relation:

$$y_{21} \approx g_m \exp(-j\,0.1\,\omega C_{gc}/g_m) , \qquad (24.2.5)$$

where C_{gc} is the gate-channel capacity.

Using this equivalent circuit, the minimum value of the field-effect transistor high-frequency noise figure is calculated to be:

$$F_{\min} \approx 1 + 2\sqrt{\alpha} f/f_1 + 2\alpha (f/f_1)^2 , \qquad (24.2.6)$$

where the cut-off frequency value is $f_1 = g_m/2\pi b C_{gc}$, and $\alpha = r_c g_m/b$. The product $r_c g_m$ is about 0.3 for the JFET and about 0.2 or even less for the

24.2. Noise figure of solid-state devices

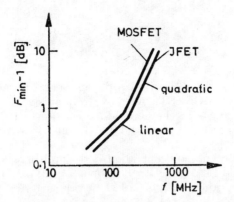

Fig. 24.7. Noise-figure frequency function of field-effect transistors on a logarithmic scale

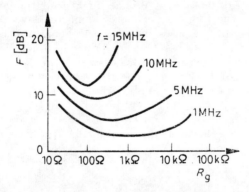

Fig. 24.8. Dependence of field-effect transistor noise figure on the generator resistance at various frequencies

MOSFET. The above approximate relation for the high-frequency noise-figure may be used with acceptable accuracy up to frequencies which are less than $3f_1$. It can be seen that the slope of the noise-figure frequency function is first 6 dB/octave (linear rise), and then 12 dB/octave (quadratic rise).

Figure 24.7 shows the quantity $F_{min} - 1$ as a function of frequency on a logarithmic scale, allowing a clear distinction between the two sections with different slopes. The MOS transistor noise figure is higher than that calculated from the constant b, and may be verified definitely by measurements.

From the equivalent circuit of Fig. 24.6, the optimum value of the generator admittance may also be determined. The real and imaginary parts of this admittance are given by

$$g_g^* \approx \sqrt{\alpha}\omega C_{gc};\qquad(24.2.7)$$
$$b_g^* \approx -\omega C_{gc}.\qquad(24.2.8)$$

Figure 24.8 shows the noise-figure dependence on generator resistance for a low-frequency MOSFET. At low frequencies, the noise figure is not sensitive for changes in the generator resistance. On the other hand, a sharp minimum may be observed in the short waveband at a given generator resistance.

The noise figure of monolithic integrated amplifiers may be calculated from the bipolar and field-effect transistor noise figures, using eqn. (24.1.2) and applying the parameter values of these transistors given earlier.

24.3. Design of input stages

In the preceding discussion, it was assumed that the generator was connected directly to the four-pole. Even if some reactive element should be needed for tuning-out purposes between the generator and the four-pole, this was assumed to be lossless, thus generating no noise. In fact, all reactances, especially inductances have losses which are taken into account by parallel resistance R_c in Fig. 24.9. In the simplest case this may be the resistance at resonance of the input parallel tuned circuit (unloaded), but it can also be the equivalent parallel loss resistance of any reactive filter. Resistance R_c generates noise and thus has an effect on the design of the input circuit. For simplicity let us assume that the four-pole is noise-free, and noise is only generated by the input resistance. In this case, the noise figure may be expressed in the following form:

$$F = 1 + R_g/R_{in} + R_g/R_c.\qquad(24.3.1)$$

Design is carried out using the design curves given in Fig. 24.10. Figure 24.10 (a) shows the noise figure as a function of the ratio R_g/R_{in} for various values of loss resistance R_c. For zero loss ($R_g/R_c = 0$) a noise figure of 3 dB pertains to the matched case when $R_g/R_{in} = 1$, and the noise figure decreases with decreasing generator resistance. For increasing losses, i.e. for increas-

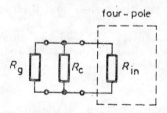

Fig. 24.9. Coupling between generator and four-pole using lossy reactive elements

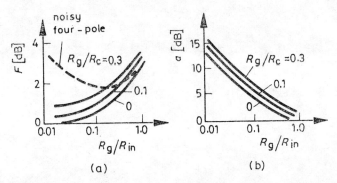

Fig. 24.10. Noise-free four-pole parameters as a function of the generator resistance, for various values of loss resistance R_c. (a) Noise figure, (b) attenuation

ing R_g/R_c values, the noise figure also increases, but the situation here is again improved by decrease in generator resistance.

However, this results in decreased power gain as a result of increasing mismatch. This can be seen in Fig. 24.10 (b), which shows the attenuation as a function of R_g/R_{in} for various values of R_g/R_c. During the design process of the input network a compromise is reached: the generator resistance is chosen to produce a low noise figure, without unduly high attenuation. The attenuation value is given from the relation:

$$a = \frac{4R_g}{R_{in}(1 + R_g/R_c + R_g/R_{in})^2}. \qquad (24.3.2)$$

In practice it is unusual to choose R_g/R_{in} lower than 0.1 because the attenuation would rise significantly, without further decrease in noise figure.

The preceding discussion is valid for noise-free four-poles. The noise figure of noisy four-poles will be higher than the values shown in Fig. 24.10 (a), especially at low generator resistance values of R_g. It has also been shown that a minimum value of noise figure appears at the optimum generator resistance value (see the dashed line in Fig. 24.10(a)).

REFERENCES

1.1 L. Esaki, New phenomena in narrow germanium pn-junctions. *Phys. Rev.*, **109** (1958) 603.
1.2 W. T. Read, A proposed high-frequency negative resistance diode. *Bell System Tech. J.*, **37** (1958) 401.
1.3 J. L. Moll, S. M. Krakauer and R. Shen, PN junction charge-storage diodes. *Proc. IRE.*, **50** (1962) 43.
1.4 S. M. Krakauer and R. W. Soshea, Hot-carrier diodes switch in picoseconds. *Electronics*, **36** 19 July (1963) 53.
1.5 H. O. Sorensen, Using the hot-carrier diodes as a detector. *Hewlett Packard J.* **17** Dec. (1965) 3.
1.6 F. L. Mergner, PIN diode and FET's improves FM reception. *Electronics*, **39** Aug. (1966) 114.
1.7 H. J. Peppiatt, Broadband AGC circuits. *Proc. IEEE*, **55** (1967) 220.
1.8 H. Mott, The harmonics produced by a PIN diode in a microwave switching application. *IEEE Trans.*, MTT-15 (1967) 180.
1.9 S. Hamilton and R. Hall, Shunt-mode harmonic generation using step-recovery diodes. *Microwave J.*, **10** Apr. (1967) 69.
1.10 H. R. Marlowe and D. E. Allen, Cut the size of your VHF attenuator. *Electr. Design*, **15** 24 May, (1967) 84.
1.11 J. Okean, Microwave amplifiers employing integrated tunnel-diode devices. *IEEE Trans.*, MTT-15 (1967) 613.
1.12 R. L. Sicotte and R. N. Assaly, Intermodulation products generated by a PIN diode switch. *Proc. IEEE*, **56** (1968) 74.
1.13 N. Cerniglia, R. C. Tonner, G. Berkovits and A. H. Solomon, Beam-lead Schottky-barrier diodes for low-noise integrated microwave mixers. *IEEE Trans.*, ED-**15** (1968) 674.
1.14 An attenuator design using PIN diodes. Hewlett Packard Appl. Note AN-912.
1.15 Harmonic generation using step-recovery diodes. Hewlett Packard Appl. Note AN-920.
1.16 Applications of PIN diodes. Hewlett Packard Appl. Note AN-922.
1.17 S. Boctor and D. J. Roulston, High efficiency conditions for nonlinear capacitor frequency multiplier. *Proc. IEEE*, **57** (1969) 688.
1.18 H. A. Watson, Microwave semiconductor devices and their circuit applications. McGraw Hill, New York, 1969
1.19 Y. Machi, Microwave frequency multiplication using hot electrons. *IEEE Trans.*, MTT-**17** (1969) 333.
1.20 V. Stachejko, Extremely high-power silicon PIN diode switch. *Proc. IEEE*, **57** (1969) 1340.
1.21 A. M. Cowley, Design and application of silicon IMPATT diodes. *Hewlett Packard J.* **21** May (1970) 2.
1.22 J. Merkelo, and R. D. Hall, Broad-band thin-film signal sampler. *IEEE Trans.*, SC-**7** Feb. (1972) 50.
1.23 G. Krause and G. Olk, PIN-Dioden als regelbare Dämpfungsglieder. *Funkschau*, **44** May (1972) 305.
1.24 Die PIN-Diode und ihre Anwendung. *Funkschau*, **45** 5. Jan. (1973) 17.

1.25 I. H. Macdonald, S. A. J. Matykiewicz and D. G. Pulley, PIN limiter for microstrip circuits. *Microwave J.*, **16** Jan. (1973) 52.
1.26 M. F. Black, Variable gain amplifier yields linear RF modulator. *Electronics*, **46** 4. Jan. (1973) 103.
1.27 H. L. Storer, H. M. Leedy and H. G. Morehead, Solid-state devices and components for mm-wave. *Microwave J.*, **16** Feb. (1973) 35.
1.28 K. K. Chang, H. Kawamoto, H. J. Prager, J. F. Reynolds, A. Rosen and V. A. Mikenas, Trapatt amplifiers for phased-array radar systems. *Microwave J.*, **16** Feb. (1973) 27.
1.29 R. J. Turner, Binary RF phase modulator switches in 3 nanoseconds. *Electronics*, **46** 26 Apr. (1973) 104.
1.30 U. L. Rohde, Zur optimalen Dimensionierung von UKW-Eingangsteilen. *Int. Elektr. Rundschau*, **27** May (1973) 103.
1.31 W. G. Matthei, State of the art of GaAs IMPATT diodes. *Microwave J.*, **16** June (1973) 29.
1.32 R. Tenenholtz, Broadband MIC multithrow PIN diode switches. *Microwave J.*, **16** July (1973) 25.
1.33 U. L. Rohe, Zur optimalen Dimensionierung von Kurzwellen-Eingangsteilen. *Int. Elektr. Rundschau*, **27** Dec. (1973) 276.
1.34 G. C. Fincke, Fast electronic tuning of high power circuits. *IEEE Trans.*, CAS-21 (1974) 313.
1.35 A. A. Sweet, Understanding the basics of Gunn oscillator operation. *EDN*, **19** 5 May (1974) 40.
1.36 R. J. Turner, Schottky-diode pair makes an RF detector stable. *Electronics*, **47** 2 May (1974) 95.
1.37 J. Hammerschmitt, Silizium Speichervaraktor und Impattdiode. *Siemens Zeitschrift*, **48** (1974) 507.
1.38 P. Chorney, Multi-octave, multi-throw PIN diode switches. *Microwave J.*, **17** Sep. (1974) 39.
1.39 J. C. McDade and F. Schiavone, Switching time performance of microwave PIN diodes. *Microwave J.*, **17** Dec. (1974) 65.
1.40 A. S. Clorfine, High-power, high bandwidth TRAPATT circuits. *IEEE J.*, SC-**10** (1975) 27.
1.41 H. Q. Tserng, D. N. McQuiddy and W. R. Wisseman, X-band MIC GaAs IMPATT amplifier module. *IEEE J.*, SC-**10** (1975) 32.
1.42 B. Siegal and E. Pendleton, Zero-bias Schottky-diodes as microwave detectors. *Microwave J.*, **18** Sep. (1975) 40.
1.43 R. Tenenholtz, Designing megawatt diode duplexers. *Microwave J.*, **22** Jan. (1979) 63.

2.1 J. Z. Ebers and J. L. Moll, Large-signal behaviour of junction transistors. *Proc. IRE*, **42** (1954) 1773.
2.2 J. Zawels, The natural equivalent circuit of junction transistors. *RCA Review*, **16** (1955) 360.
2.3 M. Das, On the frequency dependence of the magnitude of common-emitter current-gain of graded base transistors. *Proc. IRE*, **48** (1960) 240.
2.4 J. Lindenmayer, Electrical representation of the drift transistor. *Semicond. Prod.*, **4** March (1961) 41.
2.5 L. J. Sevin, Field-effect transistors. McGraw-Hill, New York, 1965.
2.6 C. A. Mead, Schottky-barrier gate FET. *Proc. IEEE*, **54** (1966) 307.
2.7 B. Reddy and F. N. Trofimenkoff, FET high-frequency analysis. *Proc. IEEE*, **113** (1966) 1755.
2.8 J. T. Wallmark and H. Johnson, Field-effect transistors, Prentice Hall, New York, 1966.
2.9 D. Koehler, The charge-control concept in the form of equivalent circuits. *Bell System Tech. J.*, **46** (1967) 523.
2.10 R. H. Crawford, MOSFET in circuit design. McGraw-Hill, New York, 1967.
2.11 E. F. Felsing, Der Einfluss von Gehäuseformen auf die Eigenschaften von UHF Transistoren. *Valvo Berichte*, **13** Oct. (1967) 71.

2.12 H. K. Gummel, Charge-control transistor model for network analysis programs. *Proc. IEEE*, **56** (1968) 751.
2.13 S. Kurtin and C. A. Mead, GaSe Schottky-barrier gate FET. *Proc. IEEE*, **56** (1968) 1594.
2.14 V. A. Dhaka, Design and fabrication of subnanosecond current switch and transistors. *IBM J. Res. Develop.*, **12** (1968) 476.
2.15 C. D. Todd, Junction field-effect transistors. John Wiley, New York, 1968.
2.16 D. Gestner, Breitbandige Hochfrequenz-Leistungstransistoren. *AEG-Telefunken Report*, 1968, p. 69.
2.17 S. M. Shannon, J. Stephen and J. H. Freeman, MOS frequency soars with ion-implanted layers. *Electronics*, **42** 3 Feb. (1969) 96.
2.18 W. Baechtold, W. Kotyczka and M. J. O. Strutt, Computerized calculation of small-signal and noise properties of microwave transistors. *IEE Trans.*, MTT-17 (1969) 614.
2.19 F. Faggin and T. Klein, A faster generation of MOS devices. *Electronics*, **42** Sep. (1969) 88.
2.20 H. F. Strom, Field-effect transistor bibliography, 1967-1968. *IEEE Trans.*, ED-**16** (1969) 957.
2.21 M. B. Das, High-frequency network properties of MOS transistors. *IEEE Trans.*, ED-**16** (1969) 1049.
2.22 J. Arandjelovic, HF MOS transistors. *Proc. IEEE*, **58** (1970) 143.
2.23 K. E. Drangeid, R. Sommerhalter and W. Walter, High-speed Gallium-Arsenide Schottky-barrier field-effect transistors. *Electr. Lett.*, **6** (1970) 228.
2.24 O. Müller, Ultralinear UHF power transistors for CATV applications. *Proc. IEEE*, **58** (1970) 1112.
2.25 R. J. Nienhuis, An MOS tetrode for the VHF band with a channel 1.5 μm long. *Philips Tech. Rev.*, **31** (1970) 259.
2.26 R. D. Josephy, MOS-Transistoren zur Leistungsverstärkung im HF Bereich. *Philips Tech. Rundschau.* **31** (1970/71) 262.
2.27 S. Kakihane and P. H. Wang, Simple CAD technique to develop high-frequency transistors. *IEEE J.*, SC-**6** (1971) 236.
2.28 R. G. Harrison, Computer simulation of a microwave power transistor. *IEEE J.*, SC-**6** (1971) 226.
2.29 J. Lange, A survey of the present state of microwave transistor modeling. *IEEE Trans.*, ED-**18** (1971) 1168.
2.30 H. J. Sigg and J. Kocsis, D-MOS transistor for microwave applications. *IEEE Trans.*, ED-**19** (1972) 45.
2.31 D. S. Jacobson, What are the trade-offs in RF transistor design?. *Microwaves*, **11** July (1972) 46.
2.32 A. B. Macnee and R. J. Talsky, High-frequency transistor modeling for circuit design, *IEEE J.*, SC-**7** (1972) 320.
2.33 J. A. Archer, Improved microwave transistor structure. *Electr. Lett.*, 8 (1972) 499.
2.34 J. A. Benjamin, New design concepts for microwave power transistor. *Microwave J.*, **15** Oct. (1972) 39.
2.35 J. R. Greenbaum, Easy-to-use HYPI program makes possible transition from nonlinear to linear transistor model. *Electronics*, **46** 18 Jan. (1973) 174.
2.36 J. Thomas and D. Gordon, Microwave small signal bipolar transistor. *Microwave J.*, **16** Febr. (1973) 43.
2.37 W. Baechtold, K. Daetwyler, T. Forster, T. O. Mohr, W. Walter and P. Wolf, Si and GaAs Schottky-barrier FET's. *Electr. Lett.*, **9** May (1973) 232.
2.38 P. C. Parekh and J. Steenbergen, The 3 keys to good transistor design. *Microwaves*, **12** Aug. (1973) 40.
2.39 H. F. Cooke and F. E. Emery, Advances in X-band bipolar transistors. *Microwave J.*, **16** Nov. (1973) 47.
2.40 A. J. Wahl, Distributed theory for microwave bipolar transistors. *IEEE Trans.*, ED-**21** (1974) 40.
2.41 P. W. Schackle, An experimental study of microwave bipolar transistors. *IEEE Trans.*, ED-**21** (1974) 32.
2.42 R. S. Ronen and L. Strauss, The silicon-on-sapphire MOS tetrode. *IEEE Trans.*, ED-**21** (1974) 100.

2.43 E. Faldella, G. Inculano, Mathematical model for transistor for VHF. *Electr. Lett.*, **10** (1974) 26.
2.44 S. Kakihana, Current status and trends in high-frequency transistor. *Microelectronics*, **5** (1974) 6.
2.45 F. A. D'Altroy, R. M. Jacobs, J. M. Macci and E. J. Panner, Ultralinear transistors. *Bell System Tech. J.*, **53** (1974) 2195.
2.46 D. M. Pietro, A new 5 GHz transistor process. *Hewlett Packard J.*, **26** Apr. (1975) 8.
2.47 H. J. Jekat, A portable 1100 MHz frequency counter. *Hewlett Packard J.*, **26** Apr. (1975) 2.
2.48 J. S. Barrera and W. E. Poole, Microwave transistor review. *Microwave J.*, **19** Feb. (1976) 28.
2.49 E. F. Belohoubek, A. Presser and H. S. Veloric., Improved circuit-device interface for microwave bipolar power transistor. *IEEE J.*, SC-**11** (1976) 256.
2.50 H. F. Cooke, Microwave field-effect transistors in 1978. *Microwave J.*, **21** Apr. (1978) 43.
2.51 E. J. Colussi, Internally matched RF power transistors. *Microwave J.*, **21** Apr. (1978) 81.
2.52 K. Honjo, Y. Takayama and A. Higashisaka, Broad-band internal matching of microwave power GaAs MESFET's. *IEEE Trans.*, MTT-**27** Jan. (1979) 3.
2.53 H. Fukui, Determination of the basic parameters of a GaAs MOSFET. *BSTJ*, **58** March (1979) 771.
2.54 J. Sone and Y. Takayama, Ku- and K-band internally matched high-power GaAs FET amplifiers. *Electr. Lett.*, **15** 30 Aug. (1979) 562.

3.1 D. A. Maxwell, R. H. Beeson and D. F. Allison, The minimization of parazitics in integrated circuits by dielectric isolation. *IEEE Trans.*, ED-**12** (1965) 20.
3.2 M. P. Lepselter, Beam-lead technology. *Bell System Tech. J.*, **45** (1966) 233.
3.3 F. A. Brand and V. Gelnovatch, Microwave solid-state devices and circuits. *Microwave J.*, **10** July (1967) 22.
3.4 J. Eimbinder, Linear integrated circuits: theory and applications. John Wiley, New York, 1968.
3.5 R. A. Hierschfeld, Linear integrated circuits in communication systems. *IEEE Spectrum*, **5** (1968) 71.
3.6 R. J. Widlar, Future trends in IC operational amplifiers. *EDN.*, **13** 10 June (1968) 24.
3.7 J. R. Hearn and D. C. Spreng, New concepts in signal generation. *Hewlett Packard J.*, **19** Aug. (1968) 15.
3.8 F. H. Lee, Dielectric isolated saturating circuit. *IEEE Trans.*, ED-**15** (1968) 645.
3.9 M. B. Vora and J. J. Chang, PIN isolation for monolithic IC. *IEEE Trans.*, ED-**15** (1968) 655.
3.10 W. C. Roswold, W. H. Legat and R. L. Holden, Air-gap isolated microcircuits. *IEEE Trans.*, ED-**15** (1968) 640.
3.11 J. E. Solomon, Trends in analog integrated circuits. *Motorola Monitor*, **6** 3 (1968) 29.
3.12 U. S. Davidsohn, F. H. Lee, Dielectric isolated integrated circuits substrate processes. *Proc. IEEE*, **57** (1969) 1532.
3.13 M. Brooksby and R. D. Pering, Monolithic transistor array for high-frequency applications. *Hewlett Packard J.*, **21** Jan. (1970) 15.
3.14 A. Boyle, Fabrication of high-frequency analog integrated circuits. *Comp. Techn.*, **4** May (1971) 22.
3.15 E. S. Narayanamurthi, New high-speed monolithic operational amplifier. *IEEE J.*, SC-**16** (1971) 71.
3.16 R. J. Plassche, A wideband operational amplifier. *IEEE J.*, SC-**6** (1971) 347.
3.17 J. Addis, Three technologies on one chip make a broadband amplifier. *Electronics*, **45** 5 June (1972) 103.
3.18 D. Jones, Taking a look at input capacitance. *Electr. Eng.*, **44** Sep. (1972) 18.
3.19 J. Kay, Mullard and Plessey IC models for use in linear CAD programs. *Comp. Aided Design*, **5** Apr. (1973) 90.

3.20 R. J. Apfel, P. R. Gray, A fast-settling monolithic operational amplifier. *IEEE J.*, SC-**9** (1974) 332.
3.21 O. R. Gray, R. G. Meyer, Recent advances in monolithic operational amplifier design. *IEEE Trans.*, CAS-**21** (1974) 317.
3.22 P. C. Davis, S. F. Moyer and V. R. Saari, High slew-rate monolithic operational amplifier. *IEEE J.*, SC-**9** (1974) 340.
3.23 J. E. Solomon, The monolithic OpAmp: A tutorial study. *IEEE J.*, SC-**9** (1974) 314.
3.24 F. L. Sanders, The bucket-brigade delay line, a shift register for analogue signals. *Philips Tech. Rev.*, **31** (1970) 97.
3.25 D. Pippenger and D. May, Put BiFET's into your linear circuits. *Electr. Design*, **26** 4 Jan. (1978) 104.
3.26 D. Senderowicz, D. A. Hodges and P. R. Gray, High-performance NMOS operational amplifier. *IEEE J.*, SC-**13** Dec. (1978) 760.

4.1 J. G. Linvill and J. F. Gibbons, Transistors and active circuits. McGraw-Hill, New York, 1961.
4.2 W. Th. H. Hetterscheid, Transistor bandpass amplifiers. *Philips Tech. Library*, Eindhoven, 1964.
4.3 G. E. Bodway, Two-port power flow analysis using generalized scattering parameters, *Microwave J.*, **10** May (1967) 61.
4.4 Transistor parameter measurement. Hewlett Packard Application Note AN-77/1.
4.5 K. L. Kotzebue, Microwave amplifier design with potentially unstable FET's. *IEEE Trans.* MTT-**27** Jan. (1979) 1.

5.1 V. Uzunoglu, Semiconductor network analysis and design. McGraw-Hill, New York, 1964.
5.2 M. S. Ghausi, Principles and design of linear active circuits. McGraw-Hill, New York, 1965.
5.3 E. M. Cherry, and D. E. Hooper, Amplifying devices and low-pass amplifiers. John Wiley, New York, 1968.

6.1 W. W. Gartner, Transistors: Principles, design and application. Van Nostrand, New York, 1960.
6.2 B. Down, Using feedback in FET circuit to reduce input capacitance. *Electronics*, **37** 14 Dec. (1964) 64.
6.3 W. Grein, M. Russell and J. Sverdrup, A solid-state 50 MHz oscilloscope. *Electronics*, **39** 25 July (1966) 95.
6.4 C. D. Todd, FET as a source follower. *Electr. Components*, **7** Sep. (1961) 831.
6.5 J. B. Compton, Designing FET's in cascade. *EDN*, **11** 14 Sep. (1966) 60.
6.6 C. D. Todd, Follower circuits combining FET's and bipolar transistors. *Electr. Components*, **7** Oct. (1966) 943.
6.7 R. W. Barkes, B. L. Hart, A very high impedance wide-band buffer amplifier. *Proc. IEEE*, **57** (1969) 244.
6.8 O. A. Horna, High-speed voltage-follower has only 1 nanosecond delay. *Electronics*, **46** 19 July (1973) 115.
6.9 K. Klatt, Die obere Grenzfrequenz der Kollektorschaltung. *Int. Elektr. Rundschau*, **28** Apr. (1974) 67.

7.1 J. R. James, Analysis of the cascode transistor configuration. *Electr. Eng.*, **32** Jan. (1960) 44.
7.2 S. H. Sen, The equivalent h-parameters for n transistors connected in cascade. *Proc. IEEE*, **53** (1965) 1803.
7.3 R. H. Riordan, The analysis of multistage transistor amplifiers. *Proc. IRE (Austr.)*, **24** (1963) 70.
7.4 D. E. Denlinger, O. A. Kolody, Simplified "Y"-parameter analysis of multistage linear amplifiers. *IEEE Trans.*, BTR-**15** (1969) 68.

8.1 S. Schwarz, A transistor video amplifier for radar display. *SGS Fairchild Appl. Rep.*, AR-96, 1963.
8.2 A. Cense, A recent development in circuits and transistors for television receivers. *Electr. Appl.*, **27** 2 (1967) 41.
8.3 P. Sturzu, Use CAD to optimize broadband amplifier design. *Electr. Design*, **22** 10 May (1974) 92.

9.1 J. Almond and A. R. Boothroyd, Broadband transistor feedback amplifiers. *Proc. IEE*, **103** (1966) 93.
9.2 S. S. Hakim, Feedback amplifier stabilization. *Electr. Techn.*, **31** Jan. (1962) 23.
9.3 S. S. Hakim, Feedback amplifier stabilization. *Industr. Electr.*, **1** Febr. (1963) 273.
9.4 S. S. Hakim, Aspects of return-difference evaluation in transistor feedback amplifiers. *Proc. IEE*, **112** (1965) 1700.
9.5 H. G. Brierley, S. S. Hakim, Transistor feedback amplifier design. *Proc. IEE*, **112** (1965) 1825.
9.6 R. F. Hoskins, Definition of loop-gain and return difference in transistor feedback amplifiers. *Proc. IEE*, **112** (1965) 1995.
9.7 G. E. Bodway, Circuit design and characterization of transistors by means of threeport scattering parameters. *Microwave J.*, **11** May (1968) 55.
9.8 O. Giust, Modulators. *Bell System Technical J.*, **50** (1971) 2155.
9.9 G. L. Fenderson, The IF main amplifier. *Bell System Tech. J.*, **50** (1971) 2195.

10.1 F. Waldhauer, Wide-band feedback amplifiers. *IRE Trans.*, CT-4 (1957) 18.
10.2 F. Blecher, Design principles for single-loop transistor feedback amplifiers. *IRE Trans.*, CT-4 (1957) 32.
10.3 J. C. Broekert and R. M. Scarlett, Transistor amplifier has 100 Mc bandwidth. *Electronics*, **33** 16 Apr. (1960) 73.
10.4 M. S. Ghausi, Optimum design of the shunt-series feedback pair with maximally flat magnitude response. *IRE Trans.*, CT-8 (1961) 448.
10.5 E. M. Cherry and D. E. Hooper, The design of wideband transistor feedback amplifiers. *Proc. IEE*, **110** (1963) 375.
10.6 P. Beneteau, A 2 nanosecond video amplifier. *Fairchild Appl. Rep.*, AR-10, 1964.
10.7 D. W. Howell, A transistor amplifier with 500 Mc bandwidth. *Stanford Univ. Rep.*, 1820/1, 1964.
10.8 S. S. Hakim, Return difference measurement in transistor feedback amplifiers. *Proc. IEE*, **112** (1965) 914.
10.9 M. S. Ghausi, A simplified analysis of the series-shunt feedback pair. *Solid St. Tech.*, **8** Sept. (1965) 17.
10.10 D. E. Harding, Wideband transistor amplifier for use in submerged repeaters. *Proc. IEE*, **112** (1965) 1869.
10.11 D. W. Howell, Transistor amplifier with 500 Mc low-pass bandwidth. *IEEE Trans.*, CT-12 (1965) 591.
10.12 W. Glathe, 150 MHz Breitband-Verstärker mit geringem Klirrfactor. *Int. Elektr. Rundschau*, **19** Okt. (1967) 261.
10.13 A. E. Hilling, A 50 to 500 MHz broadband transistor amplifier. *Electr. Eng.*, **39** June (1967) 352.
10.14 S. Sijstra, Vertical deflection amplifier for 150 MHz oscilloscopes. *Electr. Appl.*, **27** 2 (1967) 61.
10.15 S. W. Director, R. A. Rohrer, Automated network design — The frequency-domain case. *IEEE Trans.*, CT-16 (1969) 330.
10.16 J. Tuil, Transistor equipped aerial amplifiers. *Electr. Appl.*, **28** 2 (1968) 61.
10.17 K. H. Eichel, Einfache Methode zur Erzielung eines konstant Eingangswiderstandes bei Breitbandverstärkern. *Int. Elektr. Rundschau*, **27** Febr. (1973) 45.
10.18 J. Lauch, Impulsverstärker mit aktiver Verstärkungseinstellung bei konstanter Bandbreite. *Int. Elektr. Rundschau*, **27** Febr. (1973) 37.
10.19 W. U. McCalla, An integrated IF amplifier. *IEEE J.*, SC-8 (1973) 440.
10.20 R. G. Meyer, R. Eschenbach and R. Chin, A wideband ultralinear amplifier from 3 to 300 MHz. *IEEE J.*, SC-9 (1974) 167.

11.1 M. J. Gay, Design and development of linear circuits. *Component Tech.*, **2** May (1967) 24.
11.2 E. A. Evel, A voltage probe for high-frequency measurement. *Hewlett Packard J.*, **20** Nov. (1968) 19.
11.3 T. J. Van Kessel, An integrated operational amplifier with novel HF behaviour. *IEEE J.*, SC-3 (1968) 348.
11.4 J. Eimbinder, Designing with linear integrated circuits. John Wiley, New York, 1969.
11.5 J. Zellmer, High impedance probing to 500 MHz. *Hewlett Packard J.*, **21** Dec. (1969) 12.
11.6 A. J. DeVilbiss, A wideband oscilloscope amplifier. *Hewlett Packard J.*, **21** Jan. (1970) 11.
11.7 R. A. Wooley, Automated design of DC coupled monolithic broad-band amplifiers. *IEEE J.*, SC-6 (1971) 24.
11.8 R. I. Ollins and S. J. Ratner, CAD and optimization of a broad-band HF monolithic amplifier. *IEEE J.*, SC-7 (1972) 487.
11.9 G. White and G. M. Chin, A DC to 2.3 GHz amplifier using an "embedding" scheme. *Bell System Tech. J.*, **52** (1973) 53.
11.10 C. R. Battjes, A wide-band high-voltage monolithic amplifier. *IEEE J.*, SC-8 (1973) 408.
11.11 J. B. Coughlin, R. J. H. Gelsing, P. J. W. Jochems and H. J. M. van der Laak, A monolithic wideband amplifier from DC to 1 GHz. *IEEE J.*, SC-8 (1973) 414.
11.12 J. L. Sorden, A new generation in frequency and time measurement. *Hewlett Packard J.*, **25** June (1975) 2.
11.13 W. M. Sansen and R. G. Meyer, An integrated wideband variable-gain amplifier with maximum dynamic range. *IEEE J.*, SC-9 (1974) 159.
11.14 S. Christensen and I. Matthews, A new microwave link analyser. *Hewlett Packard J.*, **27** Nov. (1975) 13.

12.1 C. L. Ruthroff, Some broad-band transformers. *Proc. IRE*, **47** (1959) 1337.
12.2 W. F. Bodtmann and C. L. Ruthroff, A wide-band transistor IF amplifier. *Bell System Tech. J.*, **42** (1963) 37.
12.3 L. U. Kibler, An 80 megabit 15 W transistor pulse amplifier. *Bell System Tech. J.*, **44** (1965) 1977.
12.4 L. U. Kibler, Transistor pulse amplifier. *Bell System Tech. J.*, **44** (1965) 1983.
12.5 W. Hilberg, Einige grundsätzliche Betrachtungen zu Breitband-Übertragern. *NTZ*, **19** (1966) 527.
12.6 J. B. Payne, Impedance mismatching in wide-band transistor amplifier design. *IEEE Trans.*, CT-14 (1967) 432.
12.7 O. Pitzalis and T. P. Couse, Practical design information for broadband transmission line transformers. *Proc. IEEE*, **58** (1968) 738.
12.8 H. Seidel, H. R. Beurrier and A. N. Friedman, Error-controlled high-power linear amplifier at VHF. *Bell System Tech. J.*, **47** (1968) 651.
12.9 R. C. Hejhall, Solid-state linear power amplifier design. Motorola Appl. Note AN-546.
12.10 H. Prasad and M. Krishnaswamy, 80–160 MHz transistorized wideband power amplifier. *Electro-Technology (Bangalore)*, **13** Jan. (1969) 9.
12.11 W. H. Lambert, Second-order distortion in CATV push-pull amplifiers. *Proc. IEEE*, **58** (1970) 1057.
12.12 Transistors for single-sideband linear amplifier. *Mullard Ltd.*, TP 1337, (1972) 11.
12.13 H. L. Krauss and Ch. W. Allen, Designing toroidal transformers to optimize wideband performance. *Electronics*, **46** 16 Aug. (1973) 113.
12.14 R. G. Meyer, R. Eschenbach and W. M. Edgerley, A wide-band feedforward amplifier. *IEEE J.*, SC-9 (1974) 422.
12.15 W. Fleischhauer, VMOS Leistungs-FET's in Hochfrequenz-Schaltungen. *Nachrichtentechnik u. Elektronik*, **32** Jan. (1978) 15.
12.16 S. Ludvik, Vertical geometry is boosting FET's into power uses at radio frequencies. *Electronics*, **51** 2 March (1978) 105.

13.1 I. L. Rohde, Transistoren bei höchsten Frequenzen. Verlag Radio-Foto-Kinotechnik, Berlin, 1965.
13.2 A. Moser, Ein 140 MHz Kettenverstärker mit Feldeffect-Transistoren. *Int. Elektr. Rundschau*, **19** May (1967) 109.
13.3 K. Chen, Distributed amplifiers. *Proc. IEE*, **114** (1967) 1065.
13.4 G. Kohn and R. W. Landauer, Distributed field-effect amplifier. *Proc. IEEE*, **56** (1968) 1136.
13.5 W. Jutzi, A MOSFET distributed amplifier with 2 GHz bandwidth., *Proc. IEEE*, **57** (1969) 1195.
13.6 O. V. Alaksejev, Design of distributed amplifiers with the aid of a computer. *IEEE Trans.*, CT-**20** (1973) 702.
13.7 F. Meyer, Wide-band pulse amplifier. *IEEE J.*, SC-**13** June (1978) 409.

14.1 S. Mason, Power gain in feedback amplifiers. *IRE Trans.*, CT-**1** (1954) 20.
14.2 C. C. Cheng, Neutralization and unilaterization. *IRE Trans.*, CT-**2** (1955) 138.
14.3 A. P. Stern, C. A. Aldridge and W. F. Chow, Internal feedback and neutralization of transistor amplifier. *Proc. IRE*, **43** (1955) 838.
14.4 A. P. Stern, Stability and power gain of tuned transistor amplifiers. *Proc. IRE*, **45** (1957) 335.
14.5 L. Gohm, Neutralization über breite Frequenzbänder. *NTZ*, **18** (1960) 83.
14.6 R. Paul, Zur Stabilität von transistorbestückten Schmalbandverstärkerstufen. *Nachrichtentechnik*, **11** (1961) 295.
14.7 J. M. Rolett, Stability and power gain invariants. *IRE Trans.*, CT-**9** (1962)29.
14.8 W. Th. H. Hetterscheid, Designing transistor IF amplifiers. Philips Tech. Library, Eindhoven, 1966.
14.9 W. H. Fröhner, Quick amplifier design with scattering parameters. *Electronics*, **40** (1967) 100.
14.10 F. M. Carlson, Application considerations for the VHF MOSFET. RCA Appl. Note AN-3193.

15.1 J. Carstaedt, H. Schoen and F. Weitzsch, Nichtneutralisierte und teilneutralisierte ZF-Verstärker mit Transistoren. *Valvo Berichte*, **6** (1960) 81.
15.2 M. A. H. El-Said, Tuned transistor amplifier. *IRE Trans.*, CT-**7** (1960) 440.
15.3 G. Rusche, K. Wagner and F. Weitzsch, Flächen-Transistoren. Springer Verlag, Berlin, 1961.
15.4 D. M. Duncan, Mismatch design of transistor IF amplifier. *Proc. IRE (Austr.)*, (1962) 147.
15.5 G. G. Luettgenau and S. H. Barnes, Designing with low-noise MOSFET's. *Electronics*, **37** 14 Dec. (1964) 53.
15.6 Using Linvill-techniques for RF amplifiers, Motorola Appl. Note AN-166.
15.7 P. Show, Simplify IF amplifier design. *Electr. Design.* **14** 5 July (1966) 38.
15.8 J. G. Krijakov and J. L. Simonov, Analysis of a cascode transistor tuned amplifier. *Electr. Comm.*, **37** Jan. (1967) 40.
15.9 B. Welling, Stagger-tuned IC amplifier stages. *Electr. Design*, **15** 26 Apr. (1967) 236.
15.10 H. M. Kleinmann, Application of dual-gate MOSFET. *IEEE Trans.*, BTR-**13** (1967) 72.
15.11 J. M. Phalan, Boost FET amplifier gains. *Electr. Design*, **15** 13 Sept. (1967) 98.
15.12 O. Suominen, Series tuned amplifiers as low-noise preamplifier. *IEEE J.*, SC-**2** (1967) 116.
15.13 W. Th. H. Hetterscheid, Vision IF amplifiers. *Electr. Appl.*, **26** (2) (1966) 49.
15.14 G. Wolf, Recent developments in circuits and transistors for television receivers. *Electr. Appl.*, **26** (3) (1966) 145.
15.15 H. Kriebel, Hifi Tuner — Stand der Technik. *Funkschau*, **45** (1973) 681.
15.16 Y. Miwa, K. Okuno and T. Namekawa, High-frequency amplifier design using Nichols-charts. *IEEE J.*, SC-**7** (1972) 195.

16.1 F. Weitzsch, Einige theoretische Untersuchungen zur Leistungsübertragung und Stabilität. *Nachrichtentechn. Fachb.*, **18** (1960) 23.

16.2 C. J. McCluskey, Bandpass transistor amplifiers. *Electr. Techn.*, **38** May (1961) 183.
16.3 H. G. Cherry, The design of tuned transistor IF amplifiers. *IEEE Trans.*, BTR-**9** (1963) 48.
16.4 K. Redmond, Rapid design for a transistorized double-tuned bandpass IF amplifier. *IEEE Trans.*, BTR-**9** (1963) 52.
16.5 H. Oberbeck, Beitrag zur exackten Berechnung von breitbandigen zweikreisigen Bandfiltern. *AEÜ*, **18** March (1964) 189.
16.6 L. Blaser and E. Cummuris, Designing FET's into AM radios. *IEEE Trans.*, BTR-**10** (1964) 29.
16.7 P. E. Kolk and I. A. Maloff, The field-effect transistor as high-frequency amplifier. *Electronics*, **37** 14 Dec. (1964) 71.
16.8 D. R. Recklinghausen, Field-effect transistor for FM front ends. *Electr. World*, **74** Dec. (1965) 64.
16.9 W. M. Austin, TV applications of MOS transistors. *IEEE Trans.*, BTR-**12** (1966) 68.
16.10 D. N. Leonard, Improve FM performance with FET's. *Electr. Design*, **15** 1 March (1967) 63.
16.11 I. S. Docherty and J. L. Casse, The design of maximally flat wideband amplifiers with double-tuned interstage coupling. *Proc. IEEE*, **55** (1967) 513.

17.1 S. C. Dutta Roy, The inductive transistor. *IEEE Trans.*, CT-**10** (1963) 113.
17.2 J. Lindmayer and W. North, The inductive effect in transistors. *Solid St. Electr.*, **8** Apr. (1964) 409.
17.3 J. J. Barrett, An FM-tuner using MOSFET's and integrated circuits. *IEEE Trans.*, BTR-**11** (1965) 24.
17.4 J. Avins, Integrated circuits in television receivers. *IEEE Trans.*, BTR-**12** (1966) 70.
17.5 P. J. Whiteneir, The analysis and design of IF amplifiers. *IEEE Trans.*, BTR-**12** (1966) 75.
17.6 J. Robertson and B. Welling, An integrated RF-IF amplifier. Motorola Appl. Note AN-247.
17.7 B. Welling, Using integrated circuits in a stagger tuned IF strip. Motorola Appl. Note AN-259.
17.8 B. Welling, An IC color TV video IF amplifier. *IEEE Trans.*, BTR-**13** (1967) 24.
17.9 R. L. Weber and J. C. Prabhakar, A thick-film television video IF amplifier. *IEEE Trans.*, BTR-**13** (1967) 7.
17.10 D. A. Daly, S. P. Knight, M. Caulton and R. Ekholdt, Lumped elements in microwave integrated circuits. *IEEE Trans.*, MTT-**15** (1967) 713.
17.11 J. A. Archer, J. F. Gibbons and G. M. Purnaiya, Use of transistor-simulated inductance in broadband amplifiers. *IEEE J.*, SC-**3** (1968) 12.
17.12 N. Schater, Lumped element IC produces 1 W in S-band. *Electr. Design*, **16** Apr. (1968) 32.
17.13 R. A. Hirschfeld, Tuned circuit design using monolithic RF/IF amplifiers. *Electr. Design*, **16** (1968) 24.
17.14 M. Caulton, S. P. Knight and D. A. Daly, Hybrid integrated lumped element microwave amplifiers. *IEEE J.*, SC-**3** (1968) 59.
17.15 K. M. Johnson, X-band integrated circuits mixer. *IEEE J.*, SC-**3** (1968) 50.
17.16 R. Milton and C. Kamnitsis, RF integrated amplifiers in high power UHF broadband structures. RCA Appl. Note ST-4128, 1968.
17.17 J. A. Mataya, G. W. Haines and S. B. Marshall, IF amplifier using C_c-compensated transistors. *IEEE J.*, SC-**3** (1968) 401.
17.18 M. Caulton and W. E. Poole, Designing lumped elements into microwave amplifiers. *Electronics*, **42** 14 Apr. (1969) 100.
17.19 E. Sugata, T. Namekawa, Integrated circuits for television receivers. *IEEE Spectrum*, **6** (1969) 64.
17.20 D. K. Adams and R. Y. C. Ho, Active filters for UHF and microwave frequencies. *IEEE Trans.*, MTT-**17** (1969) 662.
17.21 G. Baskerville, A single-chip television IF system. *IEEE J.*, SC-**7** (1972) 455.

18.1 G. L. Matthaei, Synthesis of Chebyshev impedance matching network filters and interstages. *IRE Trans.*, CT-**3** (1956) 163.
18.2 H. Klink, Breitbandverstärker mit Transistoren. *AEÜ*, **18** June (1964) 350.
18.3 G. L. Matthaei, Tables of Chebyshev impedance transforming networks of low-pass filter-form. *Proc. IEEE*, **52** (1964) 939.
18.4 R. S. Engelbercht and K. Kurokowa, A wideband low-noise L-band balanced transistor amplifier. *Proc. IEEE*, **53** (1965) 237.
18.5 K. Kurokowa, Design theory of balanced transistor. *Bell System Tech. J.*, **44** (1965) 1675.
18.6 J. Lauchner, Wideband microwave transistor amplifiers. *Solid St. Design*, **6** Dec. (1965) 19.
18.7 G. Pierson, A FET operating at UHF. *Electr. Design*, **14** 29, March (1966) 48.
18.8 G. L. Matthaei, Short-step Chebyshev impedance transformers. *IEEE Trans.*, MTT-**14** (1966) 372.
18.9 R. V. Snyder, Broadband impedance matching techniques applied to design of UHF transistor amplifiers. *Proc. IEEE*, **55** (1967) 124.
18.10 T. E. Sunders and P. D. Stark, An integrated 4 GHz balanced transistor amplifier. *IEEE J.*, SC-**2** (1967) 4.
18.11 G. Sabbadini, Transistorbestückter Antennenverstärker für die Fernsehbereiche IV und V. *Int. Elektr. Rundschau*, **21** Dec. (1967) 321.
18.12 W. G. Gelnovatch, CAD of wideband integrated microwave amplifiers. *IEEE Trans.*, ED-**15** (1968) 491.
18.13 S. C. Blum, A 10 W S-band solid-state amplifier. *IEEE J.*, SC-**3** (1968) 233.
18.14 H. C. Lee, Microwave power amplifiers. *Microwave J.* **12** Febr. (1969) 51.
18.15 C. Kamnitsis, Broadband matching of UHF microstrip amplifiers. *Microwaves*, **8** Apr. (1969) 54.
18.16 T. W. Houston and L. W. Read, CAD of broadband and low-noise microwave amplifiers. *IEEE Trans.*, MTT-**17** (1969) 612.
18.17 K. Ayaki, E. Igarashi and Y. Kajinara, A 4 GHz multistage transistor amplifier. *IEEE Trans.*, MTT-**17** (1969) 1072.
18.18 F. Lüttich, Transistor Breitbandverstärker bis 1 GHz mit hoher Ausgangsleistung. *Int. Elektr. Rundschau*, **24** Apr. (1970) 105.
18.19 D. R. Bowman, 600 MHz intermediate frequency amplifier. *Electr. Eng.*, Aug. (1970) 30.
18.20 O. Pitzalis, and R. A. Gilson, Tables of impedance matching networks which approximate prescribed attenuation vs. frequency slopes. *IEEE Trans.*, MTT-**19** (1971) 381.
18.21 Improved circuit with low parasitism transistor. *Electr. Eng.*, **44** Aug. (1972) 23.
18.22 A. Presser, E. F. Belohoubek, 1−2 GHz high-power linear transistor amplifier. *RCA Rev.*, **33** (1972) 737.
18.23 J. Eisenberg and R. I. Disman, Design a 4−8 GHz FET amplifier. *Microwaves*, **12** Febr. (1973) 52.
18.24 J. Curtis, Let's simplify MIC power amplifier design. *Microwaves*, **12** Febr. (1973) 46.
18.25 P. H. Wolfert, L-band 110 W transistor amplifier, *Microwave J.*, **16** Febr. (1973) 47.
18.26 K. Richter, Predicting linear power amplifier performance. *Microwaves*, **13** Febr. (1974) 56.
18.27 G. D. Vendelin, J. A. Archer and N. G. Bechtel, A low-noise integrated S-band amplifier. *Microwave J.*, **17** Febr. (1974) 47.
18.28 P. Sturzu, Build a 12 octave hybrid amplifier. *Microwaves*, **13** June (1974) 54.
18.29 R. P. Arnold and W. L. Bailey, Match impedances with tapered lines. *Electr. Design*, **22** 7 June (1974) 136.
18.30 P. T. Chen, Design and application of 2−6.5 GHz transistor amplifiers. *IEEE J.*, SC-**9** (1974) 154.
18.31 G. Basawapatna, A 2−6.2 GHz power amplifier. *Hewlett Packard J.*, **26** March (1975) 11.
18.32 J. McDermott, The new GaAs FET's. *Electr. Design*, **26** 1 Febr. (1978) 38.

18.33 K. B. Niclas, R. B. Gold, W. T. Wilser and W. R. Hitchens, A 12—18 GHz medium-power GaAs MESFET amplifier. *IEEE J.*, SC-**13** Aug. (1978) 520.
18.34 C. M. Krowne, Extending the low-frequency range of GaAs FET broadband microwave amplifiers using microstrip transmissionl ines. *Electr. Lett.*, **15** 15 March (1979) 197.
18.35 J. Siddiqui, Synthesis technique for a 2—8 GHz 1W FET amplifier. *Microwave J.*, **22** Apr. (1979) 57.

19.1 T. M. Scott, Tuned power amplifiers. *IEEE Trans.*, CT-**11** (1964) 385.
19.2 P. Paris, Calcul et realisation d'amplificateurs VHF de puissance. *L'Onde Electrique*, **45** (1965) 328.
19.3 D. F. Page and W. D. Hindton, On solid-state class D systems. *Proc. IEEE*, **53** (1965) 423.
19.4 B. T. Vincent, Large-signal operation of microwave transistors. *IEEE Trans.*, MTT-**13** (1965) 865.
19.5 J. G. Slatter, An approach to the design of transistor tuned power amplifiers. *IEEE Trans.*, CT-**12** (1965) 206.
19.6 R. G. Harrison, A nonlinear theory of class C transistor amplifiers. *IEEE J.*, SC-**2** (1965) 93.
19.7 D. R. Lohrmann, Parametric oscillations in VHF transistor power amplifiers. *Proc. IEEE*, **54** (1966) 409.
19.8 R. Minton, Design trade-offs for RF transistor power amplifiers. *The Electr. Engineer*, **26** March (1967) 68.
19.9 P. Schiff, RF breakdown phenomenon improves the voltage capability of a transistor. *Electronics*, **40** 12 June (1967) 97.
19.10 O. Müller, Stability problems in transistor power amplifiers. *Proc. IEEE*, **55** (1967) 409.
19.11 N. O. Sokal, J. J. Sierakowski and J. J. Sirota, Use a good switching transistor model. *Electr. Design*, **12** 7 June (1967) 54.
19.12 D. M. Snider, A theoretical analysis of RF power amplifier. *IEEE Trans.*, ED-**14** (1967) 851.
19.13 D. R. Lohrmann, Exotic effects?. *Proc. IEEE*, **56** (1968) 332.
19.14 M. R. Osborne, Design of tuned transistor power amplifiers. *Electr. Eng.*, **40** Aug. (1968) 436.
19.15 H. K. Gummel, Charge-control transistor model for network analysis programs. *Proc. IEEE*, **56** Apr. (1968) 751.
19.16 J. Mulder, On the design of transistor RF power amplifiers. *Electr. Appl.*, **27** 4 (1967) 155.
19.17 O. Müller, Large-signal s-parameter measurements of class C operated transistors. *NTZ*, **21** (1968) 644.
19.18 H. Rothe, Der Bipolartransistor als Leistungsverstärker. *AEÜ* **22** Aug. (1968) 407.
19.19 S. Krishna, P. J. Kannam and W. Doesschate, Some limitations of the power output capability of VHF transistors. *IEE Trans.*, ED-**15** (1968) 855.
19.20 C. R. Zimmer, Center frequency shift in transistor class C amplifiers. *IEEE Trans.*, AES-**5** (1969) 999.
19.21 H. Hilbers, On the input and load impedance and gain of RF power transistors. *Electr. Appl.*, **27** (1967) (2) 53.
19.22 B. Reich, E. B. Hakim and G. J. Malinowsky, Maximum RF power transistor collector voltage. *Proc IEEE*, **57** (1969) 1789.
19.23 R. W. Brown, Transistor models for circuit analysis programs. *Electr. Eng.*, **41** Dec. (1969) 50.
19.24 A. H. M. El-Said, Analysis of tuned junction transistor circuits under large sinusoidal voltages. *IEEE Trans.*, CT-**17** (1970) 8.
19.25 R. L. Bailey, Large-signal nonlinear analysis of high-power high-frequency junction transistors. *IEEE Trans.*, ED-**17** (1970) 108.
19.26 G. W. Zobrist, Thinking of getting into CAD. *Electronics*, **43** 30 March (1970) 98.
19.27 T. Nygren and J. Martinson, Improved characterizing technique for microwave power amplifier transistors. *Electr. Lett.*, **6** (1970) 282.
19.28 J. Andeweg and T. H. J. van den Hurk, A discussion of the design and prop-

erties of high-power transistors of SSB applications. *IEEE Trans.*, ED-17 (1970) 717.
19.29 B. Reich, E. B. Hakim and G. Malinowsky, RF power transistor for reliable communications systems. *IEEE Trans.*, ED-17 (1970) 816.
19.30 B. K. Erickson, Temperature compensation of high-frequency transistors. *Electronics*, **46** 24 May (1973) 102.
19.31 G. K. Baxter, Thermal response of microwave transistors under pulsed power operation. *IEEE Trans.*, PHP-9 Sept. (1973) 184.
19.32 G. E. Brehm and G. D. Vendelin, Biasing FET's for optimum performance. *Microwaves*, **13** Febr. (1974) 38.
19.33 G. K. Baxter, Transient temperature response of a power transistor. *IEEE Trans.*, PHP-**10** (1974) 132.
19.34 C. B. Beuthauser, Hotspotting in RF devices. *Electr. Components*, **16** 4 June (1974) 12.
19.35 D. R. LaRosa, Hybrid-coupled amps. *Microwaves*, **14** Feb. (1975) 44.
19.36 J. Vidkjaer, A computerized study of the class C biased RF power amplifier. *IEEE J.*, SC-**13** Apr. (1978) 247.
19.37 S. R. Mazumder and P. D. Puije, Two-signal method of measuring the large-signal s-parameters of transistors, *IEEE Trans.*, MTT-**26** June (1978) 417.
19.38 F. H. Raab and N. O. Sokal, Transistor power losses in the class E tuned power amplifier. *IEEE J.*, SC-**13** Dec. (1978) 912.

20.1 R. M. Kurzrok, Design technique for lumped circuit hybrid rings. *Electronics*, **35** 18 May (1962) 60.
20.2 R. M. Kurzrok, S. J. Mehlman and A. Newton, Hybrid-coupled VHF transistor power amplifier. *Solid St. Design*, **6** Aug. (1965) 21.
20.3 R. C. French, A wideband transistorized power amplifier. *Electr. Eng.*, **38** Jan. (1966) 8.
20.4 G. H. Wood, A. W. Morse and G. R. Brainerd, Transistors share the load in a kilowatt amplifier. *Electronics*, **40** 11 Dec. (1967) 100.
20.5 J. A. Benjamin, Use hybrid junctions for more VHF power. *Electr. Design*, **16** 1 Aug. (1968) 54.
20.6 J. A. Benjamin, Build broadband RF power amplifiers. *Electr. Design*, **17** 18. Jan. (1969) 50.
20.7 R. L. Bailey, W. P. Bennett, L. F. Heckman and I. E. Martin, An all-transistor 1 kW high-gain UHF power amplifier. *IEEE Trans.*, MTT-**17** (1969) 1154.
20.8 R. D. Peden, Charge-driven HF transistor tuned power amplifier. *IEEE J.*, SC-**5** (1970) 55.
20.9 M. Ringler, Junction-gate FET RF power amplifier. *Proc. IEEE*, **58** (1970) 789.
20.10 O. Pitzalis, R. E. Horn and R. J. Baranello, Broadband 60 W linear amplifier. *IEEE J.*, SC-**6** (1971) 93.
20.11 A low-Q bandpass amplifier. *Electr. Eng.*, **44** Aug. (1972) 17.
20.12 A. H. Hilbers, Design of high-frequency wide-band power transformers. *Philips Electr. Appl. Bull.*, **32** (1) (1973) 44.
20.13 J. Mulder, Input network design for high-frequency wideband power amplifiers. *Philips Electr. Appl. Bull.*, **32** (3) (1973) 101.
20.14 K. Hupper, H. Helmrick, 50 W VHF Verstärker mit Transistoren. *Int. Elektr. Rundschau*, **27** Aug. (1973) 165.
20.15 S. Chambers, A 1000 W solid-state power amplifier. *Electr. Design*, **22** 1 Apr. (1974) 58.
20.16 J. Stammelback, Transformationsnetzwerke für transistorbestückte Hochfrequenz-Leistungsverstärker. *Int. Elektr. Rundschau*, **28** Nov. (1974) 229.
20.17 F. W. Hauer, Stop burn-out in RF power amplifiers. *Electr. Design*, **23** 4 Jan. (1975) 110.
20.18 F. H. Raab, FET power amplifier boosts transmitter efficiency. *Electronics*, **49** 10 June (1976) 122.
20.19 H. P. Graf, Ein Linear-Leistungsverstärker für Fernsehsender. *Nachrichtentechnik u. Elektronik*, **32** Febr. (1978) 48.
20.20 F. H. Raab, Get broadband, dual-mode operation with this FET power amplifier. *EDN*, **23** 20 Oct. (1978) 117.

21.1 J. S. Vogel and M. J. O. Strutt, Untersuchung der Verzerrungen und Mischvorgänge in Transistor-Stufen bei hochen Frequenzen. *AEÜ*, **16** Aug. (1972) 407.
21.2 L. W. Read, An analysis of high-frequency transistor mixers. *IEEE Trans.*, BTR-**9** (1963) 72.
21.3 E. Becher, Nonlinear admittance mixer. *RCA Rev.*, **25** (1964) 662.
21.4 S. M. Wearer, For good mixer, add one FET. *Electronics*, **39** 21 March (1966) 109.
21.5 D. R. Recklinghausen and H. H. Scott, Theory and design of FET converters. *IEEE Trans.*, BTR-**12** 1. Apr. (1966) 43.
21.6 U. L. Rohde, Transistor 2-metre converters. *Wireless World*, **72** July (1966) 357.
21.7 F. C. Fitchen and G. C. Sundberg, Conversion gain null in FET mixers. *Proc. IEEE*, **55** Jan. (1967) 101.
21.8 E. Klein, Y-parameters simplify mixer design. *Electr. Design*, **15** 1 Apr. (1967) 68.
21.9 J. T. Winkel and B. C. Bouma, An investigation into transistor cross-modulation at VHF under AGC conditions. *IEEE Trans.*, ED-**14** July (1967) 374.
21.10 J. S. Vogel, Nonlinear distortion and mixing processes in field-effect transistors. *Proc. IEEE*, **55** Dec. (1967) 2109.
21.11 G. Dijk and G. Wolf, A mixer transistor for VHF. *Electr. Appl.*, **29** (2) (1969) 39.
21.13 D. M. Miller and R. G. Meyer, Nonlinearity and cross-modulation in field-effect transistors. *IEEE J.*, SC-**6** (1971) 244.
21.13 O. Klank, Ideale und wirkliche Kennlinien von Transistoren für HF Eingangs- und Mischstufen. *Int. Elektr. Rundschau*, **24** Jan. (1972) 17.
21.14 R. G. Meyer, M. J. Shensa and R. Eschenbach, Cross modulation and intermodulation in amplifiers at high frequencies. *IEEE J.*, SC-**7** (1972) 16.
21.15 S. Narayanan and H. C. Poon, An analysis of distortion in bipolar transistors. *IEEE Trans.*, CT-**20** (1973) 341.
21.16 H. C. Poon, Implication of transistor frequency dependence. *IEEE Trans.*, ED-**21** (1974) 110.
21.17 F. Beck, V. Geissler and H. Khakzar, Prüfung des Gummel–Poon modells auf seine Brauchbarkeit. *Int. Elektr. Rundschau*, **28** July (1974) 147.
21.18 A. M. Khadr and R. H. Johnston, Distortion in high-frequency FET amplifiers. *IEEE J.*, SC-**9** (1974) 180.
21.19 H. Lindermeyer and R. Popp, Gross-Signalverzerrungen in Emitterfolgerendstufen. *NTZ*, **28** (1975) 1.
21.20 K. Breitkopf, Transistor mixer boosts up conversion gain. *Microwaves*, **14** Jan. (1975) 62.

22.1 C. B. Burckhard, Analysis of varactor multipliers. *Bell System Tech. J.*, **44** (1965) 675.
22.2 H. C. Lee and G. J. Gilbert, Overlay transistors move into microwave region. *Electronics*, **39** 21 March (1966) 93.
22.3 R. D. Hall and S. M. Krakauer, Microwave harmonic generation with step-recovery diode. *Hewlett Packard J.*, **27** Apr. (1966) 5.
22.4 B. Schiek, Frequenzverhalten von Vervielfachern mit Kapazitätsdioden. *AEÜ*, **20** Sept. (1966) 515.
22.5 R. Thompson, Step-recovery diode frequency multiplier. *Electr. Lett.*, **2** March (1966) 117.
22.6 A. P. Anderson, Circuit aspects of transistor parametric frequency doublers. *IEEE Trans.*, ED-**14** (1967) 86.
22.7 S. Hamilton and R. D. Hall, Shunt-mode harmonic generation using step-recovery diodes. *Microwave J.*, **10** Apr. (1967) 69.
22.8 R. Carlson, A frequency comb generator. *Hewlett Packard J.*, **18** (1967) 15.
22.9 J. Irwin and C. Swan, A composite varactor for harmonic generator. *IEEE Trans.*, ED-**13** (1968) 466.
22.10 K. Schünemann and B. Schiek, Optimaler Wirkungsgrad von Frequenzvervielfachern mit Speicherdiode. *AEÜ*, **22** (1968) 293.
22.11 L. I. Kuo, Step-recovery Dioden Frequenzverdreifacher. *AEÜ*, **23** (1969) 268.

22.12 K. L. Kotzebue and G. L. Matthaei, The design of broad-band frequency doublers using charge-storage diodes. *IEEE Trans.*, MTT-17 (1969) 1077.
22.13 D. K. Rytting and S. N. Sanders, A system for automatic network analysis. *Hewlett Packard J.*, **21** (1970) 2.
22.14 H. R. Bedell, R. W. Judkins and R. L. Lahlum, Microwave generator. *Bell System Tech. J.*, **50** (1971) 2205.
22.15 R. H. Johnston and A. R. Boothroyd, A high-frequency transistor frequency multiplier and power amplifier. *IEEE J.*, SC-7 (1972) 71.
22.16 Simple parametric oscillator multiplier. *Electr. Eng.*, **44** Aug. (1972) 15.
22.17 J. Racy, How to select varactors for harmonic generation. *Microwaves*, **12** Nov. (1973) 54.
22.18 H. Faller, Try a transistor in your next frequency multiplier. *Microwaves*, **12** Dec. (1973) 48.

23.1 W. A. Edson, Noise in oscillators. *Proc. IRE*, **48** (1960) 1454.
23.2 R. L. Pritchard, Discussion of matrix analysis of transistor oscillators. *IRE Trans.*, CT-8 (1961) 169.
23.3 P. J. Baxandall, Transistor crystal oscillators. *The Radio and Electr. Eng.*, **29** Apr. (1965) 229.
23.4 W. Johnston and E. Loach, Microwave oscillator. *Bell System Tech. J.*, **44** (1965) 369.
23.5 A. H. James, Crystal oscillator using integrated circuit amplifiers. *Electr. Eng.*, **38** Jan. (1966) 42.
23.6 C. MacDonald, FET replaces vacuum tube in 1 MHz oscillator. *Electr. Design*, **12** March (1966) 81.
23.7 K. K. Clarke, Design of self-limiting transistor sine-wave oscillators. *IEEE Trans.*, CT-13 (1966) 58.
23.8 R. Spence, A theory of maximally loaded oscillators, *IEEE Trans.*, CT-13 (1966) 226.
23.9 T. F. Prosser, FET's produces stable oscillators. *Electronics*, **39** 3 Oct. (1966) 102.
23.10 H. N. Toussaint, Zur Bemessung des emittergekoppelten Oscillators. *Frequenz*, **21** June (1967) 193.
23.11 D. Singh and J. Garters, Crystal controlled 164 MHz oscillator/quadrupler. *Electr. Eng.*, **39** Dec. (1967) 769.
23.12 H. N. Toussaint, Transistor power oscillator. *Proc. IEEE*, **56** (1968) 226.
23.13 G. Bethmann, Berechnung von Transistor Oscillatoren. *TH Ilmenau Mitt.*, (1968) 179.
23.14 U. Sutcliffe, Transistor LC oscillator circuits with amplitude controlled by mean current. *Electr. Eng.*, **40** June (1968) 388.
23.15 H. Ikeda, A MOS transistor RF oscillator suitable for MOS-IC. *Proc. IEEE*, **56** (1968) 1638.
23.16 G. D. Hanchett, Insulated-gate FET in oscillator circuits. Motorola Appl. Note ST-3520.
23.17 U. Bühn, Entwurf von UHF Transistor-Oscillatoren mit grossem Durchstimmbereich. *Int. Elektr. Rundschau*, **21** Febr. (1969) 33.
23.18 T. L. Grisell, I. H. Hawley, B. D. Unter and P. G. Winninghoff, Design of a third-generation RF spectrum analyzer. *Hewlett Packard J.*, **19** Aug. (1968) 8.
23.19 H. N. Toussaint, ECAP analysis of UHF transistor oscillator transient response. *IEEE Trans.*, MTT-17 (1969) 620.
23.20 W. J. Evans, Circuits for high-efficiency avalanche diode oscillators. *IEEE Trans.*, MTT-17 (1969) 1060.
23.21 D. Cawsey, Wide range tuning of solid state microwave oscillators. *IEEE J.*, SC-5 (1970) 82.
23.22 P. E. Olivier, Microwave YIG-tuned transistor oscillator amplifier design. *IEEE J.*, SC-7 (1972) 54.
23.23 R. E. Pratt, R. W. Austin and A. Dethlefsen, Microcircuits for microwave sweeper. *Hewlett Packard J.*, **22** Nov. (1970) 9.
23.24 T. J. Aprille and T. N. Trick, A computer algorithm to determine the steady-state response of nonlinear oscillator. *IEEE Trans.*, CT-19 (1972) 354.
23.25 200 MHz oscillator uses ECL. *Electr. Eng.*, **34** Sept. (1972) 25.

23.26 R. R. Hay, Versatile VHF signal generator. *Hewlett Packard J.*, **25** March, (1974) 18.
23.27 C. Hamilton, Transistor harmonic oscillator with outputs in X-band. *NTZ*, **27** (1974) 196.
23.28 G. Hodowanec, Microwave transistor oscillator. *Microwave J.*, **17** June (1974) 39.
23.29 F. Patensschat, Design RF oscillators. *Electr. Design*, **22** 20 Sept. (1974) 70.
23.30 W. Subbarao, Simplify LC RF-oscillator design. *Electr. Design*, **22** 13 Sept. (1974) 141.
23.31 J. A. Nelson, A. N. Roy and W. D. Ryan, A UHF driven subharmonic oscillator. *Int. J. Electr.*, **37** Sept. (1974) 329.
23.32 G. Bajen, Design transistor oscillators. *Electr. Design*, **24** 12 Apr. (1976) 98.
23.33 A. W. Vermis, Specifiying isolation to limit frequency pulling. *Microwaves*, **15** May, (1976) 55.
23.34 J. M. Golio and C. M. Krowne, New approach for FET oscillator design. *Microwave J.*, **21** Oct. (1978) 59.
23.35 K. M. Johnson, Large-signal GaAs MESFET oscillator design. *IEEE Trans.*, MTT-**27** March (1979) 217.

24.1 R. R. Webster, The noise figure of transistor converters. *IRE Trans.*, BTR-**7** (1961) 50.
24.2 W. Smulders, Noise properties of transistors at high frequencies. *Electr. Appl.*, **23** (1962) 1.
24.3 H. Fukni, The noise performance of microwave transistors. *IEEE Trans.*, ED-**13** (1966) 329.
24.4 W. C. Bruncke and A. van der Ziel, Thermal noise junction gate FET's. *IEEE Trans.*, ED-**13** (1766) 323.
24.5 R. Q. Lane, The comparative performance of FET and bipolar transistors at VHF. *IEEE J.*, SC-**1** (1966) 35.
24.6 G. D. Johnson, Design amplifier for low noise. *Electr. Design*, **14,** 6 Dec. (1966) 54.
24.7 J. E. Solomon, Cascade noise figure of integrated transistor amplifiers, Motorola Appl. Note AN-223.
24.8 D. C. Agourdis and A. van der Ziel, Noise figure of UHF transistors. *IEEE Trans.*, ED-**14** (1967) 808.
24.9 R. G. Meyer, Noise in transistor mixers at high frequencies. *Proc. IEEE*, **56** (1968) 487.
24.10 H. E. Halladay and A. van der Ziel, On the high-frequency excess noise. *Solid St. Electr.*, **12** (1969) 161.
24.11 F. M. Klaasen and J. Prins, Noise of field-effect transistors at very high frequencies. *IEEE Trans.*, ED-**16** (1969) 952.

SUBJECT INDEX

back-to-back diode 26
balanced modulator 327
beam-lead structure 60
biasing 302
bipolar transistor 35
 — current gain 45
 — cut-off frequency 47
 — emitter-grid structure 42
 — equivalent circuit 36
 — extrinsic element 46
 — hybrid-π equivalent circuit 83
 — interdigitated construction 40
 — intrinsic element 46
 — large-signal characterization 55
 — LOCOS structure 43
 — mesa structure 43
 — microwave structure 43
 — overlay structure 41
 — planar structure 36
 — transit time 37
 — transition frequency 37
breakdown voltage 263
bucket-brigade delay line 65
buried layer 59

CCD delay line 67
channel 49
charge storage 12
Chebyshev filter 238
class F amplifier 312
closed-loop compensation 154
coaxial amplifier 252
comb structure 40
compensation 152
conductance modulation 14
conversion 313
 — transconductance 314
coupling efficiency 172

delay lines 64
depletion layer 12
dielectric insulation 60
diffusion capacitance 12
diffusion length 12
diode
 — back-to-back 26
 — BARITT 32
 — Gunn 32

 — hot-carrier 20
 — IMPATT 30
 — junction 11
 — PIN 14
 — point-contact 11
 — step-recovery 28
 — TRAPATT 32
 — tunnel 22
 — varicap 26
directional coupler 258
distributed amplifier 175
DMOS transistor 54
dual-gate MOSFET 322
dual-mode varactor 26
duplexer 19
dynamic tuning 203

emitter
 — crowding 39
 — dip effect 39
 — follower 101
 — -grid transistor 42
epitaxial diode 14
equal-ripple response 180
error-controlled amplifier 149
Esaki diode 22
exponential stripline 256

feedback amplifier 123
feedforward method 155
four-pole
 — connection 77
 — equivalent circuit 69
 — matching 293
 — noise 364
 — parameters 80
 — power gain 72
 — scattering parameter 71
 — stability 73
frequency-comb generator 340
frequency multiplier
 — doubler 335
 — operation 329
 — — microwave 338
 — integrated circuit 338
 — step-recovery diode 340
 — tripler 336
 — — UHF 337

frequency multiplier (*cont.*)
— varactor 339
f_S cut-off frequency 48
f_T transition frequency 37

gain margin 127
group delay 180
guard bias circuit 304
Gunn diode 32

helix 235
hot-carrier diode 20
hot spot 264
hybrid
— π circuit 83
— technology 235
— 3-dB coupler 258

IMPATT diode 30
inductive output 232
interdigitated structure 40
internal matching 44
intrinsic transistor 46
iterative gain 233

JFET 49
— cut-off frequency 51
— equivalent circuit 50
— — nonlinear 56

Kirk-effect 39

large-signal transient 163
lateral transistor 58
Linvill diagram 188
— factor 75
LOCOS structure 43
long-tail pair 112
loop-gain limit 187
LSA diode 34
lumped-element hybrid 301

MAG 78
matching four-poles 293
maximally efficient gain 77
maximally flat response 180
maximum
— collector voltage 263
— stable gain 222
— usable gain 77
MESFET 51
metal-semiconductor junction 20
microwave diode 30
Miller capacitance 94
minimum stability factor 75
mismatch method 76
mixer 323
— diode 323
— integrated circuit 326
— self oscillating 325

mobility domain 32
monolithic amplifier
— fast response 231
— gain-controlled 225
— IF 229
— multistage 226
— selective 223
— two-stage 221
monolithic integrated circuit
— beam-lead structure 60
— delay line 64
— dielectric insulation 60
— high-frequency properties 61
— structure 58
— technology 59
MOS capacitor 156
MOS transistor 51
— depletion mode 53
— equivalent circuit 52
MUG 77

narrow band amplifier 274
neutralization 184
noise
— figure 363
— matching 366
nonlinear instability 266
nonlinearity 124

operational amplifier 151
optimum termination
— large-signal 291
— small signal 75
oscillators
— Clapp 346
— Colpitts 345
— crystal 360
— Hartley 344
— negative-impedance 348
— push-pull 354
— stability 349
— subharmonic 362
— transformer coupled 344
overlay transistor 41

parametric
— amplifier 30
— oscillation 268
phase
— compensation 152
— margin 127
PIN
— diode 14
— — attenuator 17
— — equivalent circuit 14
— — stripline package 16
— — switch 18
— insulation 61
planar diode 13
pole-zero analysis 85

Subject index

positive feedback 139
potentially unstable four-pole 73
power amplifier
— class F 312
— equivalent circuit 285
— guard bias 308
— hybrid technology 236
— 1 kW output power 310
— 3-dB hybrid 302
— transmission-line transformer 310
power cut-off frequency 162
push-pull amplifier
— distortion 169
— output power 282
— wideband 146
— — monolithic 158
— — power 147
— — — gain-controlled 159

Read diode 30

sampler 22
saturation voltage 262
second breakdown 264
series collector resistance 38
series-parallel-type control 229
signal-handling capability 320
slew rate 163
snap-back
— diode 28
— effect 266
SOS transistor 53
source follower 104
stagger tuning 207
static induction transistor 54
Stern-factor 74
storage-charge varactor 28
sustaining voltage 263

TED diode 32
thin-film technology 235
transducer gain 73
transient temperature peak 265
transistor
— bipolar 35
— DMOS 54
— JFET 49
— microwave 43
— MESFET 51
— MOSFET 51
— power 39
— SOS 53
— static-induction 54
— V-MOS 54
transistor configuration
— common-base 99
— common-emitter 94
— common-source 102
— emitter-follower 100

— source-follower 104
transition
— capacitance 12
— frequency 37
transmission line
— compensation 257
— transformer 164
transmission-line transformer 174
transversal filter 66
TRAPATT diode 32
tuned amplifier
— double-tuned 209
— hybrid 236
— neutralization 184
— single-tuned 193
— stability 182
tuned-out noise 365
tunnel diode 22
— amplifier 24
— equivalent circuit 23
two-branch amplifier 160
two-stage amplifier
— compensated 142
— current feedback 136
— double feedback 140
— frequency response 109
— monolithic 221
— positive feedback 140

unconditionally stable four-pole 73
unilateral four-pole 184

varactor 25
— dual-mode 26
— Q factor 26
— Schottky-barrier 30
— storage-charge 28
— tuning 26
varicap diode 26
V-MOS transistor 54

wideband amplifier
— five-stage 144
— four-stage 145
— — monolithic 158
— microwave 261
— multi-stage 106
— output power 162
— pole-zero analysis 85
— seven-stage 144
— three-stage 255
— — monolithic 157
— transfer function 86
— transmission-line transformer 168
— two-branch 160
— two-stage 142

y-neutralization 185